HUMAN INTELLECT:
OPTIMAL TUNING AND CONTROL

ASTONISHING WAY TO BECOME SMARTER

Yuri Iserlis

authorHOUSE

AuthorHouse™
1663 Liberty Drive
Bloomington, IN 47403
www.authorhouse.com
Phone: 1 (800) 839-8640

Published by AuthorHouse 07/21/2020

ISBN: 978-1-7283-6341-7 (sc)
ISBN: 978-1-7283-6340-0 (e)

CONTENTS

Dedication xi

Acknowledgments...xiii

Introduction..xv

How This Book Is Organized xxiii

Chapter 1   What is an Intellect and Intellectual abilities................................ 1

    1.1. Invisible part of human..1

    1.2. Human being wonders...1

    1.3. Some popular theories of intellect (intelligence) 3

    1.4. Intellect in Nature...5

    1.5. Intellect (intelligence), Mind and Operating system 8

    1.6. Basic functions of intellect, intellectual abilities 16

    1.7. Creativity and intuition...17

    1.8. Perception and comprehension of information in the Mind.....................21

    1.9. Subsystem of information storage and interchange...................................22

    1.10. Willpower 27

    1.11. The diversity of intellectual abilities..29

    1.12. Intellectual –psychological portrait of a person 35

Chapter 2   What level of intellect is necessary for people in the 21ˢᵗ century? 36

    2.1. People with practically no brain 36

    2.2. Distribution of people in society in terms of intellect level 37

    2.3. Contribution of smart people in the development of society and civilization.............39

    2.4. Psychological and social- psychological reasons of resistance to change 42

    2.5. Economic levers of resistance to change...45

    2.6. Prosperity or survival for a child prodigy...46

    2.7. Difficulties of overcoming the resistance to the changes 51

    2.8. Is it necessary to be ringing the alarm bells?..53

    2.9. What level of intellect is needed for a successful life?...............................55

Chapter 3   Methods of evaluation and measurement of intellectual abilities and characteristics 62

3.1. Expert evaluations and psychometric measurements 62

3.2. IQ-tests and its accuracy 64

3.3. The brain as an universal measurement device 66

3.4. Radiesthesia as subjective method of extraction of information from subconscious part of mind 69

3.5. Regimes and some rules for assessments and measurements 72

3.6. Basic scales for radiesthesia (dowsing) 73

3.7. Access codes for subconscious part of mind 74

3.8. Quality of measurement by radiesthesia methods 82

Chapter 4   Where are intellectual abilities hidden 84

4.1. Tasks formulation 84

4.2. Cell Intellect 85

4.3. A little bit about human brain from anatomy and physiology 91

4.4. Neuroplasticity of the brain 101

4.5. Some specific features of consciousness and mind 103

4.6. Abdominal brain and heart intellect 110

4.7. Super complexity of the human cerebral system. Expanded definition of the human intellect. 112

4.8. Extrasensory perception from contemporary science viewpoint 113

4.9. Automatic writing, channeling, mediumship, etc. from psychology and esoteric knowledge viewpoints 116

Chapter 5   Intellect and multi-dimensional biofield 119

5.1. The energies in the human body 119

5.2 The human organism is like as an orchestra 121

5.3. The human biofield and its components 125

5.4. Aura and subtle bodies 129

5.5. Human soul from a good sense perspective 131

5.6. The assemblage point as energy center in the human body 134

5.7. Emotions as important attribute of intellect 137

5.8. Love and intellect 140

Chapter 6   Factors that affect a power of human intellect 143

6.1. Levels of influence on intellect 143

6.2. Genetic level 144

6.3. Age and intellectual abilities 145

6.4. Man under the influence of external fields 146

6.5. Influence of biological organisms and other people 151

6.6. Influence of feeding 154

6.7. The magic role of aromas, music, colors and art 156

6.8. Influence of education and training 159

6.9. Creative environment and inspiration 161

6.10. Boundaries of artificial influences on intellect...................................... 162

Chapter 7   Some features of human thinking 164

7.1. New concepts of thinking 164

7.2. About technologies of the thought processes 165

7.3. Creativity during the sleep 168

7.4. Sequential and parallel thinking mind 169

7.5. Thought-forms 170

7.6. Dominant motivation, mental blocks and mental barriers 172

7.7. Mirror mind 173

7.8. Mindset and Mind tuning 174

Chapter 8   Problems and some methods of enhancing intellectual abilities.......... 177

8.1. Classical individual methods of mind activation.................................... 177

8.2. Compensation of homeostatic imbalance... 179

8.3. Mind liberation from mental blocks and barriers 180

8.4. Accelerated learning 181

8.5. Acceleration of information input into the brain................................... 183

8.6. Memory training 185

8.7. Changing the position of the assemblage point 185

8.8. Meditation 186

Chapter 9   Intellect tuning.. 190

9.1. Intellect optimization as unused opportunities for enhancing quality and power
      of brain functions... 190

9.2. The goal of optimal mind tuning. Criteria and Target (Goal) functions.... 192

9.3. Means and parameters for the mind optimal tuning 194

9.4. Math methods and results of optimal tuning an intellect by G-factor 198

9.5. Optimal tuning the intellect components 206

9.6. Optimal tuning a willpower 207

9.7. Measurement of vital energy index 208

9.8. Aging and optimal tuning 213

9.9. Some words about random errors of measurements............................... 215

Chapter 10 Intellect and mind control (management)... 216

10.1. The necessity for the mind (intellect) control 216

10.2. Mind Technology for enhancing the brains performance 220

10.3. Hygiene and prevention of the intellect and brain diseases 225

10.4. Mindset management 227

10.5. Control the subconscious mind by means of information influence 229

10.6. Cognitive control 234

10.7. Creativity control. Algorithms and methods 237

10.8. Control emotions 241

10.9. Brain training as a part of intellect control 244

10.10. Collective intellect and Apollo Syndrome 245

Conclusion 249

About the author 251

References 253

You cannot hope to build a better world without improving the individuals.
                                                                    - Marie Curie

DEDICATION

To my wife Lyuba Iserlis with love and to all members of my family with boundless gratitude

ACKNOWLEDGMENTS

I would like to thanks Eugene Aseykin for his help in hatching and shepherding the idea of book; Lyubov Iserlis, Nadegda Zakhidova, and Mark Shklyar participated in several experiments, Dr. Wayne Baird who also taught me some methods of diagnosis with Applied Kinesiology and treatment of such diseases as allergies, arthritis, gout, etc. based on methods of energy exposure without the use of any medication, Alex Yunerman who help me in the design some R-scales, especially my colleagues and friends who spent their time for reading and commenting a manuscript: Mark Shklyar, Wayne Baird, Dr. Vadim Gritsus, Irene Zak.

A lot of interesting information I received attending meetups of the Dowsers Association on San Jose, headed by Wayne Hoff.

Some ideas of book was discussed and tested in the health club Flower of Live (San Jose) with participation of Boris and Molly Shedrin, Valentina Walz, Nadegda Zakhidova, Lyudmila Gritsenko, Vladimir and Lyubov Troussov, Nadia Nelson, Dalia and Roman Neshmit, Zina and Peter German, etc.

I would like to thanks the sponsors, who helped me in crowdsourcing for publishing, Valentin Malin, Sofia and Eugene Rubinstein, Alexander Tuchinsky, etc.

I am very grateful and indebted to workers of publishing company AuthorHouse for patience and creative support during the process.

INTRODUCTION

Studies in recent years have shown that the certain animals, insects, and even some mushrooms have a mind. Studies show also that the man could win a dominant position on our planet in the civilization to which we belong, because his mind was much better. It has not been possible for ants that created social order allowing them to survive in difficult conditions, and nor dolphins and some cetaceans, that have the brain capacity close to the man.

The word intellect we often use in everyday life. However, few people can give this a scientific definition, primarily because these definitions are quite a lot. Intellect is a term usually used in studies of the human mind, and refers to the ability of the mind to come to conclusions about what is true or real, and about how to solve problems.

The analysis shows that a man became the king of nature for two reasons: first, through the personal intellect, and secondly, because in a community, people were able to strengthen their intellect, creating a so-called social intellect, which allowed them to solve more difficult problems of survival in times of planet Earths cataclysms.

What is intellect really, and where is it located? Whether it can be seen or touched? How the level of the intellect can be connected with success in the life? Can a man of average ability to develop his intellect to the intellectual level of well-known human geniuses, such as Leonardo Da Vinci, Amadeus Mozart, Nikola Tesla and Albert Einstein? These questions concerned humanity for a long time. Many theories have already been created to explain intellect, and many methods for its evaluation, improvement and development have been created too. Unfortunately, still now here are many questions left, for example, not clear what structures of brain responsible for intellectual abilities, if they are creating not only in the brain, then where? It is also not completely clear what factors influence the functioning and development of intellect.

Our experience shows that the intellect is not a program or an organ of the human body, generally this is a poorly described system of thinking for humans or animals. In addition, it is poorly described because there is still no complete clarity how the brain works, what it can and what cannot, and why? Using modern scientific language, our brain is rather a system to control the person; most likely, this is a multi-level and distributed system whose efficiency depends on many factors that are different for different people. For example, for best results of mind activity, one person should be hungry; the other should have a good night's sleep; the third - to be in a state of sexual satisfaction, etc. One needs silence around him, but the other cannot live without the noise or music. One always has a solution for every situation, the second needs a long time to think, the third looks for an advice. For some people, intellect is associated with intuition, others never believe intuition.

Intellect and intellectual abilities are not constant things. They can be developed in the different rate and can change during lifetime. According to many experts, every healthy child is born with many intellectual possibilities, some of them depending on the circumstances realized to a greater extent, others hardly implemented. First become talented people or even geniuses (no matter recognized or unrecognized), others join the ranks either "average Joe (or Jane)" or foolish and ignorant people.

Until now, here are plenty other questions concerning a human intellect. What are intellects signs, which could be evaluated in the appearance of person? For example, whether we can consider the genius the person who has a high forehead as Socrates, who prefers to sit as the Thinker of Rodin, or who smiles as Mona Lisa?

Maybe the size or weight of the brain is important, the more the better? However, there are examples of geniuses with the sufficiently small size of the brain. Several Nobel laureates have a lower brain volume of 900 cm3, and some - of more than 2000 cm3. Definitely, intellect depends on genes, but how? What kind of a text written in DNA determines the level of intellect? On the other side, what are the best criteria for measurement or evaluation of intellect?

Many experts also try to find answers to the question: why the development of intellect needed for many years of training. Why athletes can be in a relatively short period to train to produce outstanding results, and attempts to do the same with scientists give the worst effects?

We do not know an answer on the question: a man is a product of evolution, or he was created by God (by some entity or civilization). If the God or somebody other created man in his own image and type, why he did not supply him with detailed instructions for training and use of intellect?

Amid many problems today, it is not clear how to interact the subconscious mind because like as in case of iceberg, most of the operation of our minds remains out of sight. It is not clear what is intuition, how memories work, why some people are logical thinking, and others – via imaging, one need inspiration, others can work as much as necessary in any situation (workaholics), etc.

The 20th century brought significant changes in the understanding of the intellect. What is most importantly, our knowledge is increasingly moving from a description of encountered cases of brains oddity and psi-phenomena, accepted before as quackery, to scientific knowledge, approved by many experiments.

In modern medical science, there are several disciplines, which are engaged in the study of anatomy and functions of the brain, as well as the treatment of its diseases: neurophysiology, neurology (or neuropathology), psychiatry and psychology. Indirectly, the work of the brain is the subject of pathological anatomy and pathophysiology.

As well to the end of the XX century was created a plenty scientific disciplines about human history, behavior, including such disciplines as anthropology, gerontology, biochemistry, sociology, social psychology and so on.

Ironically, there is not a discipline that would study entirely all functions of the intellect. It looks like as main function of the brain was divided between several disciplines. This explains why the brain and other neural structures located, for example, in the area of the intestine and the solar plexus, have been the least studied from point of view of the entire control of mind and body.

We need to say several words about the level of complexity of the problem of studying the brain. According to the theory of complex systems, the complexity of any system depends on the number of elements of this system and the number of links between these elements.

On average, the human brain has about 100 billion cells, and connections between them are much more. For description of operating processes for each cell can be used a system of a large number of chemical or physical equations. Therefore, the integrated complexity of our mental system is huge, and it is similar and compare with the complexity of the Universe.

Theoretically, any system can be described using mathematical models. But this can be done only with condition of the accuracy of the description of all variables accounted for the operation of this system, including records of all initial and boundary conditions under which the system exists and operates. For the brain, this task is doubly difficult today, because some unknown variables and functions of the brain exist. For example, it is unknown how the brain controls all the body's cells, which exchanged between themselves not only by chemical and electrical signals, but and wave information. It is unknown how the brain creates thought forms and sends them (or perceived)signals instantly for any distance, what physical nature of clairvoyance, a forecast of the future, psychokinesis, etc...

This means that the problem cannot be solved only by the professional psychologists and other medical scientists. Here is a requirement to bring together experts in various fields of knowledge with the arsenal of new research methods developed in the different disciplines.

The United States Congress declared the ninety years of the last Century as decade studying the human brain. The initiative quickly became international. Now all over the world the best laboratories and institutes employ the study of the human brain.

In different countries successfully are functioning a plenty specialized research centers and institutes such as The Brain Research Institute and Helen Wills Neuroscience Institute at the University of California at Berkeley in California (USA), The Center for Brain and Cognition (CBC) at the University of California, San Diego, Duke Institute for Brain Sciences in North Carolina (USA), INC (Institute for Neural Computation), CARTA (Center for Research and Teaching in Anthropogony), the Institute of the Human Brain of the Russian Academy of Sciences, St. Petersburg, Russia, and others.

At present, brain and mind studies become the objects of the Big Science. Many areas of research have grown so complex and expensive that large-scale collaboration is the best method for moving forward. At last century, Big Science has long been the norm in physics, where probing nature's frontiers require massive particle accelerators such as the Large Hadron Collider at CERN near Geneva. In biology, Big Science had its debut in 1990 with the Human Genome Project, a 13-year, roughly $3-billion (in 1991 dollars) effort co-founded by the National Institutes of Health and the Department of Energy to sequence human DNA. At present, we see examples of Big Science collaboration in The Human Brain Project (2012- 2022), funded by the European Union, The White House BRAIN Initiative (Brain Research through Advancing Innovative Neuroethologies), etc.

Probably, the existed knowledges about brain, mind and intellect, and the results of the researches described above will lead to the creation of a separate scientific discipline devoted to the description of the mind and intellect in all its diversity of manifestations. The first attempt in this direction was made by the Russian scientist Konstantin Sheremetyev in the book "Intellectica.

How do your brain works "(Sheremetev, 2013). In this book, practically unknown in the USA, author discovers how the intellect changed in the process of human evolution, how the male intellect differs from the female one, and so on. In English-language literature are known a book written by Ian J. Deary "Intelligence: A Very Short Introduction" (2001) and many books devoted description of many parts of this huge item (cited in this book).

Obviously, in order for a new scientific discipline to appear, much more needs to be done. The term Intellectica is used by the Association for research on the science of Cognition for a biannual peer-reviewed academic journal of cognitive psychology that covers research in a broad range of subjects and publish articles in French and English.

Pretty long period of time in my life, I have worked in field of computer science, math (optimizations problems) and control systems for industrial objects. In the early 90s, I created from scratch a company "Intellectual system" in St. Petersburg, which was one of the first in Russia developed intelligent add-in for monitoring and control systems for gas pumping stations for gas mains. At about the same time as a result of cooperation with General dietologist of S. Petersburg (Russia) Prof. Loiko V. I was able to formulate the optimization problem of the individual therapeutic treatment of patients with the help of mathematical methods (Iserlis, 1981).

All my life I was interested in the problem of creativity and other abilities of the human mind, why smart people committed and continue to commit unforgivable errors, what kind of relationship between wise and stupid people can be useful for any society. Fruits of my reflection culminated in an essay "The formula of happiness and why fools rule the world" (Iserlis, 1996- in Russian).

The saved-up experience in this area, in the field of the system analysis, and in the field of artificial intelligence allowed me to consider on the one hand human intellect as something special in the world or the Universe, and on the other hand, to try to formulate a problem of intellectual control for complex systems, optimization of such systems in the real time and management man-machines systems. In 2009, in the USA based on my experience I wrote the book "Artificial Intelligence Around Us» [Iserlis, 2009].

In 2010, I created the health club "Flower of Life", the purpose of which was to study the human health from a variety of positions, including the positions of alternative medicine.

At last decades, I witnessed the rapid introduction of artificial intelligence systems in many industries, economics and even art (7D hologram technologies). Artificial intelligence systems, including the latest generations of robots and three-dimensional printers, over the past decades reached unprecedented success. What previously seemed to be the pinnacle of human creativity — the game of chess, Go game or checkers, the recognition of visual images and sound patterns, the automatic synthesis of new technical solutions — in practice turned out to be not so difficult.

A famous futurologist Kevin Kelly authored the New York Times best-selling book "The Inevitable" (Kelly, 2016) predicted that the most prominent powered by artificial intelligence technologies likely impact society over the next three decades. He proposed that AI of future might be not human-like. It means that humanity may invent many new types of thinking that do not exist biologically today and that these types are not like human thinking. The future will show whether it is possible.

This means that artificial intelligence may not only accelerate technological progress, but also heal diseases, destroy hunger and extend the lives of future generations. On the other side, such superintelligence may destroy life on the planet also.

If you look at the world experience of using artificial intelligence in science and industry during last 50 years, easy to make sure that we until now may imitate only a small part of the capabilities of human intellect.

At present, no one artificial intelligence system can be like a person with a developed intuition, because it cannot solve incompletely defined tasks in accordance with the goals set. Here are many questions in robotic ethics, for example for self-driving cars. Today, the ways of perceiving and transmitting information between people in time and space remain unknown. At present, some people can enhance their intelligence in ways that machines are not capable to do, for example, with the help of emotions. On the other hand, the existing laws of the development of society may impede or limit the development of the human intellect for part of this society in order to prevent the possibility of acting to the detriment of society with established traditions or in harm of existed of regime controlling society. According to the Noam Chomsky theory in many developed countries today exist and are successfully used - "10 Strategies of manipulation of the population through the media" (https://www.ccsd.ngo/10-strategies-of-manipulation/).

Therefore, for further development of artificial intellect, we need to understand more about the many outstanding problems of humans, animals, or insects' intellect.

At one moment, I realized that my experience with technical intellectual systems and some knowledges and studies in the field of medicine could allow understanding the physical and information processes in the brain, from the standpoint of cybernetics and Informatics. I understood also that the further development of artificial intellect systems is impossible without expanding our knowledge of the human intellect[1]. This inspired me to research this item seriously and based on results of 5 years' experience to write a book on a topic that interested me during the lifetime. To be honest, lacking professional psychological and medical education, I hesitated whether I can do this very complex work one circumstance convinced me the right choice.

Unfortunately, today universal science about human not exists. Historically, many problems of human studies have been solved by people with non-medical and non-humanities education, for example, such as: a physicist, mathematician and philosopher of science Roger Penrose; biophysicist Maxwell Cade; Jeff Hawkins, the man who created the Palm personal computer and many application programs for the intellectualization of its work (Hawkins, 2004), and then plunged into the study of the work of the subcortical brain.

At the contrary, some untechnical scientists tried to create a kind of robots. For example an American-born British neurophysiologist William Grey Walter, author of a number of brain waves discoveries using his EEG machines at Burden Neurological Institute in Bristol. He was the first to determine by triangulation the surface location of the strongest alpha waves within the occipital

---

[1] I was right. Now leading experts and analysts see signs of impending stagnation in the development of artificially intelligence systems. For example, here is what Heikki Ailisto writes in the article "Third AI winter is coming unless we change course", published on the website of VTT on 23/07/2019: " We are heading towards a third AI winter. AI winter refers to a period of time during which companies, researchers, research funders and the general public are disappointed in artificial intelligence and the results achieved through it. This, in turn, leads to the freezing of funding and investments as well as to the stagnation of development. Researchers and experts move on to other technologies or start calling their activities by other names".

lobe (alpha waves originate from the thalamus deep within the brain). William Walter is known also as a creator of some of the first electronic autonomous robots.

It is obvious that the problem of the human mind, in general, and intellect, in particular, are interdisciplinary, on the one hand, close to medical sciences, and on the other hand – close to physics, mathematics, cybernetics, the theory of complex systems, control theory to name just a few.

Within the framework of our civilization, humanity accumulated the huge amount of knowledge about the Universe and our environment, including the nature of the Earth and the creatures that inhabit it. But there is a paradox, our knowledge is increasing annually, but their introduction in science disciplines is delayed due to the existence of so-called official stereotypes that unfortunately live longer than new knowledges. So, many facts about telepathy, telekinesis, subtle bodies, I am not talking about the soul and spirit remain outside the scope of the review of the contemporary science.

One of the world's leading physicists in Torsion-Physics research Dr. Gennady Shipov, who devoted his life to the theory of vacuum and torsion fields, in the report on the Conference on torsion fields in 2009, said, "From the analysis of the current state of micro-, macro and mega physics is obvious that this science needs a deep revision of its foundations. This, of course, can only go on the strategic expansion of our understanding of the world. Such an extension is impossible without a generalization of fundamental concepts of physics, such as space-time, the principle of relativity, the principle of inertia, reference system, mass, charge, quantization, etc. (Shipov,2009).

By analogy with computer systems, it can be assumed that human intellect is an operating system that has as minimum two control centers. This concept allows to use a systematic approach to describe and control the processes of thinking in order to optimize them using various criteria.

It is easily can be proved that the human brain equipped sense organs can work as the universal and precision measuring tool. Unfortunately, this instrument does not have an indicator device (panel) and results of measurements usually hidden in the depth of subconscious part of mind. To extract these results of measurements at first is needed to find an access code for this information, secondly, to deduce this information in a convenient for perception form, and after that to decipher it. Based on this approach a new method of direct measurement of intellect characteristics was used for appraisal such characteristics of intellect and mind as creativity, intuition, willpower, vital force, etc. Using algorithms of multi-parameter optimization, the method allows increasing the level of intellect and its components in several times practically for everyone. Measurement of willpower and stress opens the prospect for many people to maintain their health and activity at the proper level throughout life.

In the book are collected some rules and methods allowing to support intellectual abilities of the mind on an optimum level by means of mindset management, control of the subconscious mind, cognitive control, and control emotions.

The book presented below is an attempt to advance not only in understanding the work of human intellect, but also in its management capabilities, on the one hand, based on the achievements of artificial intelligence, and, on the other hand, based on some new discoveries in the field of psychology, sociology and other related scientific disciplines.

In the book used a plenty information from researchers, practicing medical doctors, scientist and so on, working in different fields of knowledge whose are "just" unbiased persons seeking the truth, wherever it may lead.

Below, we will try to discover some certain functions of the cells, brain and other parts of the human body related to intellect. In this case, we will use two approaches: the principle of the black box and causal analysis with main goal to identify as possible fully all the characteristics of the intellect. In some cases, we will only show these characteristics, in others cases -their relation to the causes or sources, and in the third, - description methods for improving the functioning of the brain in order to maximize intellect. But keep in mind, the ultimate goal of our study - to identify brain puzzles and try to find them an explanation which does not contradict current scientific paradigms.

This book is for students and teachers, employees and employers, art masters and any kind writers, sportsmen, basically for anyone who wants to get more out of their time, develop superhuman powers, and live a more successful life.

This book is for young and adult people, who lead, manage, supervise, consult, counsel, train, coach or teach, who want to have personal improvement or team performance as well.

The book is written for anyone who seeks to master the methods of solving intellectual problems of the highest complexity that require the human maximum mobilization of all his intellectual resources. The book is also for people who want to keep their brain functioning at its peak throughout mature years.

This book is written so that it would be possible to use the obtained results in the life of any person. Unfortunately, the volume of the book did not allow describing many techniques such as relaxation, meditation, memory training and many other things. If the reader understood the ideas outlined in the book, he will find these techniques himself.

Not all scientists will agree with every finding in this book. The information here is based largely on researches of many authors from different field of knowledge and from our experience.

In advance, I apologize to those authors who is not sited and who has also contributed a great deal to the field of intellect (intelligence) science.

In the process of writing the book, artificial intelligence systems were used in the form of different kind of word processors, style editors, grammar editors and even translators. For this reason, some inaccuracies discovered by the reader can be explained by the imperfection of these systems.

HOW THIS BOOK IS ORGANIZED

In Chapter 1 reader will receive common information about intellect in nature, human intellect, its types, and categories. He met hypothesizes that the intellect is a real-time operating system of human body, the mind of man comprises intellect and knowledge base, the mind is governed from a certain command centers by the aid of will force or force of willpower.

Based on the theory of László Polgar we can assume that the innate human intellect has at the beginning of life elements of all types of intellects, but the conditions of education and training of the child in the first years of his life determine which side of his talents obtained at birth will be significantly developed.

In Chapter 2 reader will learn that in any societies can live people practically without a brain, and rest of them have very different mental abilities from idiots to geniuses. This diversification can give the completely unexpected manifestations when the leading positions in the society occupy the fools, who oppress or destroy clever people. In this case, the size of society does not have a value. This can be family, corporation's staff or the country population. On the other side, this variety can give soil for the conflicts: psychological, social or socio-psychological, and in these conflicts do not always the best person or team win.

According to the wave theory of Alvin Toffler, 20th and 21th centuries brought new demands to workers everywhere. Work is becoming increasingly diverse, less fragmented, each worker performing a larger, rather than shallower task. A flexible schedule and free pace replace the former need for mass synchronization of behavior. Workers have to cope with more frequent changes in their work, as well as product changes and reorganizations. The under such conditions developed intellect is necessary for each person first to survive, and secondly for the realization of desires and ideas.

In Chapter 3 is proved that a brain together with the organs of the senses may be an amazingly accurate, versatile, reliable and easy-to-use meter that people use in real practice. This meter allow to measure changes and increments many intellectual function with sufficient accuracy. The proposed measurement method belongs to the class of physical measurements.

In Chapter 4 is described, where can be buried intellect. Collected information allowed to assume that hardware for intellect system can include all cells of a body, which form Corporeal Mind, the brain structures in the skull, the abdominal brain and one additional brain in the heart. This makes possible for us to come at the conclusion that human organism has united brain or distributed brain structures, therefore intellect is real-time distributed operating system, ensuring the work of this super complex cerebral system. Any parts of this system can have either same functions, or doubled, tripled, or holographic functions, or different overlapped functions. Here are represented also general information about brain states, conscious, subconscious, unconscious and superconscious parts of the mind.

Chapter 5 devoted description of the functional architecture of mind from points of interdisciplinary views. Here are described phenomena of aura and subtle bodies, other human bio fields, soul and assembly point, what are emotions and how love connected with intellect,

In Chapter 6 are analyzed different internal and external factors affecting human intellect, like genes, age, personality, willpower, emotions, ability to focus attention to name just a few.

Chapter 7 provides an overview of the functional models of mind, regimes of thinking, stages of the creative process, thoughts forms, mindset and mind tuning, dominance factors, mental blocks, and barriers.

Chapter 8 includes some achievements in the world experience in enhancing mind abilities. Here are described some brain-boosting technologies, transcranial brain stimulation, brainwave entrainment methods, problems of mind ability optimization, sound and light mind machines, techniques for enhancing some mind characteristics, brain and mind training.

Chapter 9 includes a description of the new method of intellect optimal tuning, which allows increasing G-factor of intellect with help of Audio-visual entrainment (AVE) systems, enhancing some intellect features such as creativity and intuition by dosed classical music hearing, enhancing Vital energy and willing power also by dosed revolutionary song music hearing. Experiments showed that optimized parameters of intellect: G-factor, creativity and intuition- can retain their values for a long time.

Chapter 10 is devoted to description of Intellect and Mind control which must include several parts: mindset management, control of the subconscious mind by means of information influence, cognitive control including creative control, and control emotions.

What came out of it, you will learn below chapter behind chapter.

CHAPTER 1

What is an Intellect and Intellectual abilities

We may have three main objects in the study of truth: first, to find it when we are seeking it; second, to demonstrate it after we have found it; third, to distinguish it from error by examining it.

Blaise Pascal

All human beings, all persons who reach adulthood in the world today are programmed biocomputers. No one of us can escape our own nature as programmable entities. Literally, each of us may be our programs, nothing more, nothing less.

Dr. John C Lilly (2014)

1.1. Invisible part of human

Each person receives at birth is something that impossible directly to see, touch or smell, something, which in general, not available for the detection of our main five senses. Ones of that is the soul and the subtle bodies, the other ones – all, that relates to the intellect. For a long time, it was recognized that intellect belongs only to human, animals and other alive creatures do not have that. Today, scientists are not so categorical. Studying dolphins, apes, parrots, and crow, on the one hand, a variety of insects (for example, bees and ants), and even fungi, on the other hand, scientists forced to reconsider their views.

Until now many manifestations of intellect are not yet fully understood and unraveled by modern science. Let's consider some of them.

1.2. Human being wonders

The human body - like a small universe. It would seem that our body is the most studied, but many facts proved that there are still many undiscovered opportunities. It is known, that some individuals could gains strength to lift a multi-ton weight, to jump over his head, to swim in ice temperature water, to walk on hot coals, and so on.

According to numerous documented facts, on our planet lived and now live people who can either generate high voltage static electricity; or have the ability to attract to body enough heavy metal and non-metallic objects; or may allocate willpower heat, or can move objects at a distance, without touching them. Also, there is a so-called contactless fight, when opponents strike each other blow without touching. In this case, the defender feels physical contact and impact forces. Today nobody knows why these people can do it from the point of view modern physics and other science disciplines.

Here are several other examples of human being phenomena unexplainable by modern science:

1. At last decades, many investigations were devoted to study several types of human physical and psychic phenomena, including precognition (knowing future events before they happen); pyrokinesis (creating fire with the mind, and telepathy (describing things at a remote location).
2. Mirin Dajo was the pseudonym of a Dutch named Arnold Gerrit Henskes. He became famous for radically piercing his body with all kinds of objects, for example pierced by rapiers through the heart, lung and kidneys, and did not bleed or feel anything (see Wikipedia).
3. The scientific literature describes people with the ability to remember an abnormally large number of their life experiences in vivid detail. Brad Williams is an American man from Wisconsin, who have one of the best memories amid people on the our planet. He can remember almost every day of his life, easily naming the day of the week, date, month, and year of innumerable personal and public events. Persons with such type of memory (about 25 people in the world) remember as well as recall any specific personal events or trivial details, including a date, the weather, what people wore on that day, from their past, almost in an organized manner. The World Memory Championships is an organized competition of mental sports in which competitors memorize as much information as possible within a given period of time. Dominic O'Brien is 8 times World Champion of this competition. He has a record in the Guinness Book (2002) committing to memory a random sequence of 2808 playing cards (54 packs) after looking at each card only once.

4. Another phenomena –magnetic people. Liew Thow Lin of Malaysia is known as the "Magnet Man" because he has the ability to stick metal objects to his body (see Wikipedia). He was able "to carry weights of up to thirty kilograms on his chest, including bricks stacked on metal irons". Using a chain hooked to an iron plate placed on his stomach," he has apparently pulled a car for over fifty meters. His two sons and two grandchildren also have the magnetic-like ability. Scientists from Malaysia's University of Technology found no magnetic field in Lin's body, but did determine that his skin exhibits very high levels of friction, providing a "suction effect". His two sons and two grandchildren also have the magnetic-like ability.

Since all these people are made of the same materials as ordinary people, and in all other functions almost do not differ from them, it can be assumed that any person can have a huge memory, read thoughts, transmit or perceive commands or energy, etc. if to find and use special non-invasive or invasive methods of intellect tuning.

1.3. Some popular theories of intellect (intelligence)

During the last 100 years different researchers have been proposed a variety of theories to explain the nature of intelligence.

Within the framework of the test paradigm in psychology two directly opposite in their theoretical results lines of interpretation of the nature of intellect are observed: one is connected with the recognition of the general factor of intellect represented at all levels of intellectual functioning (K. Spearman), the other line is related to the denial of any common beginning of intellectual activity and the assertion of the existence of many independent intellectual abilities (L. Thurstone).

A pioneer of factor analysis an English psychologist Charles Spearman in his article in American Journal of Psychology in 1904 year (Spearmen. 1904) suggested a hypothesis of "two-factor" intelligence, according which all mental performance must be evaluated by a generalized intelligence factor g (general factor), associated with mental processes, and specific abilities that he called "s" factors, such as verbal, mathematical, and artistic skills. Spearman noted that while people certainly could and often did excel in certain areas, people who did well in one area tended also to do well in other areas. Unfortunately Spearman could not find a method for measurement both of these factors.

Initially, in the framework of Louis Thurstone's multifactor theory of intelligence, the possibility of the existence of a general intellect was rejected. Having correlated the results of 60 different tests performed by the subjects to identify various aspects of intellectual activity, Terstone received more than 10 group factors, 7 of which were identified by him and called "primary mental abilities":

In contrast to this theory, a psychologist at Harvard University, Howard Gardner proposed the theory of multiple intelligences (Theory of Multiple Intellect - TMI), which in many positions is diametrically opposed to the concept of IQ (see section 3.2), and today is gaining more and more supporters. According to Gardner (H. Gardner, 2000), there are several different kinds of intellect that are independent of one another and acting in the brain as kind of modules, each with its own rules. Based on his observations, he concluded, the people have many different intellectual abilities found in various combinations. Gardner distinguishes the following main types of these modules:

- Natural intellect that determines a person's ability to feel components of nature including animals and plants
- Musical intellect as the ability to distinguish pitch, rhythm, timbre and tone. This type of intellect allows us to feel, create and play music
- Logical-mathematical intellect, which helps to solve complex mathematical and logical problems
- Existential intellect of philosophers

- Interpersonal intellect as the ability to understand and communicate effectively with others.
- Bodily-kinesthetic intellect as the ability to perceive their own and someone else's body and manage them. Athletes, dancers, surgeons and another doctors, massagists, chiropractors, etc. should have a well-developed bodily-kinesthetic intellect
- Linguistic intellect in the first place, as the ability to learn foreign languages, and secondly, as the ability to accurately interpret thoughts and concepts in foreign language
- Personal intellect as the ability to understand themselves as well as own thoughts and feelings, and to use this knowledge in the process of life planning.
- Spatial intellect as the ability to navigate in space: in the ocean, in the woods or in the desert, to be oriented in three dimensions.

This theory more than all others allows us to evaluate the qualitative sides of intellect.

According to Robert Sternberg's proposed the triarchic theory, intelligence has three aspects: analytical, creative, and practical (Sternberg, 1985).

Among the modern models of the intellect, it is necessary to mention a prominent psychological theory of human intelligence coined as he Cattell–Horn–Carroll theory (CHC- model). CHC-model of intelligence is a synthesis of Cattell and Horn's Gf-Gc model of fluid and crystallized intelligence and Carroll's Three Stratum Hierarchy (Sternberg & Kauffman, 1998).

Raymond Cattell and John Horn proposes that general intelligence can be divided into two broad sets of abilities, namely:

- fluid ability (G-f) as a capacity to reason and solve novel problems, independent of any knowledge from the past
- crystallized ability (G-c) as an ability to use skills, knowledge, and experience.

These intelligences must be different because crystallized intelligence increases with age, whereas fluid intelligence tends to decrease with age.

Carroll's Three Stratum Hierarchy means a three-layered model where each layer accounts for the variations in the correlations within the previous layers. The three layers (strata) of this model are defined as representing narrow, broad, and general cognitive ability (Keith, 2010).

Iconoclast Reuven Feuerstein, working with disadvantaged children in Israel, challenged the prevailing notion of a fixed intelligence with his theory of cognitive modifiability. Feuerstein believes that intelligence is not a fixed entity but a function of experience and mediation by significant individuals (parents, teachers, caregivers) in a child's environment.

This modern theory underlies a fresh view of intelligence as modifiable; it contends that intelligence can be taught, that human beings can continue to enhance their intellectual functioning throughout their lifetimes, and that all of us are "gifted" and all of us are "retarded" simultaneously (Feuerstein, et al., 1980)

Hungarian chess teacher and educational psychologist László Polgár came to the conclusion, that in the civilized countries approximately 80 percent of all children at the age of one year — potential geniuses! To three years a quantity of geniuses is reduced to 60 percent. To the five- to 50. In their 12 years remain 20 percent, and to 20 years — only five.

This means that if to teach and to educate children from the earliest age, then it is possible to make from them geniuses. But it is known that in the course of time, capabilities for self-realization can attenuate, if we them, of course, do not develop.

He proved this theory on the example of his daughters Susan (Zsuzsa), Sophia (Zsófia), and Judit, whom he raised to be chess prodigies. Polgár's experiment with his daughters has been called "one of the most amazing experiments…in the history of human education".

Susan Polgar according to on the July 1984 FIDE Rating List, at the age of 15 became the top-ranked woman player in the world, and remained ranked in the top three for the next 23 years. For a time, Sophia Polgar ranked as the sixth-strongest female player in the world. Judit Polgar was the number 1 rated woman in the world from January 1989 up until the March 2015 rating list, when she was overtaken by Chinese player Hou Yifan; she was the No. 1 again in the August 2015 women's rating list, in her last appearance in the FIDE World Rankings.

The theory of László Polgar shows that all usually innate types of intellect, described by Gardner, can belong practically (in 80 % cases) to each person, and therefore some of these types can be developed, and some cannot. Based on this theory we can assume that the innate human intellect having at the beginning elements of all described above types of intellects is capable to be developed one or several types of them in accordance with conditions and methods of nurture and training.

Described above theories have their supporters and opponents. The debate about which theory is best still does not subside. Therefore, we will not discuss them.

We can compare the intellect with a diamond. A brilliant is a diamond cut in a particular form with numerous facets to have exceptional brilliance. Usually value and price of brilliant depends not only on its size, color and cleanliness, but also on quality of cut-faceting. We often think of a brilliant's cut as shape, but a brilliant's cut grade is really about how well its facets interact with light. As the brilliants the gifted people must be by the best way adapted to the application of their abilities and knowledge where they can be claimed. There is no rule without exception. Some geniuses like as Leonardo Da Vinci or William Shakespeare are not made or created, they were born.

It is very important to bear in mind that as the brilliants in darkness can be equaled cheap stones, the gifted people can be equaled talentless in the wrong place and in the wrong time.

1.4. Intellect in Nature

For many years, it was believed that among the inhabitants of the earth only man able to think and make decisions, only man has intellect. The existence of reasonable behavior observed in the living world, considered as a manifestation of instincts, where latter is the inherent inclination of a living organism towards a particular complex behavior. Among the most vital instincts were: foraging instinct, the instinct of reproduction, the protective instinct of self-preservation, the migratory instinct.

The conventional concept of instinct, which determines the behavior of the animals, doesn't allow to correctly explain the behavior of animals in extreme situations for them. If anyone does not believe this, let them try to create an algorithm for the hunting of any mammal predator on those creatures who do not want to be eaten, different for the lion and antelope, the fox and mouse; or algorithms for finding food for herbivorous animal or bird in a strange place.

Cleve Backster, Jr. best known for his experiments with plants using a polygraph instrument, developed his theory of "primary perception" where he claimed that plants "feel pain" and have extrasensory perception (ESP).

In 1966, Cleve Backster attached polygraph electrodes to a Dracaena cane plant, a houseplant that he kept in his office. To his astonishment, Backster found that human intentions, thoughts and emotions caused reactions in the plant, which could be recorded by a polygraph instrument (popularly referred to as a lie detector test). He coined the plants' sensitivity to thoughts "Primary Perception", and published his findings from the experiments in the International Journal of Parapsychology in 1968.

In 2000, Toshiyuki Nakagaki, the professor at Future University Hakodate, Northern Japan, together with a group of researchers has discovered that small mushroom can solve the problem, that requires serious mental effort. Was taken by a tiny piece of yellow mold fungus Physarumpolycephalum and placed at the entrance of a small maze - a 30-centimeter copy of the labyrinth which is usually used to test the intellect and memory in mice. At the other end of the maze researchers put a sugar cube (Jabr Ferris, 2012).

Under natural conditions, this a monocellular being is growing under natural conditions on the leaves and stones. If it from a distance could smell the sugar, mushroom began to send his sprouts in the search. In the experiment, this brainless organism bifurcates at every intersection of the maze, and those who fell into a dead end, turns around and starts to look for the way to other destinations. Within a few hours of mushroom filled the aisles of the maze and at the end of the same day, one of them found their way to the sugar.

After this was taken a little piece of gossamer fungus, which participated in the first experiment, and placed at the entrance of the same labyrinth, also with a sugar cube on the other end. The result was unusual. In the first moment of gossamer was divided into two: one thin and precise process of paved its way directly to the sugar without any extra turn. The second cobwebs climbed the wall of the maze and labyrinth crossed in a straight line across the ceiling, straight to the goal. Mushroom gossamer did not only remember the road but also changed the rules of the game. The experience was repeated again and again with different mazes. (Toshiyuki Nakagaki, 2000). For comparison human mind and this mold mind, since ancient times, people have used a ball of thread to exit an unknown maze.

Let's see if there is kind of mind and intellect in some insects. Ants form staggeringly complicated groups, organized in a hierarchical manner. They use the division of labor. They fight and pay prisoners to "slavery." In some species, there is even subsistence agriculture: they breed edible mushrooms and herding aphids secreting a sweet substance that the ants eat.

The brain of an ant consists of only 250,000 neurons. For this indicator, they are worse not only bees but even cockroaches that have a million neurons. Through research an American biologist, Edward O. Wilson ants transmit information to each other with the help of odorous chemicals - pheromones.

In 1991, the features of the behavior of ants attracted the attention of the Belgian mathematician Marco Dorigo and his colleagues. Observation and analysis of real ants behavior was a source of inspiration for the design of novel algorithms for the solution of optimization and distributed control problems. Ant Colony Optimization theory and some algorithms of optimization are described in his book (Dorigo, 2004).

As researchers found, birds have the intellect also. A speed of perception and reproduction of the sounds of birds in the tens of times higher than that of any other kind animals. If you listen to the birds singing in the slow embodiment, the amazing beauty of the melody is heard. It turns out that each bird in its own way is a composer. At the sounds of the birds present the same tunes that were invented by mankind. It can be assumed, through their trills and singing in different keys, bird talks, and some people can define in the tone of birds singing their mood (Narby, 2006).

The family of hawks is considered as the most intelligent large birds. Some of them use not the beak to break eggs of other birds, but a stone previously firmly clamped it in its beak. It was also found that, quite often, they put the egg on a hard surface, and then with a sniper's precision throw a stone at him, while continuing to soar in the sky.

A crow is another example of the smart bird. For the experiment when in a transparent bottle have been put pieces of food, a crow could not get food, then scientists have put next to bottle a number of fragments of wire, nails and all kinds of small items. Crow after curiously inspected the pile of things, grabbed a piece of wire and made a hook, which deftly picked up food. How much intellect must have birds to almost instantly figure out what to do? (Ackerman, 2016).

In Africa, land mines kill thousands every year, but African pouched rats have been trained to save people lives by sniffing out the scent of explosives.

At January 2015, many newspapers reprinted interesting information from American city Seattle about a black Labrador named Eclipse who if wants to get to the dog park, take a bus alone. When her owner takes too long to finish his cigarette, and their bus arrives, she climbs aboard solo and rides to her stop — to the delight of fellow Seattle bus passengers. The dog and her owner, Jeff Young, live right near a bus stop. In Young's words, "She's a bus-riding, sidewalk-walking dog." Young says his dog sometimes gets on the bus without him, and he catches up with her at the dog park three or four stops away.

Swiss scientists who conducted the research abilities of animals have found that the level of intellect of dolphins is on the second place after the man. Elephants were third and apes took only fourth place.

Dolphin brain stores a large amount of information obtained in different ways. But in addition to its accumulation, dolphins are able to use it. In many situations, they use research approach. They quickly assessed the situation and made fit for their behavior.

Dolphins learn to execute commands sometimes even faster than dogs. Dolphin is enough to show the trick 2-3 times, and he could easily repeat it. In addition, dolphins show creativity. Dolphins have a distinct personality, a strong sense of his "I", and can think about the future.

Dolphins - are social animals, living in packs. American scientists have concluded that each dolphin in the pack has his name. Dolphins have a distinct personality, a strong sense of his "I", and can think about the future.

Practitioner of Cranio sacral therapy and founder Upledger Institute International John Upledger in his book "Cell Talk. Transmitting Mind into DNA" described how dolphins assisted him in healing patients with serious diseases by contacts and distantly (Upledger, 2010).

Chimps display a remarkable range of behavior and talent. They make and use simple tools, hunt in groups and engage in aggressive, violent acts. They are social creatures that appear to be capable of empathy, altruism, self-awareness, cooperation in problem-solving and learning through example and experience. Chimps even outperform humans in some memory tasks (Wilford, 2007).

Drs. Allen and Beatrix Gardner, researchers from the University of Nevada could teach a chimpanzee Washoe not only to brush her teeth, dress herself, and play games, but and use American Sign Language (ASL) to communicate with human. Washoe even became a teacher – she taught some 50 ASL signs to her adopted son, Loulis, without any assistance from researchers.

Many people who have owned a pet will swear that their dog or cat or other animal has exhibited some kind of behavior they just can't explain. How does a dog know when its owner is returning home at an unexpected time? How do cats know when it is time to go to the vet, even before the cat carrier comes out? How do horses find their way back to the stable over completely unfamiliar terrain? And how can some pets predict that their owners are about to have an epileptic fit?

These intriguing questions about animal behavior convinced world-renowned biologist Rupert Sheldrake that the very animals who are closest to us have much to teach us about biology, nature, and consciousness (Sheldrake, 2011).

1.5. Intellect (intelligence), Mind and Operating system

When two dozen prominent theorists were asked to define intelligence, they gave two dozen somewhat different definitions (Sternberg & Detterman, 1986).

According to analysis of American psychologist Prof. Linda Gottfredson, amid many theories and definitions, we can find a broader definition of intelligence with which 52 prominent researchers on intelligence agreed: "Intelligence is a very general capability that, among other things, involves the ability to reason, plan, solve problems, think abstractly, comprehend complex ideas, learn quickly and learn from experience. It is not merely book learning, a narrow academic skill, or test-taking smarts. Rather, it reflects a broader and deeper capability for comprehending our surroundings—'catching on', 'making sense' of things, or 'figuring out' what to do. Intelligence, so defined, can be measured, and intelligence tests measure it well" (Gottfredson, 1997).

Why is there such diversity of theories and hypothesizes about how the intelligence and mind works? Why is there so much complication? Perhaps it is because each of scientists only sees part of the picture from point of view of his science: biology, physiology, psychology and so on.

Many of us may remember the old Indian parable about the six blind men and the elephant. Six blind men, having never been near an elephant, suddenly had an opportunity to touch one.

"Hey, this is a pillar," said the first man who touched his leg.

"Oh, no! It is like a rope," argued the second after touching the tail.

"Oh, no! It is like a thick branch of a tree," the third man spouted after touching the trunk.

"It is like a big hand fan" said the fourth man feeling the ear.

"It is like a huge wall," sounded the fifth man who touched a side of the elephant.

"It is like a sharp spear," said the sixth man who touched one of his tusks.

They began to argue about the elephant and every one of them insisted that he was right. At this time, the elephant's owner was awakened by the noise and came out to quiet the blind men. After each blind man explained his case, the owner responded: "The elephant is a very large animal. Its side is like a wall. Its trunk is like a snake. Its tusks are like spears. Its legs are like trees. Its ears are like fans. Its tail is like a rope. Therefore, you are all right. However, you are all wrong too because each of you touched only one part of the animal. To know what an elephant is really like, you must put all those parts together." (Backstein, 1992).

The same is with intelligence. It is not easy to find a right definition of intelligence because this is a systemic concept, i.e., it is connected with the description of the complex systemic functions from the positions of not only psychology but also other sciences: biology, genetics, physiology, neurology and other disciplines including mathematics, physics and chemistry.

According to Mariam -Webster dictionary, "intelligence" is:

a. The ability to learn or understand things or to deal with new or difficult situations,
b. Secret information that a government collects about an enemy or possible enemy; also a government organization that collects such information. Here and below we will consider only the first meaning of the term.

In my opinion, to exclude dualism in the definition of the "intelligence" more conveniently use the synonym of this term "intellect". The latter like as its synonym derives from the Latin word "intellectus", which means comprehension, understanding, meaning, reason, discerning.

Many scientists are sure, that a man by his anatomy, physiology and psychology is the most sophisticated system in the world. Based on the human study were created computers, intelligent control systems and robots, for example: automobile engine controllers, industrial intellectual control systems, space robots, automatic and automated production lines and plants, and so on.

Vice versa, creation and development of computers and systems of artificial intelligence made it possible to understand much better as the brain and mind work. And what is more, in philosophy, psychology, cybernetics, and studies on artificial intelligence appeared an opinion, which emphasizes the analogies between the functioning of human brain and computers systems. According to the extreme version of this view, the brain is compared with a computer, and consciousness — with a computer interface and programs.

Let consider a basic organization of the brain from point of view a Control theory. In computer system analogy, the brain equals the Central Processing Unit (CPU) of the body, which performs most of the processing inside a human body. The human brain as super complex system can be divided in to following basic functional subsystems (call them units). They are:

- Input/output Interface Units
- Operating system Unit
- Memory Unit
- Body Control Unit
- Psychological Control Unit.

Input-output Interface Units provide communication links between brain system and sensors of human body for input information from sensor organs and speaking, gestures, mimic and other subsystem for information output.

Idea of existence in our brain the Internal operating system a long time hangs in the air and used in some works (Kancharla, 2014, Socki, 2019, etc.). This idea has some roots in the computation theory of mind which was proposed in its modern form by Hilary Putnam, and developed by the MIT philosopher and cognitive scientist Jerry Fodor (Putman, 1961; Fodor, 1983).

Term "operating system" is used in the computer science for description of an integrated set of specialized programs used to manage overall resources and operations of the computer. The three most common operating systems for personal computers are Microsoft Windows, Mac OS X, and Linux.

For different types of control system are used real-time operating system like as VxWorks, OS Lynx, Wind River Linus, and so on. What means real-time? The information processing by some control system must correspond with real environment as in real life, i.e. the scale of time of the controlling system and medium controlled by it must coincide. By this reason, such type of operating system named the real- time operating system.

The Operating system Unit generates the key commands (for executing Operating functions) to run the human. The mind executes these commands at a lower level under the control of two following functional subsystems of brain: Body Control Unit (BCU) and Psychological Control Unit (PCU).

A central place in human physiology takes the concept of homeostasis, which is the relative constancy of the dynamic parameters of the internal environment, as well as the stability of the basic physiological functions and processes in the human body. Homeostasis is a prerequisite for the normal functioning of the body, and cannot exists without the special human body control systems or subsystems, because when these systems cannot support of the homeostatic condition, an individual falls into a state of discomfort or illness. American physiologist Walter B. Cannon in 1932 in his book "The Wisdom of the Body" proposed this term as the name for "the coordinated physiological processes, which supports the majority of the steady states of organism" (Cannon, 1932).

In literature to describe system which ensure homeostasis are uses many terms, such as effortful control, willpower, executive control, self-discipline, self-regulation, self-control (Bukowsky, 2014) and so on. Providing a state of homeostasis is the main task of controlling the body with the help of a Body Control Unit.

Practically all functions of the Body Control Unit are not accessible to consciousness. Obviously BCU is also a part of the autonomic nervous system (the vegetative nervous system), that acts largely unconsciously and keeps on normal level such functions as the regulation of body temperature at 98.6 degrees Fahrenheit, heart rate, regular digestion, respiratory rate, urination, etc. It is not easy, because it maintains a balance among the hundreds of chemicals in billions of cells so that all organs and body systems work in complete harmony most of the time. Only well-developed control system can do it.

The other functional subsystem of the brain Psychological Control Unit (PCU) provides a person's abilities for the regulation of behavioral, emotional, and attentional impulses in the presence of the arising pleasant temptations or the appearing external influences. Using PCU people regulate thoughts, trying to concentrate or removing from thought stream. By the same way, they use to regulate emotions and even moods. They may regulate different kind of impulses of wish, temptations, and so on. Finally, they can regulate any performances, changing accuracy, volume of used factor and speed or time of execution. It can be assumed, that PCU mostly locate in conscious part of mind, mentioned below.

Hypothesis of existence of BCU and PCU corresponds with discovery of Dr. Iosipovic from New York University and his colleagues, assumed that the cerebral cortex divided into two interconnected systems: an "extrinsic" system that respond more to external stimuli and tasks, and an "intrinsic" system composed of brain areas that respond less to external stimuli and tasks (Iosipovic, 2012).

One of the proofs of the existence of PCU can be a mental illness like dissociative identity disorder (DID) described by the American writer and producer Sidney Sheldon in his novel "Tell Me Your Dreams". DID, according to the definition, used by NAMI (National Alliance of Mental Illness), previously referred to as multiple personality disorder, is a dissociative disorder involving a disturbance of identity in which two or more separate and distinct personality states (or identities) control an individual's behavior at different times. When under the control of one identity, a person is usually unable to remember some of the events that occurred while other personalities were in control. The different identities, referred to as alters, may exhibit differences in speech, mannerisms, attitudes, thoughts and gender orientation. The alters may even present physical differences, such as allergies, right-or-left handedness or the need for eyeglass prescriptions. These differences between alters are often very striking.

A person living with DID may have as few as two alters (i.e. two PCU) or as many as 100. The average number is about 10. Often alters are stable over time, continuing to play specific roles in the person's life for years. Some alters may harbor aggressive tendencies, directed toward individuals in the person's environment or toward other alters within the person.

William Stanley Milligan, known as Billy Milligan, was the subject of a highly publicized court case in Ohio in the late 1970s. After having committed several felonies including armed robbery, he was arrested for three rapes on the Ohio State University campus. In the course of preparing his defense, psychologists diagnosed Milligan with multiple personality disorder. His lawyers pleaded insanity, claiming that two of his alternate personalities committed the crimes without Milligan's being aware of it. He was the first person diagnosed with multiple personality disorder to raise such a defense. Billy Milligan had about 10 personalities. In 1988, after ten years in a psychiatric hospital Milligan was recognized recovered and released.

By brain-as-computer analogy, dissociative identity disorder looks like an embedded similar to computer processors in some parts of brain with some a switch. It means that people with DID must have several Physiological Control Units.

It is possible to assume that for a normal person these two subsystems (BCU and PCU) interconnected for example when organizing the work of muscles in accordance with the tasks of the brain, and they can be parts of one to my knowledge not yet discovered by science multi-level distributed command center of the person.

Memory Units are an essential and very important part of brain. Memorization of all information and ability to fast recall are two main functions of the mind. This information storage does not include kind of brain short-term memory similar RAM (Random Access Memory) used in many computer operating systems, which is the memory structure needed for the operating system itself.

For all I know, the term Body Operating Unit (BOU) was proposed by Veera Raghavaiah Kancharla in his article (2014). The work of this Body Operating Unit had programmed genetically for realization three main strategies of the human life: survival, procreation and pleasure as a reward for the implementation of first two strategies (functions).

Knowing the functional structure of the brain, we now have the opportunity to answer correctly on the questions of what are intelligence and mind for the brain from point of view of theories of control and complex systems.

"Our computer – like brain obviously had to have a pretty wonderful-and wonderfully mysterious- "operating system! "- wrote guru in the field of the brain neuroplasticity Dr. Michael Merzenich in his book "Soft-Wired: How the New Science of Brain Plasticity Can Change Your Life" (2013).

Let's try to find some arguments in favor of hypothesis that Intellect is operating system of the human brain.

For comparison goal in the Table 1.1. is collected main functions of Computer or real-time Operating systems (RTOS) and Intellect.

Table 1.1

Number	Function	Computer or RTOS	Intelligence
1	Ensuring the activity of all subsystems by using languages that are understandable for each subsystem for interaction	It is like a "translator" between the hardware of the computer and its software.	It creates and distributes to all systems of the body generated in the brain commands. Similarly, it collects and analyzes information from all organs and body systems
2	Coordination of all subsystems included in the system	It is able to coordinate in time the actions of all computer systems and programs	It is able to coordinate the actions of all physiological and mental functions of the body
3	Automatic execution of some tasks	The operating system performs some tasks independently and does not even inform its user about this, except in cases where it is impossible to do without human intervention.	Automatically analyzes information from the sense organs and body control system, automatically responds to changes in the external situation, except in cases where it is impossible to do without human psychic

Number	Function	Computer or RTOS	Intelligence
4	Information flows management	The system allows access and management of data and programs available in a computer or a control system	Provides access to existing long-term and short-term memory systems, programs human actions and ensures the execution of these programs
5	Providing multitasking control mode	The operating system allocates each task a certain amount of memory, analyzes the results and switches from one task to another.	Provides multitasking mind control
6	Energy Management	Performs the role of a resource manager.	Regulates the ability of thinking, protects the brain from overwork
7	Ability to learn	Has the ability to add or tune some programs	Allows learning and improving quality most of functions
8	Basic intellectual abilities	Use special programs of artificial intellect	Had a complex programmed genetically programs for realization 3 main strategies of the human life: survival, procreation and pleasure.

If compare the basic features of the human intellect and computer or real-time operating systems, we will notice that they are very similar (7 functions from 8) by functionality with one exception. The main distinction of both analyzed systems is that the computer or real-time operating systems do not have basic intellectual abilities such as creativity, situation comprehension, intuition, and so on. However, if to think about it properly, it can be understandable that the operating system of a living organism should be different from the operating system of any machine, because any living organism should always solve three important tasks: how to survive, how to produce offspring and how to enjoy. Intellectual abilities allow solving these tasks with priority.

Perhaps the human operating system has undergone the most changes in the process of human evolution, and it should have become the smartest and most reliable subsystem of not only the brain but also the entire body. We can assume that thanks to this our civilization was able to conquer all the living space on earth. When people were hunters, they had to be smarter than all the surrounding animals in order to ensure their own food, when people mastered agriculture; they had to understand the laws of the plant world and how to use its fruits. In order to reproduce themselves, people had to learn how to attract attention to them and have a fan through not only usage of jewelry or makeup, but also by use various ways of contacting each other. Practically from here are roots of all intellectual abilities of a person, here are the reasons for the appearance of the arts. Hence, there is also a reason to consider the intellect as the Operating System Unit of a person as a living organism. Somebody can object to me: what about different kind of fools (i.e. people without any intellectual abilities), they should not have any intellect. If we recall that even for fools, there is enough intellect to ensure their survival, reproduction and pleasure. This argument once again confirms the correctness of the proposed hypothesis.

Here are also other not so important distinctions between computer (or real-time operating system) and human operating system. For example, memory in a computer or in computerized system has access by the address, but access to human memory is via content. Function of computer can be changed by changing software application. Today is not exist methods to download in the brain any application, although we cannot deny the existence of psychological methods of influence. Generally, these differences human brain and computer or control system are not important, because human brain function much more complex than computer or control system functionality.

Intellect works with information with and without vocabulary content, which can be encoded into various forms for storage, transmission and interpretation (for example, information may be encoded into a sequence of signs, or transmitted via a sequence of signals).

Unlike a computer, human intellect does not have a graphical user interface (GUI), which lets use special devices (mouse, joystick or touchscreen) to input or output information to or from screen using a combination of graphics and text. However, beside of sense organs we have a kind of input/output interface developed in different degree for different people. For some people, these are encrypted feedback signals in the form of a gut feeling, a shiver, a bad mood, tears, or a bowel sensation. Acupuncture points are also an interface. For people with developed sensitivity, this may be information in the form of pictures or films in a dream or reality, or verbal information communicated by an internal voice.

Therefore, from computer science viewpoint we can assume as a hypothesis that the intellect is a real-time operating system of the human brain with all the ensuing consequences from this definition.

The proposed hypothesis is consistent with the Spearman theory and, in general, with Gardner's theory, only in the latter case Gardner intelligences will be equal intellectual capabilities like as musical, logical-mathematical, interpersonal and so forth, developed to one degree or another.

On the other side, this hypothesis has some common features and differences with the Internal Human Operation System' hypothesis of Norbert Soski. The author of this hypothesis in his book "Only Human: Guide to our internal Human Operating System (iHOS) and Achieving a Better Life" (Soski, 2019) defined this system as "a soul and body software". To explain the operation of its iHOS, Norbert Soski introduces the concepts of two souls inside of human: Primitive Soul and Angelic Soul and corresponding them the control loops.

Ones else important argument in favor of this hypothesis is one of the postulates of the Indian traditional system of medicine Ayurveda according to which one of the root causes of a human disease is a mistake of human intellect ("prajnaparadha").

And finally, here is the last argument. According to studies carried out in recent years in genetics, among the main components of the gene there are elements (codons) that have ambiguous significance, for the correct interpretation of which they require knowledge of the context and some logical, and maybe mathematical operations (Shcherbak, 2003). Proceeding from here, Russian scientists Gariaev (Gariaev, 2009) and Puchko (Puchko,2010) proposed a hypothesis of the existence of biological computers inside cells and even inside of subtle bodies. It is obvious, that a biological computer without units of energy, storage, interface, etc. is above mentioned operating system.

Amid plenty definitions of a word "mind" let choose 2 meanings: 1. A functional system of human, that can think, imagine, feel, remember, take a decision, etc.; 2. Process that includes thinks, feels, wills, perceives, judges, etc. Below we will speak only about the first meaning. Intellect as an operating system works for mind needs. Many scientists equate the concepts of mind as a system and concept of intellect, but in fact, it is wrong, because intellect is a subsystem of the mind.

The assumed above definition of the intellect allows easily answer the following question: what is the difference between mind and intellect?

According to studies of Russian psychologists "mind and intellect have a similar functional structure, which includes thought processes" (Selivanov (Селиванов), 2017). Therefore, the answer is very simple: **mind combines intellect and knowledge, organized in the subsystems of information storage and interchange (Memory Units).**

It does not mean that the mind is a computation system according to the computational theory of mind. It means that intellect supports all functional brain systems. Based on this assumption, the intellect can be considered as the internal functional structure that selects, organizes and transforms information. By analogy with the computer, the intellect processes the information coming to the person from outside and from his body for the purpose of the realization of one or another intention of personality or internal "I".

Obviously that the more intellect or the more knowledge, the better mind. For many persons knowledge in the brain may replace the abilities of the intellect or compensate its disadvantages. Knowledge in any discipline allows a fool in our century looks wiser than some geniuses in last centuries. Very often a person with advanced intellect and having no knowledge cannot have advantages in comparison with the stupid person, who are well educated and know how to use those or other knowledge bases, for example, such as the Wikipedia or Google. In this context we need to remember words of the famous Czech writer Karel Čapek: "One of the greatest troubles of civilization - scientific fool".

Theoretically, all mentally healthy people in terms of the mind could be divided into three main groups with blurred boundaries:

- smart (including geniuses, gifted, talented, and wise people),
- normal people with common sense (like as average Joe or plain Jane),
- fools (here is a set of synonyms like as idiot, dumb, asshole describing different rate of mental retardation from mentally deficient persons to ones who are deficient in judgment, understanding, etc.).

Intellect of a talented person distinguishes from intellect of genius. The philosopher Arthur Schopenhauer said it best:" Talent hits a target no one else can hit: genius hits a target no one else can see".

1.6. Basic functions of intellect, intellectual abilities

Commonly intellect is considered as a set of the certain interconnected mental functions: perception, comprehension, comparison, abstraction, concept formation, judgments, conclusions, and so on.

Using information of previous section, we can consider that intellect comprises the following basic subsystems, which ensure the following functions of human being:

- Basic kernel services of physiological control for the main functions of human body
- Communication with external world, including nature, people and machines
- Linguistic performance (term of Noam Chomsky to describe both the production, and the comprehension of language)
- Brain-body parallel processing organization
- Cognitive control, including attention and concentration abilities, i.e. the support of the channel of perception. Attention is ensured through the blocking not important channels of perception and by emotions.
- Perception that turn into comprehension, which somewhere processed (critically reviewed, analyzed from different perspectives, based on existed knowledge, compared with the existing models or means, based on the chains of association), and after that transferred in new knowledge.
- Generation of desires
- Thinking as generations of thoughts (imaginative, verbal, or image-verbal) in association of events, knowledge, and emotions
- Validity of thinking - to confirm or double check the accuracy of judgments and conclusions; critical thinking – to refuse initiated actions, if they are contrary to the requirements of the task;
- Generation of emergency reactions on external or internal dangerous problem
- Using emotions as the source of checking the correctness (it pleases or not to please) or as the amplifier of activity (in the case of success) or weakening (in the case of failure).
- Generation of new ideas or problem solution (creativity) on the base not only Intuition or instincts, but and logical reasoning, rules of analysis and synthesis
- Storage of information as knowledge base
- Using associative centers for remembering and ideas creations
- Transmission of solution into command for the executive control center located in frontal lobes of the brain, checking results of operations
- Capability for adaptation and external instruction
- Generation and support of willpower
- Task scheduling
- Implication and empathy, empathy – the capacity to share and understand emotional states of others in reference to oneself – plays a critical role in human interpersonal engagement and social interaction.
- Predictions, or anticipations of events
- Curiosity - to know "what is it" in many material respects

The intellect may include also other functions, some of which will be described below in one form or another.

Generation of new ideas or problem solution is one of the most important functions of intellect. Quality of this function defines very important characteristic of mind – mind flexibility realized by definition of Eduard de Bono in the diapason from vertical to lateral thinking (de Bono, 2010). Vertical thinking is distinguished as something that is linear, while on the other hand lateral thinking can be seen as non-linear. By another words lateral thinking is non-formal, non- standard or non-dogmatic thinking.

There is definitely a correlation between intellect (G-factor or other functions) and sense of humor. Albert Einstein said that without a sense of humor, his mental abilities would not be so brilliant. According to the researchers, this is because to understand someone else's humor and create their own jokes both cognitive and emotional abilities require. The analysis of data has shown, that for fun-loving people the verbal and nonverbal intellect is higher, due to what they usually are less inclined to negative and aggressive mood.

Intellectual abilities are broad term. We will define this term as a set including all or part specified above functions. These functions may be developed to varying degrees by forming different kind and level of intellect.

Let consider some basic functions which form the most important intellectual abilities.

1.7. Creativity and intuition

Since time immemorial, the problem of creativity and intuition agitates humankind because in a greater degree than other functions they characterize the intellect. During centuries, here were many questions: How do define these? Can creativity and intuition be learned? How to think like a genius? Where does creativity or intuition come from? How can we nurture this by yourself? For today, some of these questions have answers, some – do not have. Let consider these functions from the perspective of their management capabilities.

Creativity and intuition have many similar traits:

- Both are needed for solving decision-making problems,
- Both are mostly congenital
- Are fundamentally interdependent and sometimes interchangeable.
- The specific ways in which both abilities actually influences decisions remain poorly understood
- Both can be developed with help of special training
- Both cannot be measured by traditional methods.

Creativity. Generation of new ideas or problem solution is one of the most important functions of intellect. Creativity is defined as the capacity to produce either new and original ideas, or new approach, or a new method, or a new technique.

Creativity is very complex function, including many components (Gautam, 2012). Usually creativity has been associated with a wide range of behavioral and mental characteristics, including associations between semantically remote ideas and contexts, the application of multiple perspectives, curiosity, flexibility in thought and action, rapid generation of multiple, qualitatively different solutions and answers to problems and questions, tolerance for ambiguity and uncertainty, and unusual uses of familiar objects. Simplest criteria of the creativity are, for example, the number of written pages of the author's text for the writer, or notes for composer, or pictures for the artist, etc.

The entire creative process– from preparation to incubation, illumination, and verification– consists of many interacting cognitive processes (both conscious and unconscious) and emotions. Depending on the stage of the creative process, and what you're actually attempting to create, different brain regions are recruited to handle the task.(https://blogs.scientificamerican.com/beautiful-minds/the-real-neuroscience-of-creativity/)

In 2012, Adobe Systems Inc. has just published a global study on creativity for National economies of 5 developed countries USA, UK, Germany, France and Japan. The research revealed that for these countries 8 in 10 people feel that creativity is key to driving economic grow and nearly two-thirds of respondents feel creativity is valuable to society, in opposite to a striking minority – only 1 in 4 people believe they are living up to their own creative potential (State of Create global benchmark study (www. Adobe.com, 2012))

At present, there are many theories of creativity and many heuristic methods of its development and application in use. Three of the best-known creativity techniques was developed:

- by Alex Osborn's - the author of brainstorming (1941) and founder the Creative Education Foundation,
- by Creative Problem Solving Institute; really by Genrikh Altshuller- author of Theory of Inventive Problem-Solving (TRIZ, 1956);
- by Edward de Bono, who originated the term "lateral thinking", wrote the book "Six Thinking Hats"(1999), and was a proponent of the teaching of thinking as a subject in schools.

A huge contribution in the theory of creativity was made by the American psychologist, Dr. Ellis P. Torrance, creativity experts and author of many books Michael Michalko, business theorists Clayton M. Christensen, just to name a few.

Mentioned in the several sections above (1.7, 5.7, 8.2) creativity is very complex function of intellect, including many components: ability of idea generation, flexibility as ability to generate different categories of idea, originality as a degree of novelty, fluency as total number of ideas per unit of time, elaboration as detailing degree of any idea (Gautam, 2012), ability for visualization of ideas (Muchalko, 2011).

Amid creative people exist people for whom it is enough to create a new idea, in contrary with perfectionists, who need to develop a new idea including all details.

Creative abilities in many disciplines can manifest themselves in different ways and creativity itself can have different names: in industry it is called innovation, in business it is 'entrepreneurship', in mathematics it is often equated with math problem solving, and in music and art it is creativity.

In 60th of last century American psychologists Joy Guilford and Ellis Torrance constructed several psychometric tests to measure creativity. For example, the Torrance Tests of Creative Thinking include some simple tests of divergent thinking and other problem-solving skills, which were scored on:

- Fluency – the ability to produce a number of different ideas,
- Originality – the ability to have a new or novel idea.
- Elaboration – the ability to extend ideas.

The Creativity Achievement Questionnaire, a self-report test that measures creative achievement across 10 domains, was described in 2005 and shown to be reliable and valid when compared to other measures of creativity and to independent evaluation of creative output.

All these tests have advantages and disadvantages that limit their wide application.

A study by George Land reveals that people are naturally creative and as they grow up they can lose this ability. In 1968, George Land conducted a research study to test the creativity of 1,600 children ranging in ages from three-to-five years old who were enrolled in a Head Start program. This was the same creativity test he devised for NASA to help select innovative engineers and scientists. He re-tested the same children at 10 years of age, and again at 15 years of age. The results were astounding. Test results amongst 5 year olds: 98%, amongst 10 year olds: 30%, amongst 15 year olds: 12%, Same test given to 280,000 adults: 2% (Land, 1993).

Today many scientists believe, that a skill of creativity has several parts- one from birth, other as results of training process by experimenting, exploring, questioning assumptions, using imagination and synthesis of information, and third by using special tools based on TRIZ and similar techniques.

Training process and tools based on TRIZ and similar techniques can be successful if to have a method of measurement of creativity described below.

Intuition. According to Encyclopedia Britannica, "Intuition is the power of obtaining knowledge that cannot be acquired either by inference or observation, by reason or experience. Usually intuition referred in science as "paranormal abilities" or "Extra Sensory Perception" (ESP) for ability to obtain information about past, present or future events without using the "normal" five senses. Albert Einstein considered intuition one of the main components in making discoveries. Although many people and world-famous geniuses have testified that they have given solution to difficult problems by following their intuition, official science did not consider the existence of intuition reliable. On the other hand, there are many scientifically substantiated facts that speak out in favor of intuition existence for many people. W.E. Cox, in a well-known study (1956), analyzed the number of tickets sold for 28 passenger trains that crashed between 1950 and 1955. He found that the trains that crashed always had fewer people than similar trains on the same day of the previous week. The situation has changed in recent years. The study of team of Joel Pearson, an associate professor of psychology and director of the Science of Innovation Lab at the University of New South Wales in Australia showed that intuition does, indeed, exist and that researchers can measure it. They found evidence that people can use their intuition to make faster, more accurate and more confident decisions (Lulfityanto, 2016).

Today, more and more scientists and philosophers suggest that the logical, materialistic thinking does not provide us successful, happy and safe life. Life experience demonstrate us very often, that man with intuition always gets an advantage over other people in any case. Professor John Michalasky of the New Jersey Institute of Technology conducted intuition tests on the group of CEOs. He discovered that: out of 100% of CEOs with high intuition scores, 81% doubled their business in 5 years, but out of 100 % of CEOs with low intuition scores, only 25% managed to double their business in 5 years. These results proved a strong correlation between intuition and CEO ability, meaning that "gut feeling" may play an important role in making successful decisions.

At present there are many hypotheses explaining what intuition is, amid them we can distinguish four main ones:

- It is an unconscious human experience that accumulates throughout life, "written" in the subconscious and suddenly manifested as insight (illumination)
- Intuition is a normal logic. The process of finding a solution in the consciousness or subconscious, in which intermediate logical sequence lost, ie, the result appears as if by itself
- Intuition is knowledge derived from one or more egregores ("group mind"), or from the knowledge base that Carl Jung called the collective unconscious, which represents a kind of "archive", which stores the experience of mankind
- It is knowledge derived from the Universal (cosmic) Mind.

For some people intuition is that guiding "inner voice" which always knows what is ultimately best for any individual, in all situations. Today more and more regular people are awakening their intuition to achieve tremendous success in all areas of their lives, including wealth, relationship, career, and much more. Intuition comes in several forms:

- sudden flash of insight, visual or auditory
- predictive dream
- spinal shiver of recognition as something is occurring or told to you
- sense of knowing something already
- gut feelings
- snapshot image of a future scene or event
- knowledge, perspective or understanding divined from tools which respond to the subconscious mind (Hetherington,2013).

It is undesirable to confuse intuition and insight, because in contrast to intuition, insight involves a period of incubation of the problem before the recognition of a solution (McCrea, 2010).

Dr. Raymond Trevor Bradley, who spent a lot of time in research to understand fundamental processes of communication and group function in social collectives, described some features of intuition, which probably can be useful for the future a quantum-holographic theory of intuition:

- the transmission of intuitive information between persons is provided without electromagnetic waves;
- quality of the transmission of intuitive information does not decrease over distance and is not affected by location;
- intuitive foreknowledge about a future event is not limited by the normal causal relations of time;

- intuitive perception is related to the degree of emotionality of a person;
- both the heart and the brain (and possibly other bodily systems) are involved in intuitive perception of future events (Bradley Raymond Trevor, 2006).

It is known, that the intuition can be enriched by self-training and learning to perceive and understand the information passed by our emotions, or organs, for example, from the heart, a skin and so forth.

Below we describe a methods of measurement and optimization of intuition.

1.8. Perception and comprehension of information in the Mind

For assessment of the level of development, the ability to understand (perception and comprehension) sometimes more important than the ability to find solutions for complex problem.

A function of human perception is a holistic reflection of real world objects, including the sensation of his own body with the help of the senses. The process of perception occurs in close association with other mental processes: thinking, speech, memory, attention, will (to organize the process of perception), directed motivation, emotions.

Perception is a complex process that involves monitoring skills, decoding (different for understanding speech or drawing), a dictionary or knowledge in the needed areas), the skills of building a reflection model of the object with a certain correction of errors in the perception of information, saving results in mind.

Usually objects and phenomena of the surrounding world are perceived by our senses not as separate objects, but are integrated into complete images.

Depending on the purpose of human activity are allocated involuntary and voluntary perception. Involuntary perception may be caused by the peculiarities of the surrounding objects, and these objects matched the interests and needs of the individual. Arbitrary perception involves goal setting, application willpower, a deliberate choice of the object of perception.

Comprehension is the action or capability of understanding the meaning or importance of the information perceived via special channels; video, audio, tactile and so on.

Comprehension is a complex cognitive process of the brain that searches relations between perceived information from a given object or its attributes and other objects and their attributes in the long-term memory, and establishes a representational model for the object or attribute by connecting it to appropriate clusters of memory. It is obvious that before any information can be possessed and processed, it should be comprehended properly.

The ability of perception and comprehension can be varied depending on:

- Types of perception. Everyone healthy person owns all kinds of simple perception like as visual, auditory, tactile, but sometimes one of these systems generally developed better than others. One person understands better by reading, other by hearing, next with help different kinds of pictures
- speed of thinking, flexibility, rigidity and inertia of thinking, ability to change the point of view

- Ability to learn something, like a new language, a new discipline, or art,
- Ability to communicate and interact with people.

There are people who have a lot of stereotypes in thinking or behavior, and people who have a minimum amount of these stereotypes or not at all. In this case, the stereotype is a preconceived attitude towards the events, actions, deeds, etc. All varieties of patterns can be divided into three broad categories:

- Stereotypes of response
- Stereotypes of perception and evaluation
- Stereotypes of thinking.

Some of these stereotypes based on some opinion of a respectful teacher, doctrine, or position taken on faith in the absolute truth, unfailing in all circumstances, i.e. what is called dogma. Dogmatism can manifest itself in various forms.

1.9. Subsystem of information storage and interchange

Without the involvement of memory, most of the mental processes cannot be realized, without memory the functioning a full-fledged personality is impossible also. Memorization and recall are two amid main functions of the intellect.

Memory units in one or another part of the brain may influence on the many functions of intellect: communication with external world, perception, generation of new ideas or problem solution, to name just a few.

Until now the intellects subsystem of information storage and interchange (ISI) is very complex, and still has many unknown in the field of structure, functions and its capacity.

The brain's exact storage capacity for memories is difficult to calculate, because nobody knows how actually to measure it. There are a few theories. For example, for one of them human brain has the capacity to store almost unlimited amounts of information, for other one storage capacity equal 100 terabytes of info, for next other – about 2,5 petabytes.

ISI is multilevel system located not only in the brain, but and in the cells of the other organs. It called a memory, but in reality, it must have for normal functionality such functions as attention, concentration, filtration of information, fast access to information and so on. Today, it is not known how information is transmitted not only within brain parts and the body, but also between 3 levels of mind.

Thanks to studies of many scientists now are exist many models of memory work. Here are only very popular two.

According to multi-store model of Richard Atkinson and Richard Shiffrin (Atkinson,1968) information in the brain stores information in one of 3 states of memory: the sensory, short-term and long-term stores:

- Sensory is a structure in which sensory information is stored for a short time (less than 0.5 seconds for visual information and 2 seconds for sound)

- Short-term - a structure in which a small amount of information(7+/- 2 pieces) is stored for less than 20(30) sec
- Long-term - the structure of a large volume, which can store information during a long time.

In 1974, Alan Baddeley and Graham Hitch proposed a model of working memory (Baddeley, 2015). According to their hypothesis, working memory splits short-term memory into multiple components: a visual-spatial sketch, a phonological loop, episodic buffer and central executive element.

Long-term memory has three main functions: acquisition, consolidation, and recall. Acquisition occurs when we are introduced to new information; consolidation involves the stabilization of this information, and recall is the ability to access the information later on. Since the early neurological work of Karl Lashley and Wilder Penfield in the 1950s and 1960s, it has become clear that long-term memories are not stored in just one part of the brain, but are widely distributed throughout the cortex.

Short-term memory holds a small amount of information (typically around 7 items or even less) in mind in an active, readily available state for a short period (typically from 10 to 15 seconds, or sometimes up to a minute). Short-term memory should be distinguished from working memory, which refers to structures and processes used for temporarily storing and manipulating information.

Working memory acts as a form of mental workspace, providing like as a stage for thoughts. It is usually assumed to be supported by attention ability, and to be linked with resources within short-term and long- term memory (Miyake, 1999, Chun, 2007)).

Nicola Tesla had phenomenal working memory, he could keep in this memory a huge volume of information (Tesla, 2014).

Short-term memory can go in the long-term memory: for example, the phone number if we constantly use is fixed in the long-term memory.

Dr. Endel Tulving, an Estonian-Canadian experimental psychologist and cognitive neuroscientist, suggested an existence of episodic, semantic and procedural memory. Episodic memory is responsible for storing information about events, semantic memory- for storing our knowledges, procedural memory - for knowing how to do things, i.e. memory of motor skills (Tulving, 1972).

Very seldom, are born people with extremely rare kinds of memory called an eidetic memory and a photographic memory. Eidetic memory referring to the ability to recollect an image so vividly that it appears to be real. Photographic memory referring to the ability to recall page or picture in great details. The history knows many people with outstanding eidetic memory:

- Antonio Magliabechi - Italian bibliophile and librarian of Florence, who knew to the end of his life by heart almost all the books that he had read from own library (40,000 books and 10,000 manuscripts).
- Arturo Toscanini, Italian conductor. He memorized about 100 operas and over 100 symphonies. Toscanini could memorize poems after a single reading and could pick out on the piano the songs and arias he had heard people singing

- John von Neumann - Hungarian mathematician, was able to memorize a column of the phone book at a single glance. He could recite exactly word for word any books he had read, including page numbers and footnotes—even those of books he had read decades earlier
- Laurence Kim Peek - an American savant- was able to scan two pages of any book, one with left eye and other with right eye, and all those information got stored in his memory. According to an article in The Times newspaper, he could accurately recall the contents of at least 12,000 books.

An autistic British architectural artist Stephen Wiltshire is known for his ability to draw from memory a landscape after seeing it just once. He is an example of man with photographic memory. In May 2005 Wiltshire produced his panoramic picture of Tokyo on a 32.8-foot-long (10.0 m) canvas within seven days following a helicopter ride over the city. Since then he has drawn Rome, Hong Kong, Frankfurt, Madrid, Dubai, Jerusalem and London on giant canvasses.

More often there are people with an emotional or flashbulb memory who have unusual form of eidetic memory to remember as well as recall any specific personal events or trivial details, including a date, the weather, what people wore on that day, from their past, almost in an organized manner.

Associative memory is a content-addressable structure that allows the recall of data, based on the degree of similarity between the input pattern and the patterns stored in memory. Associative centers are parts of the associative memory that usually is defined as the ability to learn and remember the relationship between unrelated items. Associative memory is very useful for creative goals. Intensive language studies in a formal setting may improve associative memory performance. In any training is very important the dosage of practice (Mårtensson. 2011). Associative memory can be such types as emotional, auditory, visual, gustatory, and combined.

Unfortunately a long-term memory is not constant. In 1885, Hermann Ebbinghaus published the hypothesis of the exponential nature of forgetting (Thorne, 2004). A typical graph of the forgetting curve purports to show that humans tend to halve their memory of newly learned knowledge in a matter of days or weeks unless they consciously review the learned material. The speed with which we forget any information depends on a number of different factors:

- How difficult is the learned material
- How easy is it to relate the information with knowledge base of person
- How is the information represented
- Under which condition are a person learning the material.

Famous American psychiatrist, one of the founders of the field of transpersonal psychology and a researcher into the use of non-ordinary states of consciousness for purposes of exploring, Stanislav Grof suggested hypotheses of COEX principle- systems of COndensed EXperience. A COEX system can be defined as 'a specific constellation of memories (and related fantasies) from different life periods of the individual. The memories belonging to a particular COEX system have a similar basic theme or contain similar elements and are associated with a strong emotional charge of the same quality. COEXs may be positive or negative depending on whether or not the emotional experiences were pleasant (Grov, 1973).

Adopting terminology of computer science we can assume that our long-term memory is knowledge base that may make the processing of incoming information from all sensory organs, including products of mind –thoughts and emotions at any given moments.

According to Emmanuel Kant here are two worlds of reality, which he called the nouminal and the phenomenal worlds. The nouminal world is the objective external world. The phenomenal world is the internal perceptual world based on the conscious experience, which is a copy of the external world of objective reality constructed in our brain. (Lehar, 2013). Each person has and remember own phenomenal world not only in the description or representations of facts (which he saw, heard or felt), but and a structured model of this world, that includes all knowledges, which person received about this world during his life. Structured model means that this model divided on the parts having associative links and pointers.

The difference between the phenomenal model of the world and the real world largely depends on the individuality of our senses. To better understand this, let's remember about synesthesia - the ability to unite different sensations into one. These people are not so rare, the estimated occurrence of synesthesia ranges from rarer than one in 20,000 to as prevalent as one in 200 (Palmary, 2006). Of course, people with synesthesia represent the world around them in their brain differently than ordinary people.

Adopting computer science terminology, we can assume that human long-term memory is a complex system which includes the following interconnected parts:

- Knowledge base of events representing facts about world through a person experience
- Analytical knowledge base, which must have an inference engine that can reason about those facts and use rules and other forms of logic to deduce new facts or highlight discrepancies.
- Belief system, which helps to explain everything what we don't know or cannot explain by another way.

A belief system is an ideology or set of principles that helps us to interpret our everyday reality. This could be in the form of religion, political affiliation, philosophy, or spirituality, among many other things. These beliefs are shaped and influenced by a number of different factors. Our knowledge on a certain topic, the way we were raised, and even peer pressure from others can help to create and even change our belief systems. The convictions that come from these systems are a way for us to make sense of the world around us and to define our role within it. For some people their current life situation is the reflection of their belief system. Beliefs get stronger when more evidence supports them and they become weaker when something appears to be contradicting with them.

These integral knowledge base must have control subsystem allowing fast remember and recall any data. The all knowledge bases and belief system have to include the most important parts of our personality- traits, goals, values, beliefs, etc.

All parts of this integral knowledge base is not a static collection of information, but a dynamic resource that may have a capacity to learn.

Logical analysis of intellect subsystems shows, that memory units in one or another part may affect on the following functions of intellect:

- Communication with external world
- Perception of information
- Generations of thoughts
- Generation of new ideas or problem solution on the base not only Intuition or instincts, but and logical reasoning, rules of analysis and synthesis
- Task scheduling,
- Etc.

Some from main components of memory are attention and cognitive filters.

Attention is controllable, selective, and limited. Attention is the most essential mental resource for any organism needed to select relevant information. It determines which aspects of the environment we deal with, and various automatic, subconscious processes to make the correct choice about what gets passed through to our conscious awareness. The human brain has evolved to hide from us those things we are not paying attention to. In other words, we often have a cognitive blind spot: We do not know what we are missing because our brain can completely ignore things that are not its priority at the moment— even if they are right in front of our eyes. Cognitive psychologists have called this blind spot various names, including inattentional blindness.

Today are created many hypotheses and theories about how attention may control the flow of thoughts in consciousness. Here are the idea of bottleneck and filters that restricts the rate of flow (model of Donald Broadbent (1958), multilevel selection of Anthony and Diana Deutsch and Donald Norman model (1968), Daniel Kahneman's attentional resource theories (Kahneman,1973), and so on).).

Two of the most crucial principles used by the attentional filter are change and importance. The brain is exquisite change detector. The second principle, importance, can also let information pass through. Here is what is personally important.

Cognitive control system together with attention filter allows shorting a volume of information in our consciousness, and increasing a speed of access to our mind.

Attention can be exogenous when a trigger comes from outside the brain, or endogenous, when a trigger located inside of brain (Hopfinge, 2012). It is possible to distinguish sensual and intellectual attention. The first usually is connected with emotions and selective work of sense organs, and the second — with concentration and an orientation of thoughts. A sensual attention has a place when in the center of consciousness there is any sensual impression; intellectual attention is needed when an object of interest is the content of one or several thoughts. On the other side, attention can be involuntary as automatic reaction on changing something around and voluntary. Psychological studies of attention, as a rule, devoted the study of voluntary attention - its volume, sustainability, distribution and to be switched ability. The volume of attention is measured by quantity of objects, which are perceived at the same time. Sustainability of attention means preservation of attention regardless of external influences. Distribution of attention is expressed in the ability to perform simultaneously multiple actions or monitor many processes. Ability of attention to be fast switched for the other action (a kind of cognitive flexibility) is very important in real life, when we needed to participate in the several projects practically at the same time. It is known that this ability depends on temperament type of the person.

Memory and attention are the interconnected functions of a person thinking. Without skill to concentrate on object of memorizing and emotional feedback from studying this object it is impossible to keep in mind a necessary material qualitatively. As a matter of fact, attentiveness and interest of the person are primary factors for his short-term memory. Significant impact on the success of the remembering provides health, vitality, the mood of the person.

It is known, that if a person motivates his efforts towards a particular goal, his steady mental concentration and strong willpower would accelerate specific activities in his brain and give rise to manifestations of mental potentials of particular types. Each one of us can achieve such targets by internal desire, dedicated attempts, perseverance and focusing of the mind. With help of attention and willpower we can stop a stream a thoughts to work with one thought or group of connected thoughts.

1.10. Willpower

A driving force for PSCU (see section 1.5) is based on the willpower, the supreme psychic functions giving a person's ability to make informed decisions and implement them. Not without reason still William Shakespeare wrote in his tragedy Othello about it:" Our bodies are our gardens, to the which our wills are gardeners…"

The will and willpower as ability to self-organization and self-government in developments and realization of the purposes is necessary for ensuring work of intellect in full volume. It manifested in overcoming the obstacles encountered on the way to achieve any goal consciously. Persistence is a very useful way of exerting willpower to get what a man wants. Willpower and volition are colloquial and scientific terms for the same process.

Harry Carpenter suggested a very interesting approach to understanding what is it willpower. He wrote, "So, the conscious mind has the will and the subconscious mind has the power. When the conscious mind and subconscious mind are in harmony, you have the willpower. You are single- minded. But when the conscious mind and subconscious mind are in conflict, there is no willpower. You are "double-minded." Your conscious mind cannot directly overpower your subconscious mind and "will" it to do something (Carpenter, 2011).

Willpower is the inner manager in order to take decision for any problem, and execute any task until it will be done regardless of outer and inner resistance, difficulties and discomfort. It helps human beings to overcome temptations, negative habits, and laziness and to carry different actions, even when it requires effort and unpleasant experiences. Willpower is able to hold back any desire or impulsive actions that will have effects on human beings in the future (Law, 2013).

According to Kelly McConigal, willpower includes three forces which may be called, respectively: "I want", "I will", and "I won't" (McGonigal, 2012).

This definition can associate with famous "The Serenity Prayer"- the common name for a prayer authored by the American theologian Reinhold Niebuhr. The best-known form this prayer is: "God, grant me the serenity to accept the things I cannot change, the courage to change the things I can, and the wisdom to know the difference." First part of this prayer means "I won't", second part – "I will", and the third part, defined by wisdom, means "I want to do what I can".

The willpower in conscious mind can stimulate the subconscious mind into action, change habits, reverse negative thinking patterns, improve our physical and emotional health, and our conscious mind can even influence our involuntary functions.

Roy Baymeister discovered that willpower has some traits as muscles, because "becomes fatigued from overuse but can also be strengthened over the long term through exercise"(Baymeister, 2011). Now there is a plenty the proven methods that we can use every day to boost your willpower, eliminate instant gratification, and get ready to reach goals and dreams. Meditation can be helpful to create ironclad willpower.

Here is one example what do willpower can do. When teacher Helena Tangiyev said to young Mikhail Baryshnikov (one of the greatest ballet dancers in history) that he needs to grow at least 4 centimeters, otherwise it will never happen to be a soloist in any solid ballet theatre, he was able to stretch themselves on these 4 centimeters, i.e. he brought the body to the needed condition. In the 1975 tragedy struck - Baryshnikov broke his leg. Gypsum, crutches. Five knee operations, the eternal ice after the performance, hobbling gait. He overcame all and again began to dance. PSCU is a crucial element in man's ability to achieve his goals. The level of self-control is defined as a congenital genetic characteristics, culture traditions and psychological skills of a person. Self-control is opposed to impulsiveness — inability to confront personal short-term wishes.

One of the most replicated findings about willpower is that "it seems to be finite value—it means, that is, we only have so much and it runs out (become weak) as we use it." This reason can explain why the same person can have today the strong, but tomorrow a weak willpower. It means that for self- control we need to have a skill to restore willpower (McGonigal, 2012).

About hundred years ago famous American psychologist and psychotherapist Orison Swett Marden, who rightfully considered to be the person changing the lives of millions of people with his books and ideas, which became proponent in the "New Thought Movement" in the late 19th century, wrote:"The way to learn to run is to run, the way to learn to swim is to swim. The way to learn to develop will-power is by the actual exercise of will-power in the business of life" (Marden, 2008).

Self-discipline (or self-control) is the companion of willpower. It endows the stamina to persevere in whatever one does. It bestows the ability to withstand hardships and difficulties, whether physical, emotional or mental. Willpower is a power for a single action. Self-discipline includes a set of rules allowing to achieve the intended goals.

"Self-control also proved to be a better predictor of college grades than the student's IQ (see section 3.2) or SAT score" –wrote Roy Baumeister and John Tierney in their book "Willpower: Rediscovering the Greatest Human Strength" (2011).

PSCU controls the following very important brain functions:

- Censorship of the psyche
- Generation and ranging needs from bodies impulses and temptations
- Executing needs
- Evaluation of any actions effectivity
- Common behavior of person, individually or in society.

The description of all these functions is beyond the scope of this book. We can mention only one thing, important for description of intuition below. PSCU in many cases must act in uncertainties originated from incompleteness of information about environment properties.

1.11. The diversity of intellectual abilities

List of intellectual abilities can include for example such as the ability to listen and hear; to highlight the most important (general and excellent) things; to compare, systematize, and summarize variants, to deduce and substantiate the conclusion; to establish and support any relationships, to perceive and build a chain of judgments, analyze any situation, find the necessary arguments in the discussion and so on.

In the process of human life, the intellectual abilities of a person are not constant: at the beginning of life they develop and maturate during 10-15 years, and at the end of life as a result of aging they may degrade to one degree or another.

The Swiss psychologist J. Piaget identified four main stages of the development of intellect during the period from infancy to adolescence:

- The stage of sensorimotor intellect (from birth to 2 years). Its content is the mastery of the child operations with real objects (toys and things). The child learns to manipulate them, thereby developing "manual" intelligence.
- Stage pre-conceptual intellect (from 2 to 6-7 years). The main achievements of this stage are the mastery of speech and the formation of preconceptions with its help. They are distinguished from concepts by the incompleteness of the ideas embodied in them.
- Stage of specific operations (from 6-7 to 10-11 years). At this stage, which coincides with the beginning of schooling, the child begins to master simple mental operations with objects (classification, serialization, finding correspondences between objects).
- Stage of formal operations (from 10-11years to adolescence). At this stage, the teenager is already mastering logical operations. He learns the reverse operations (addition - subtraction, multiplication-division), thanks to which he masters the ability to track the course of his thinking and check the results obtained (Piaget, 1983).

Diversity of intellect abilities can be covered with help of the simple labels. Any folks can be divided by giftedness into the following groups: average citizens, prodigies, talents and geniuses (Briggs, 2000). Representation of these groups in different nations is different.

Theoretically, intellect as a complex system with many functions can be described with the help of complex mathematical models. Unfortunately, the volume of book and wish to give a description of this system to be understandable as minimum for students of high school don't allow using this approach.

Because the basic functions of intellect and memory described above do not allow to get a complete picture of all intellectual abilities for any person, below we will try to evaluate a diversity of the intellects characteristics with the help of the different classification indicators.

The human intellectual abilities may vary according to:

A. role of personality
B. number of developed talents
C. using the different forms of information (signals, symbols, words, etc.) for perception, storing, comprehension and presenting information

D. diversity of used senses

E. information processing method

F. depending on the so-called creative environment

G. by way of social dependence.

Let us consider these variation.

A. Role of personality

Each person has his own personality formed from birth (including identical twins) and as a result of the processes of education and training. Through nurture and training in family, in educational institutions or in the process of communication with other people, this person acquires a culture that largely determines his behavior in society (Bender, 2016).

In recent decades, psychologists have proposed a number of theories explaining personality. Some, such as Friedman and Rosenman, focus on an individual's observable behavior. Other theories, such as the five-factor model, takes a trait theory approach seeking to understand personality in terms of specific attitudes and types of behavior.

In psychology, the five-factor model (FFM) considers five broad domains that are used to describe human personality (Digman, 1990). The five factors are openness, conscientiousness, extraversion, agreeableness, and neuroticism:

- Openness is a kind of developed curiosity, easy and active perception of art, unusual ideas, new knowledge and experience
- Conscientiousness is a tendency to be organized and dependable, to have self-discipline, act dutifully
- Extraversion means to have assertiveness, energy, positive emotions, surgency, sociability, and so on
- Agreeableness is a trait to be compassionate and cooperative towards others in interpersonal relationships
- Neuroticism means to have a degree of emotional instability.

One of the most researched and supported personality tests is developed by psychologists Katherine Cook Briggs and daughter Isabel Briggs Meyers (The Myers-Briggs Type Indicator (MBTI)). According to MBTI all population can be divided on 16 psychological types, each with their own values around a huge variety of applications, including career choice, relationships, communication, spirituality, conflict resolution, study habits, leadership, relationship needs, just to name a few.

A lot of studies has been devoted to describing the role of individuality in shaping intellectual abilities (Personality and Intelligence, 1994; Maltby, 2017). Intellect and personality have some common features; for example, they both follow a relatively stable pattern throughout the whole individual's life.

Numerous attempts to build a model of the influence of personality traits and temperament of human on the intellect have not finished until now.

It should be obvious that intellect and personal human traits are interdependent. Moreover, this dependence may change as with age, so and with the development of intellectual abilities.

Though the traditional view in psychology is that there is no meaningful relationship between personality and intellect, it's easy to see that this is not true, because positive personality traits should strengthen intellectual abilities, and negative ones - to reduce. Among the positive traits you can specify for example the following:

- Being honest and taking responsibility for actions,
- Adaptability and compatibility that can help you get along with others,
- Compassion and understanding,
- Patience,
- Courage to do what's right in tough situations.

Among the negative traits you can specify for example the following:

- A propensity for lying,
- Being rigid and selfish,.
- Being full of laziness,
- An inability to empathize with others,
- Being quick to anger.

B. According to the number of developed talents

Intellect abilities can be manifested like unilateral or multilateral.

Unilateral ability manifested in any one developed hypertrophic ability, for example, in philosophy, music, drawing or mathematics. Amid people with unilateral abilities were many famous geniuses: Aristotle, Van Gogh, Euler, and others.

Unilateral abilities have also people with Savant syndrome. Savantizm (from fr.savant-"scientist") - a rare condition in which persons with developmental disabilities (including autism) have a so-called "island of genius" - outstanding ability in one or more areas of expertise, contrasting with the general limitations personality.

For example, autistic from the UK Daniel Tammet hardly speaks, does not distinguish between left and right, cannot insert the plug into the outlet, but it is easy to mentally perform complex mathematical calculations, operating with numbers, consisting of more than one hundred characters. He also proved that during a week can learn a completely unfamiliar language.

"I represent the numbers in the form of visual images. They have color, texture, shape, - says Tammet. - Numerical sequences appear in my mind as landscapes. As paintings. In my head, as if there is a universe with its fourth dimension."

People with multilateral abilities usually are geniuses in two or more fields. They are represented by such geniuses as a Persian polymath Ibn Sina (Avicenna) who is regarded as one of the most significant thinkers and writers of the Islamic Golden Age and was not only philosopher, but and physician, naturalist, poet, author of a treatise on the theory of music; Galileo Galilei - an Italian physicist, engineer, astronomer, mathematician and philosopher, which has had a significant impact on the science of his time, Leonardo Da Vinci, who had greatest skills in painting, drawing, mathematics, anatomy, engineering, architecture, etc. People with multilateral intellect usually are much more likely to be found in the ability of the multilateral framework of one branch of art. For example, a German composer of the early classical period Christoph Willibald von Gluck was composer, reformer of opera, singer, cellist, violinist, organist, conductor; a prolific and influential composer of the Classical era Wolfgang Amadeus Mozart was also harpsichordist, violinist, conductor; Romanian musician George Enescu was composer, conductor, violinist, pianist and teacher, and so on. The owners of multilateral artistic abilities manifested themselves in different branches of art: Russian painter Karl Briullov was artist, teacher, cellist, pianist; Italian operatic tenor Enrico Caruso was also a sculptor, film actor, cartoonist; the English actor Charles Spencer Chaplin was also a director, film actor, screenwriter, composer.

C. According to using the different forms of operating information (signals, symbols, words, etc.) for perception, storing, comprehension and presenting information

Depending on the used forms of information encoding the following types of intellect can be distinguished:

- Verbal based intellect - the ability to speech production, including mechanisms responsible for the phonetic (speech sounds), syntax (grammar), semantic (meaning) and pragmatic components of speech (use of speech in different situations).
- Images based intellect - the ability to see in consciousness any object in two or three dimensions
- Music based intellect - the ability for the generation, transmission, and understanding of meanings associated with the sounds, including the mechanisms responsible for the perception of pitch, rhythm and timbre.

Most people have verbal based aptitude. Above mentioned Nicola Tesla had images based aptitude.

People with music based intellect like to transmit essence of the events, moods and feelings through music. Not only composers may have musical intellect, but also the performers, working in the different genres.

D. According to diversity of used senses

By analogy with computer a human body has sense organs like as equipment used to provide data and control signals to an information processing system (intellect). The conventional list of five senses doesn't really give our bodies credit for all of the amazing things they can do. There are a more different things we can sense with the help of special recently discovered indicators (sensory cell types).

Usually, the senses are frequently divided into exteroceptive and interoceptive. First of all, exteroceptive senses include the traditional five: sight, hearing, touch, smell and taste, as well as thermoception (temperature differences) and possibly an additional vibroception - the sensitivity to waves or oscillations, which includes seeing, hearing and touch.

Exteroceptive senses allows also an organism to sense body movement, direction, and acceleration, and to attain and maintain postural equilibrium and balance.

Interoceptive senses are senses that perceive sensations in internal organs. In the special literature you can find references about some of these senses:

- The sense of thirst: This sense more or less allows your body to monitor its hydration level and so your body knows when it should tell you to drink.
- The sense of hunger: This sense allows your body to detect when you need to eat something.
- The times sense: This one is debated as no singular mechanism has been found that allows people to perceive time. However, experimental data have conclusively shown that many people have a startling accurate sense of time.

It makes sense to indicate more about the extraordinary abilities of some people to feel what most people do not sensate. People who experience taste with greater intensity than the rest of the population are called supertasters. These people have the mushroom shaped bumps on the tongue that are covered in taste buds. Of the five types of taste, sweet, salty, bitter, sour, and umami, a supertaster generally finds bitterness to be the most perceptible. People with absolute pitch are capable of identifying and reproducing a tone without needing a known reference. Tetrachromacy is the ability to see light from four distinct sources, for example from the red, green, blue, and distinguish about 100 million colors. Surprisingly some humans like bats are also capable of using echolocation. Synesthesia is a neurological phenomenon in which stimulation of one sensory or cognitive pathway leads to automatic, involuntary experiences in a second sensory or cognitive pathway. The people with profound synesthesia are able, for example, hear or taste colors, feel noises, and smell pictures. In grapheme-color synesthesia (the most-studied form of synesthesia) an individual's perception of numbers and letters is associated with colors.

It must be obvious, that the more senses have a person, and the more quality of information from sensors, the more information he or she can receive. Because for many people a main parts of intellect are perception and comprehension, it means that with increasing number of senses people will have more intellectual abilities.

E. According to the number of main information flows in the brain

Information flows in the human brain can be:

- Sequential
- Parallel
- Combined.

Sequential flows of information is when we for example recognize what we see, after that what we hear, etc.; parallel processing of information is when we recognize at the same time an information from all the senses. Combined processing is a combination of the sequential and parallel processing.

Accordingly, during processing the information we can think sequentially, parallel or in combination. Similarly, we can perform the entire circle of activity: from collecting information to action, based on this information.

Well known, that Gay Julius Caesar - in ancient Rome's most famous figure was able to dictate different letters to several secretaries at the same time without losing the line of thought for each dictation. Theodore Roosevelt like as Napoleon Bonaparte could multitask in extraordinary fashion, dictating letters to one secretary and memoranda to another, while browsing through a book. Ambidexterity - a surprising phenomenon where a person can work equally well by the left and right hands. People that are naturally ambidextrous are uncommon, with only one out of one hundred people being naturally ambidextrous.

A young Chinese artist Song Syaonan became a sensation on YouTube after it posted a video in which he also paints portraits of Morgan Freeman and Tim Robbins both hands. More than 20 years, he easily draws two separate portraits with both hands simultaneously. Song Syaonan paints portraits of his friends and celebrities, spending on each picture for 2-3 hours, regardless of whether he is working with one hand or two.

The chess players have the parallel processing brain, which capable of conducting simultaneous games on many boards. Chess player can hold in memory not only current position of the pieces, but the whole history of the game. In July 1933, famous Russian-French chess player Alexander Alekhine played 32 people blindfold simultaneously at the Century of Progress Exhibition in Chicago (World's Fair), winning 19, drawing 9, and losing 4 games in 14 hours.

F. Depending on the so-called creative environment

Different people have a maximum of their creativity or productivity under special conditions. These conditions usually called creative environment and here it is possible to indicate the following:

- Special internal condition for inspiration, for example, sufficient meal, sufficient sleeping time. and so on;
- Boundary conditions like silence, temperature in the room, maybe special scent and so on;
- Influence of other people, like supervisors, favorite women or man
- And so on see subdivision

The more content about this item can be found in Chapter 6 (see subdivision 6.11).

G. By way of social dependence

Social dependence is defined as needing to be surrounded by others. Many people cannot function well unless they are with other people all of the time. This makes them socially dependent, first of all in the work of intellect. It means that people in the group can create a collective intellect.

Collective intellect emerges from the collaboration, collective efforts, and competition of many individuals and appears in consensus decision making. In some cases, groups are remarkably intelligent, and are often smarter than the smartest people in them. For the best results any group must have an intellectual manager.

The main conditions for a group to be intelligent are diversity and independence of any members, and centralization or decentralization of management depending on the size of the group. Information aggregation functionality is needed. The right information needs to be delivered to the right people in the right place, at the right time, and in the right way.

1.12. Intellectual –psychological portrait of a person

To describe the diversity of intellectual abilities and their connection with personality traits for a person can be used an idea of building of psychological portrait based on the existed psychological theories, which allow to predict what and why this person as an individual or in the team will feel, think, and do in the different circumstances in his/her future.

Intellectual–psychological portrait of a person is not only sole way to differentiate one person from the others, but also can be used as a method to evaluate his/her suitability to ensure success of this person in his/her activities for any given situation, and goal.

Psychological portraits have been used in the works of Lithuanian sociologist Ausra Augustinaviciute (Аугустинавичюте, 1998), her followers (Bukalov, 2000), Russian psychologists from Tomsk polytechnic university and Tomsk state university (Kiseleva, 2016), and so on. For example, team of Russian psychologists (Kiseleva, 2016) composed the psychological portrait of the client from three facets which are based on the following theories: the theory of temper, socionic theory (Аугустинавичюте, 1998), and the theory of the types of thinking (perception of the world).

Unfortunately these psychological portraits had very limited possibility to take in account many intellectual abilities.

Using a computer technique for building and analyzing an intellectual-psychological portrait, any expert can evaluate in any convenient form all the necessary intellectual and psychological characteristics of a person, needed for any goal. The quality of the assessment will depend on method of evaluation or measurement of these characteristics. The described in the next chapters methods for the numerical measurement of such abilities as creativity, intuition, and others can greatly facilitate the task of constructing such a portrait, and its use for the purpose of self-improvement or training to obtain the necessary for success qualities.

CHAPTER 2

What level of intellect is necessary for people in the 21st century?

Where ignorance is bliss, 'tis folly to be wise.

Thomas Gray

2.1. People with practically no brain

Dr. John Lorber, neurology professor at the University of Sheffield in the United Kingdom, collected research data about several hundred people who functioned quite well with practically no brain. This doctor described in his publications some of the subjects as having no "detectable brains". Amid the patients cases collected by Dr. John Lorber was socially completely normal student of university, who gained a first class honor degree in mathematics. Dr. Lorber discovered that the student had only a thin layer of mantle measuring about a millimeter and his cranium was filled mainly with cerebrospinal fluid. It was very interesting that this subject was purportedly not even aware of his condition until the initial brain scans were performed. (Levin, 1980).

Dr. Lorber addressed a conference of pediatricians the paper entitled "Is your brain really necessary?" This paper led the scientific world to rethink its knowledge of the brain activity. There are many facts that this article has not lost its relevance today.

Let see some information. Everybody has met people who do not like to think and only do it in critical situations. They know society's rules and live according to them. We can call them "people whose brains are turned off". Bernard Shaw wrote about this: "Two percent of the people think; three percent of the people think they think; and ninety-five percent of the people would rather die than think." In this regard, today we can state very similar question: what kind of intellect is necessary for the real life?

In the usual sense this word, high level of the intellect is practically not necessary to a cattle drover, food or pizza delivery person, store clerk, loader, sorter, harvest collector, or other workers with primitive jobs. These jobs usually occupied by people without education, who very often need a few minutes or hours of instruction to get started and receive a salary. However, it is difficult to understand the results of research from Moscow's (Russia) consulting company, Detech

(Simonenko, 2012). They discovered the fact that high-level intellectual abilities are not a defining factor for a manager's success. According to this results, "the data obtained in the research shows that managers with low intellectual ability are likely to be less successful in their professional activity, while persons with high intellectual ability do not always have a competitive advantage over the ones from groups with average and above average levels of intellectual development".

These data allow formulating a question: if people practically without a brain or ones that turned off their brain can exist in our society, do the rest of society's members really need a good intellectual abilities? If managers with a high level of intellectual abilities do not have advantages over managers with a lower level of intellectual abilities, why should we force children to study 12 or more years in various educational institutions? How much money each family could have extra without this expensive education?

2.2. Distribution of people in society in terms of intellect level

In their excellent book "The Bell Curve" Richard J. Herrnstein and Charles Murray statistically have proved an existence differences in intellectual capacity among people and groups (Hernstein, 1996). According to authors, intellect is predominantly an inherited genetic attribute on which are superposed external factors such as nutrition, education, family composition, and so on.

All people have different levels of intellect. Today there are various classifications of intellect levels, including 7 or 12 levels.

According to Wechsler Intelligence Scales, a society can be divided into several groups with different levels of IQ (see section 3.2). IQ range is not absolutely correct, but according to the law of large numbers (the quantitative laws of mass phenomena are clearly manifested only in a sufficiently large number of them) it allows to make here and below qualitative analysis[2]. From the Table 2.1 below we can see that a society on average has 7 different levels of intellect from "mentally retarded" to "very superior intellect". Usually scores over 140 are considered in the genius range. It is obvious, that people with high level of intellect is tiny part of society (1 to ten millions according to Efroimson (Efroimson, 2002); geniuses resemble grains of gold in the gold-bearing sand.

---

[2] Here and below we don't refer an source of information, when that easily can be found, for example, in the Google.

Table 2.1

IQ Range ("deviation IQ")	IQ Classification	Actual Percent Included	Theoretical Percent Included
130+	Very superior	2.6	2.2
120-129	Superior	6.9	6.7
110-119	High Average	16.6	16.1
90-109	Average	49.1	50
80-89	Low Average	16.1	16.1
70-79	Borderline	6.4	6.7
Below 70	Mentally Retarded	2.3	2.2

The number of defective people in the last group is higher than that of geniuses. Sometimes it is difficult to distinguish between the aforementioned two because there are people who can be classified as genius but are not mentally healthy. For example, Van Gogh, Čiurlionis and Strindberg were schizophrenics; Newton, Pascal, Gogol, Maupassant, Hoffmann, Kafka, and Vrubel suffered from various mental disorders; Dostoevsky, Flaubert, Julius Caesar, and Napoleon were epileptics. This does not include people with savant syndrome, a condition in which a person with a mental disability, such as an autism spectrum disorder, demonstrates profound and prodigious capacities or abilities far in excess of what would be considered normal.

For the purposes of our analysis, used in the section 1.6 the simplest classification of level of mind including 3 groups with low, middle and high mind will suffice for intellect too.

If unite some groups from the Table 2.1. with low and high level of intellect, we can conclude that these new 3 groups comprise about 25 percent of people with low level of intellect (IQ less than 90), about 50 percent with an average level of intellect (IQ in diapason from 90 to109), and about 25 percent with high level of intellect (IQ more than 109).

In democratic societies and in societies with an authoritative regime, mass media allow uniting specified above group, transforming them in a conglomerate of like-minded beings.

In everyday life people do not always use whole power of their mind. Concrete analysis shows that, seemingly, most people think no more than 10% of the time. Even creative people do not think 100 % of the time. People quite often do nothing not because of a lack of mind at all, but because they get tired of thinking as the work of the brain is connected with the high expenditure of energy. The brain can require up to 20-25 percent of the energy needed for the entire body in the process of intensive work and about 10 percent in condition of the rest and relaxation.

Practically all modern Western societies have stratification, i.e. structure with a hierarchy of levels or "stratum" on a variety of dimensions, which may include education level, power, wealth, social status, social standing, and many others (Gabrenia, 2003). Social stratification typically is distinguished as three (sometimes more) social classes: the upper class, the middle class, and the lower class. It is obvious that in each stratum of society their own distributions of people with different intellect levels can be.

Usually people in the top stratum enjoy power, prosperity, and prestige that are not available to other members of society; people in the bottom stratum endure penalties that other members of society escape. It is easy to assume that the higher the class, the higher percentage of intellectual people. Unfortunately, in real life can be some exemptions. For example, a smart banker amassed a huge fortune through a tremendous amount of work and intellect. After his death, his estate can be divided between his children. According to the theory of probability, not all of his children can be smart, but each of them can be a banker with its own part of the inherited capital. Because of this, new dumb bankers will be included in the upper class. Sometimes they may unfortunately lose their fortune and move down to a lower class.

On the other hand, in humanity's history is known many cases when talented people were born into extreme poverty but escaped it. Due to this, it is obvious that any strata of the society contains a mixture of clever people and fools, possibly carrying different cultures.

The complex composition of society and the complexity of the interactions between its members cause many problems in the way of the advancement of new ideas emerging in the depths of society.

2.3. Contribution of smart people in the development of society and civilization

Innovations in mechanics, computing technology, medicine, and business practices has driven economic growth, raised wages, and helped people lead longer and healthier lives.

In the 20th century, was invented many things without which we cannot live now: airplane, television, penicillin, computer, copy machine, holography, autonomous robot, Internet, mobile phone, and so on. The each innovation was connected with a name of one or several smart persons or geniuses.

It is obvious, that without Alexander Graham Bell there would not be the telephone, without Thomas Edison there wouldn't be the electric light bulb, without the Wright brothers there wouldn't be the airplane, without Henry Ford there would not be mass production of not only cars but motorcycles, ships, airplanes, and so forth.

Women have also made many inventions in various fields of knowledge that changed the world. Amid them here are Josephine Cochrane, who invented the dishwasher (patent of 1886 year); Marie Curie - the first woman to win a Nobel Prize – she won it twice – jointly with her husband, and Henri Becquerel, for their research on radioactivity in 1903 and then on her own in 1911; Margaret Knight (nicknamed as "the lady Edison) that received 27 patents in her lifetime for inventions including shoe-manufacturing machines, a "dress shield" to protect garments from perspiration stains, a rotary engine and an internal combustion engine; Katharine Blodgett- the first woman to receive a Ph.D in physics at England's Cambridge University and the first woman hired by General Electric –invented non-reflective glass, which was initially used for lenses in cameras, movie projectors, in wartime submarine periscopes; and finally Evelyn Berezin- founder of Redactron Corporation and an innovator of the first computer-driven word processor.

Most people took innovations for granted. For example, the introduction of the blackboard into our schools had a revolutionary effect on teaching methods. The exact origin of the chalkboard remains rather unclear. Nonetheless, the idea of writing on a classroom wall may well have originated with James Pillans (1778-1864), a teacher in Edinburgh, Scotland, who documented his adoption of a chalkboard for use in geography classes. Beginning in the early nineteenth century, the chalkboard steadily gained prominence in the United States, especially in the Northeast.

In the beginning of the 20-th century, German chemist, Nobel Prize in Chemistry Fritz Haber innovated technology of the ammonia synthesis necessary for the production of fertilizers and explosives. It is now estimated that Haber's innovation helped to survive one-third of the Earth's population.

Even though Louis Pasteur and Joseph Lister were to first to discover bacteria, Alexander Fleming found the first effective antibiotic, penicillin. The discovery of antibiotics saved millions of lives by killing and preventing the growth of harmful bacteria.

Without a doubt, the great inventions changed the quality of life and increased life expectancy of people. According to statistics, the life expectancy for last 100 years increased for many developed countries approximately twice.

The 21-st century brought no less serious discoveries like the iPhone, iPad, Human Genome Project, Driverless Car, Electric-Car Charging Stations, Google Android, the Tesla Roadster, Cold fusion reactor (Rossi), etc.

Now, no one is surprised, that robots are used in the automotive industry manufacturing. There are already fully automated plants that can, for example, assemble televisions. Robots are appearing in more and more new areas of human activity. A separate issue - is fighting robots for military-operators. People in the US and Europe admire new models of cars that park automatically or can drive without a driver.

Today the effects from discoveries, inventions and innovations in computer equipment, modern communication systems, ways of receiving and transforming energy, methods of genetic engineering and selection are already estimated in hundreds and thousands of billions of dollars (Reamer, 2014).

Because of the technical progress affected the entire world the number of people living in extreme poverty has reduced from 1.9 billion to 1.2 billion over the span of the last 20–25 years.

It is impossible to estimate in any material units what ingenious composers, playwrights, poets gave. However, if they were not, labor productivity of other people connected with production of material values would be much lower.

The above mentioned proves that innovators and geniuses are necessary. On the other hand, the complexity of new machines and computers increased the demand of knowledge and resulted in a need for a more capable user. Idiots cannot drive cars stuffed with electronic devices, use smart phones, or laptops. Technical progress changes the demands to quality of mind process, including cognitive abilities allowing to work quickly and readily with big massive of information. Fortunately, more and more people beginning to understand that intellect and knowledge are needed for everybody the more the better.

However, this is not as simple as it seems. Unfortunately, the history of mankind shows sometimes other results. For example, people with high levels of intellect cannot easily find jobs, realize their innovation, or receive needed funding. History has preserved numerous dramatic stories of widely known people fighting for their ideas and inventions – first proving their advantages, then obtaining copyrights on them. Geniuses such as Hector Berlioz, Andre-Marie Ampere, Nikolai Lobachevsky, and Paul Gauguin died in poverty and obtained recognition and glory after death. At nowadays, in fact Steve Jobs, who together with Steve Wozniak founded Apple Inc. in 1976 and built and sold the world's first commercially successful personal computers at a time when mainframes were the norm, was fired in 1985 year, when about 2.1 million Apple's computers had sold. An internationally known otolaryngologist and inventor Alfred Tomatis, founder of Audio-Psycho-Phonology treatment method, was banned for life from the French Medical Council in 1977 (a sentence that is extremely rare) and subsequently convicted for "illegal practice of medicine" in 1990. Now Tomatis Method is wide used, Tomatis training courses and centers created in many countries. The innovator of new methods of the rehabilitation of stroke and other neurological disabilities Dr. Edward Taub was involved in the Silver Spring trial, where he must be charged with 119 counts of animal cruelty and failing to provide adequate veterinary care.

Explanation of opposition to the acknowledgement of the geniuses and different kind of innovators can be based on inertia of society as a complex dynamic system, developing by its own laws. It also connects with culture and tradition. In addition, most people have inherent conservatism and inertia, which greatly complicate the perception of any innovation.

Classical examples of some expert opinions about known innovations can be found on the Internet:

- "This 'telephone' has too many shortcomings to be seriously considered as a means of communication. The device is inherently of no value to us". – from the document of Western Union in connection with the telephone of Bell (1876).

"Airplanes are interesting toys but of no military value".- Marshal Ferdinand Foch, French military strategist, 1911. He was later a World War I commander.

"Radio has no future".- Lord Kelvin (1824-1907), British mathematician and physicist, ca. 1897.

History shows that as a rule great discoveries and innovation made by highly creative persons. Their creativity is motivated by the task rather than by external rewards. They are usually nonconformists, risk takers, persistent, sometimes stiff-necked but flexible. They do not give up when they encounter frustration, instead they keep at it. They become very absorbed in their work; use intuition as well as logic to make decisions and produce ideas. They can find order in confusion and discover hidden meanings. Here are some examples of life of three great innovators.

Practically everybody knows of the saxophone, but just a few people know that it was invented by Adolphe Sax, Belgian-French maker of musical instruments. In 1845, Adolphe Sax patented a family of brass instruments he called saxhorns, and later the saxophone family, ranging from bass to sopranino. Sax overcame many difficulties such as being the victim of crooks and slanderers, money-lenders, jealous competitors, venomous critics, and mediocre musicians. He declared bankruptcy three times: in 1852, 1873 and 1877 respectively. Emperor Napoleon III, his former admirer, saved him from a fourth bankruptcy. Having lived a long life, Sax did not live until the jazz period and died in poverty.

Another example is Charles Babbage, a mathematician, philosopher, inventor, and mechanical engineer. Many computer users know practically nothing about Babbage, "father of the computer", who created the concept of the programmable computer. In the 19th century, Babbage designed and partly built the analytical engine, the prototype of the computer. The basic parts of the Analytical Engine resemble the components of modern computers. Babbage's ideas were far ahead of his time. It is known that a scientist to earn money to make his machine wrote the novels, trying to get elected to the Parliament of the British Empire, even at one time played the lottery.

The inventions and discoveries that Nicola Tesla made over his lifetime, particularly in the late 1800s, are the basis for much of our modern lifestyle. Here are rotating magnetic field (1882), AC motor (1883), Tesla coil (1890), radio (1897), robotics, laser, wireless communications to name just a few. Nikola Tesla perhaps had thousands of other ideas and inventions that remained unreleased and suppressed by big industry, bank systems or other unknown power. Nikola Tesla's most famous attempt to provide everyone in the world with free energy was his World Power System, a method of broadcasting electrical energy without wires, and through the ground. His Wardenclyffe Tower was never finished. Arguably, one of the greatest scientific geniuses of all time, Tesla faced poverty, slander and persecution throughout his lifetime.

2.4. Psychological and social- psychological reasons of resistance to change

As we know, a society needs talented people and geniuses who provide ideas for its development, but on the other hand, society can fights with such individuals. Very often, these people faced with resistance to change, which may exist in many forms, dependent on the quality and quantity of changes. Resistance to change can have different roots (diversity of society by cultures, age, professions, mindset, etc. and reasons: psychological, social, economic, and others).

Every society is diversely composed up of smart people and fools, honest and mean people, people who are engaged in business, and those who want to thrive at the expense of others. Because of this, between people inside the parts or stratums of society can originate misunderstanding, antagonism and even fighting, starting like conflicts.

Unfortunately, the psychology of man is more conservative than other abilities for adequate reaction under influence of changing environment. There is reason for internal and external psychological conflicts and stresses in contemporary society, first of all as results of improper development, nurture and education of the children and youth.

Differences in intellect and mind can create psychological antagonism between people. As a rule, people with the high level of intellect do not like people with low level of the intellect, and vice versa. Arthur Schopenhauer wrote: "For the fool society of other fools incomparably more pleasantly than society of all great minds, together taken". Ones more reason for antagonism described Bertran Russel: "The whole problem with the world is that fools and fanatics are always so certain of themselves, but wiser people so full of doubts".

Let us try to describe some of typical cases of resistance to change having in the root difference in level of intellect and other reasons.

Resistance due to misunderstandings. A major obstacle when proposing a new idea has to deal with close-minded or dogmatic experts. This makes it impossible to understand the essence of new ideas despite the numerous attempts at explaining the essence of the invention. Another reason is that in order to understand a new idea, any expert needs to have enough knowledge and time to understand a concept, for which author could spend many years or life. Very often the experts does not have this time and knowledge.

Immunity to change was discovered by two Harvard researchers, Robert Kegan, and Lisa Laskow Lahey (Kegan, 2009). This phenomenon, in short, is the inertia of behavior for new opportunities. On one hand, immunity to change exists to protect us from the psychological trauma and danger that sudden changes can bring losses.

On the other hand, the reason of Immunity to change comes from the fear either of the unknown or fear based on previous negative experiences. As an example, here is the known historical fact – Luddites rebellion. The Luddites were textile workers in Nottinghamshire, Yorkshire and Lancashire in the Great Britain, skilled artisans whose trade and communities were threatened by a combination of machines and other practices that had been unilaterally imposed by the aggressive new class of manufacturers that drove the Industrial Revolution.

Other example from nowadays is anti-GMO movement. GMO (a genetically modified organism) is any organism whose genetic material has been altered using genetic engineering techniques. An anti-GMO movement includes groups like GMO Free USA, Millions Against Monsanto, Just Label It, and so on. These groups organize genetically modified food controversies, which disputes over the use of foods and other goods derived from genetically modified crops. The disputes involve consumers, farmers, biotechnology companies, governmental regulators, non-governmental organizations, and scientists.

Sometimes, people are not afraid of negativity, but they are afraid of fame and success – all that pulls people from their traditional way of life.

Resistance due envy and laziness. Two incredibly common human psychological traits such as envy and laziness can cause resistance and many conflicts as well. It is human nature to envy those who are stronger, smarter, or better than they are. We experience envy not only over other peoples' intellect or talents, but appearances, relationships, and of course bank accounts. If a person is smarter than everyone else in the group is, that person will have to constantly defend their right to a higher status. The group may work together to undermine the status of the envied person, and make things more "equal". As James Fenimore Cooper said, "the tendency of democracies is, in all things, to mediocrity."

Conflict of interests. This is a situation in which the personal interest of the individual, company, or society can influence the decision-making process. For example, a researcher conducts a clinical trial which is sponsored by a person or organization with a financial interest in the results of the trial. The conflict of interests is actually problematic for both private business and public servants.

Resistance from seemingly clever people. At the end of the 20-th century, psychologists isolated two psychological laws or principles:

A. the Dunning–Kruger effect is a concept where relatively unskilled people suffer from illusory superiority. They mistakenly assess their ability to be much higher than it really is. Dunning and Kruger attributed this bias to a metacognitive inability of these people to recognize their own ineptitude and evaluate their own abilities accurately (Kruger, 1999);

B. the Peter Principle is a concept of management theory, which was formulated by Laurence J. Peter. In this principle, any person who moves from one job position to another higher position early or late destinies to the level of his/her incompetence (Peter, 1969).

Based off the afore mentioned, it is easy to explain why every society has people who fight against things they cannot understand and assume. Sometimes they count their self as scientists, sometimes argued to common sense. These types of people assume roles of the judges: what sciences have to consider reliable and what are under doubt. For examples, James Randi has an international reputation as the world's most tireless investigator and demystifier of paranormal and pseudoscientific claims. The James Randi Educational Foundation (JREF) offered a prize of US $1,000,000 to anyone who could demonstrate a supernatural ability under scientific testing criteria agreed to by both sides. Nobody knows why JREF didn't pay a cent to many real sensitives who are now known thanks to the books, television, and internet. The Committee for Skeptical Inquiry (CSI) was founded in 1976 in USA as a bimonthly American magazine "Skeptical Inquirer". Stephen Barrett, M.D., a retired psychiatrist and sportsmen founded and supported the website Quakwatch (www.quakwatch.com). The future will show what harm these people have brought upon society.

Thanks to people likes James Randy, dowsing, extra-sensory perception, psychokinesis, and clairvoyance are considered to have no scientific basis and should rather be considered an example of pseudo-science.

Demographic factors, traditions and culture. Very often choices in human lives are driven by from traditions, whether religious traditions, rituals, cultural taboos, or what people learn from elders and their families. Traditions are important because they contain best knowledge collected by many generations over the years. As many cultures throughout history have found, traditional ways can help enhance rather than undermine sustainable life choices.

American society is the "melting pot", welcoming people from many different countries, races, and religions, all hoping to find freedom, new opportunities, and a better way of life. That means that here melted together different cultures when a heterogeneous society becoming more homogeneous, with a common culture. Reaction of such society on innovation in field of industry, or art depends on the distribution of clever and fools in the society, level of their education and who of them have more strong power.

Sometimes in different countries differently innovations collide with the restrictions of traditions and actions of "traditionalists," who are against any kind of newness and govern people's mentality by inculcating them with outdated concepts of ethical and religious values. In the process of evolution, society has learned how to effectively use people with different intellect and how to accept innovations of technical progress.

Artificially organized resistance. Robert Proctor together with linguist Iain Boal, coined in 1995 a new term "Agnotology". Agnotology means the study of willful acts to spread confusion and deceit, usually to sell a product or win a competition (Proctor, 2008). In his articles and books, Proctor explains that ignorance can often be propagated under different guises. Based on the ideas and tools of agnotology, mass media can sow doubt and confusion in the battle for dissipation of harmful products or technologies such as tobacco products and wrong medical treatments.

Dialectics of development in west society demonstrate not only cases about the difficulties connecting with new discoveries, but also examples how easily it is to produce incorrect and sometimes harmful scientific and unscientific theories concerning medicine, products use, and so forth.

Daniel J. Levitin wrote about it in his book (Levitin, 2015): "Many of us find we don't know whom to believe, what is true, what has been modified, and what has been vetted. We do not have the time or expertise to do research on every little decision. Instead, we rely on trusted authorities, newspapers, radio, TV, books, sometimes your brother-in-law, the neighbor with the perfect lawn, the cab driver who dropped you at the airport, your memory of a similar experience.... Sometimes these authorities are worthy of our trust, sometimes not".

All above mentioned allows understanding why some revolutionary ideas met resistance from different kind people. British science fiction writer Arthur C. Clarke, formulated Clarke's Law of Revolutionary Ideas: Every revolutionary idea – in science, politics, art or whatever – evokes three stages of reaction. They can be summed up in three phases:

- "It is completely impossible – don't waste my time."
- "It is possible, but it is not worth doing."
- "I said it was a good idea all along."

2.5. Economic levers of resistance to change

Any invention or new idea for its implementation requires economic support to overcome resistance in a given sphere. There is a set of economic levers to advance the invention or any new idea and not smaller quantity of economic levers to prevent its introduction.

It has been proven many times in practice that if a dollar is needed for the innovation of an idea, then thousands upon thousands more are necessary to develop it. This is reason why in modern times innovators need the help of investors, managers, and marketing specialists to realize their idea.

Many things depend on the state policy. If the state helps to create favorable investment climate, the economic and financial conditions in a country affect whether individuals and businesses are willing to lend money and acquire a stake in the businesses operating there. In such climate, for example, startup companies, with the aid of special grants immediately leads to an increase in similar type of companies within different fields of industry. In general investment climate is affected by many factors, including: political instability, regime uncertainty, national security, rule of law, property rights, government regulations, etc.

If innovation threatens to decrease the profit of one or several corporations like Big Pharma, Big Oil, Media conglomerate, etc., their financial power can be used to fight the innovation through various methods.

Today is appeared change management companies (for example, TBO International), that assists public and private sector organizations to overcome resistance to change.

When creating new things or technologies it is very important to understand ways of technical progress and a possibility of a competition. Famous corporations like Kodak, Sony, and Hewlett

Packard lost a massive fraction of their market cap by ignoring radical innovations in digital photography, digital music, tablets computers, and smartphones, respectively.

In 2002, professor of psychology emeritus at Princeton University Daniel Kahneman first in the history won the Nobel Prize in economic science. In a number of accurate scientific experiments, Daniel Kahneman proved that the majority of people in the everyday life are not guided by common sense and take irrational solutions very often. Usually they do not want to think. For example, to make purchases they are guided not by logic, but by emotions, random impulses, what they heard on TV, or from a neighbor, established prejudices, advertising, etc. It means they cannot use their intellectual abilities in real life (Kahneman, 1973, 1982, 2013).

Any errors in the evaluation of an effect from innovation will have an economical outcome. There are two possible types of errors: the false-positives (false acceptances) and false-negatives (false rejections). The false positives are usually more costly than false negatives (Bohus, 2001).

2.6. Prosperity or survival for a child prodigy

In many countries, when a child prodigy grows up in a family, many experienced parents find it difficult to answer the question of what his/her future may be. May a boy or a girl expect prosperity or survival? "Only a very few of the gifted become eminent adult creator"- wrote Ellen Winner in her book "Gifted Children: Myths and Realities" (Winner, 1996). Here is two from many examples. First: American child prodigy with exceptional mathematical and linguistic skills William James Sidis, who graduated Harvard at the age of 16, had never achieved anything of significance despite his talent in the country of great opportunities. With the intellect to win a place at Oxford at the age of 13 because her remarkable aptitude for mathematics, Sufiah Yusof decade later has been exposed as a £130-an-hour prostitute (Dolan, 2008).

There may be many reasons for this phenomenon. It is interesting to consider the impact on these children and adolescents from people around them in the light of their relationships in society.

A research professor at the Free University of Brussels Frencis Heylinghen write:

"Highly gifted people have a number of personality traits that set them apart, and that are not obviously connected to the traits of intelligence, IQ, or creativity that are most often used to define the category. Many of these traits have to do with their particularly intense feelings and emotions, others with their sometimes awkward social interactions. These traits make that these people are typically misunderstood and underestimated by peers, by society, and usually even by themselves. As such, most of their gifts are actually underutilized, and they rarely fulfill their full creative potential"(Heylinghen).

In English, there is an expression "Ignorance is bliss", and there are people to whom this slogan is pleasure. Sometimes we can meet such expression: "My ignorance is just as valid as your knowledge." Very often, the ideology of ignorant, stupid, or at best, people with average abilities wins primacy in society or even in the country. For many, it may seem strange and as a rule, this is not written in the newspapers and not discussed in television programs, but unfortunately it's true.

Jonathan Swift wrote: "When a great genius appears in the world you may know him by this sign; that the dunces are all in confederacy against him". The famous American historian Richard Hofstadter showed that throughout all of time, an antagonism between the clever and the foolish has existed and coined a name for this phenomenon anti-intellectualism. His Pulitzer Prize-winning book "Anti-Intellectualism in American Life" was as the bomb explosion when it was first published (Hofstadter, 1966).

Anti-intellectualism is conception including mistrust of developed intellect and even hostility towards bearer of developed intellect and everything connected with intellectual pursuits: education, philosophy, literature, art, and science, counting this impractical and contemptible.

According to Hofstadter and other authors, anti-intellectualism believes that:

- academics or experts (even in their fields of expertise) aren't worth listening to because they lack "common sense" or are "out of touch"
- academics are "others" and have little concern for the common people
- academics promote "sinfulness" or moral degeneracy
- the arguments of experts are impossible to comprehend.

In the history of humanity, anti-intellectualism started from time of French, Spanish and Portuguese Inquisition in the Middle Ages, and later was used by totalitarian dictatorships to oppress political dissent. In the Soviet Union one of dissipated form of anti- intellectualism were denunciations and cavils to KGB departments. According to several data, in Soviet Union was written about 4 - 5 million these ones. Behind every denunciation was ether a death or broken fate of many innocent persons. In some European countries campaign of anti-intellectualism was leaded under anti-Semitic slogan. Such a policy led to an increase in the emigration of talented people and outstanding scientists to Western countries. For example, like in 1940 year in the Germany, in the Soviet Union to end of the 80s practically all of the outstanding mathematicians, who "made the weather" in science, left country.

Perhaps appearance of anti-intellectualism's most extreme political form occurred during the 1970s in Cambodia under the rule of Pol Pot and the Khmer Rouge, when people were killed for being academics or even for merely wearing eyeglasses in the Killing Fields. Sixty-five million were murdered in China – starved, hounded to suicide, shot as class traitors, 2 million in North Korea, and 1.7 million in Africa.

In the political arena of many countries there are radical groups especially the far-right and far-left, that often take up the mantle of anti-intellectualism also. Fortunately, in American history, anti-intellectualism does not relate to a sea of blood as it have place in Germany, Soviet Union, China and Cambodia.

Sometimes people in power hate smart people. For some form of state control it's easier to keep a stupid, uneducated population in check, because they won't question their rulers. It's in the government's best interest to dumb down the public education system (while always claiming to improve it, of course), to insist that people who are proud of their intellect are elitist, to encourage the breeding of the stupidest and weakest members of society through welfare programs which subsidize those too dim or lazy to care for themselves.

History has plenty examples of the political resistance transformed to conflicts and wars between the new and the old, the good and the evil, new scientific theories and old dogmas, and so forth.

Social and psychological reasons of conflicts can be closely interconnected. According to different theories of conflicts (Galtung, 1958-1973), social conflict connected with the confrontation of social powers. Roots of social conflicts can be: changes in the social systems; the polar opposition of interests; insufficiency of resources, religion and so on.

Social conflicts can have the enormous destructive force. Several generations in Russia were victims and the involuntary witness of realization in practice one idea offered by Karl Marx for a certain period of the class fight and introduced by Bolsheviks in Russia as a form of governing (from 1917 to 1961). This idea, to be exact, the concept, called "Dictatorship of the proletariat" gave the absolute power to communists, which have been used not only for destroying the existing political system, and also to wipe out physically the huge groups of the population supporting old system.

In 1917, the Russian Empire was one of the largest empires in world history, stretching over three continents. Throughout all the history this country remained mainly agrarian country, that had have instead of proletariat mainly lumpens and marginals with low levels of education and intellects There is evidence that among hundreds of thousands of people who came to the power in the country in October, 1917, only 13 thousand were competent people. All other contingent of people having real power was semiliterate or at all illiterate (Svetun'kov, Светуньков, 2015).

During the period of dictatorship, Bolsheviks generated a wave of anti-intellectualism like hostility and mistrust toward intellectuals as well as intellectual pursuits in the areas of education, philosophy, literature, art, and science. Intellectual pursuits were seen as impractical and contemptible which meant that those who manifested kind of intellect, must be repressed or killed.

According to one of latest investigations made by Pr. Nikolay Koposov (2007), during the period of mass repression since 1917 until 1953 in the Soviet Union each fifth adult person suffered from the terror. R. J. Rummel, a professor of political science at the University of Hawaii, has recently calculated that 61.9 million people were systematically killed by the Soviet Communist regime from 1917 to 1987 (Rummel, 1990). Millions of intellectuals emigrated from Soviet Union (3 waves). These numbers don't consider the number of unborn people and the impact of repressions and emigration of best people on Russian Gene pool. If a nation loses people with high levels of intellect, for the next several generations, degradation will fatally occur.

For the sake of completeness, ethnic, religious-based, conflicts cannot be ignored. Recent violence in South Sudan, Mali, Nigeria, Syria, Iraq, Yemen, Turkey, etc. focuses attention on conflict as sources of contemporary global political instability. These conflicts, ranging from immigrant-native tensions to civil wars, can go unsettled for decades.

Stupid and ignorant people tend to see violence as the best means to solve a conflict. However, clever people believe that freedom and justice can be achieved through peaceful means. Results of activity of Mahatma Gandhi, Dr. Martin Luther King and Nelson Mandela are prominent examples of how freedom and democratic rights can be won without violence.

The root of all problems, with which innovators of different level collide, it lies in the fact that the innovation especially such as computer, smart phone and internet, usually disrupts the kept balance or customary order of things in the society. This causes opposition and resistances described above.

Therefore, to realize some ideas or innovations is necessary to know who has real power in a society and how overcome this power. Let us begin with a very old parable. Once upon a time, a king wanted to go fishing. He called the royal weather forecaster and inquired as to the forecast for the next few hours.

The weatherman assured him that there was no chance of rain in the coming days. So the king went fishing with his wife, the queen. On the way he met a farmer on his donkey.

Upon seeing the king the farmer said, "Your Majesty, you should return to the palace at once because in just a short time I expect a huge amount of rain to fall in this area".

The king was polite and considerate, he replied: "I hold the palace meteorologist in high regard. He is an extensively educated and experienced professional. And besides, I pay him very high wages. He gave me a very different forecast. I trust him and I will continue on my way." So he and his retinue continued their journey. However, a short time later a torrential rain fell from the sky. The King and Queen were totally soaked, and their entourage chuckled upon seeing them in such a shameful condition. Furious, the king returned to the palace and gave the order to fire his palace meteorologist.

Then he summoned the farmer and offered him the prestigious and high paying job of royal forecaster. The farmer said, "Your Majesty, I do not know anything about forecasting. I obtain my information from my donkey. If I see my donkey's ears drooping, it means with certainty that it will rain." So the king hired the donkey. Since then it began the practice of hiring dumb asses to work for the government and paying them outrageous salaries to occupy its highest and most influential positions (http://www.mybowlsclubs.com.au/jokes.php).

In the process of human evolution was created many methods that lead to fools power. One way to do this is through democratic elections. Because in many countries a nice and cozy relationship exists between big governments, mass media, and big business, next way for fools to be fire in the top structures is corruption. Some fools leave traces of their activities in the form of state laws. It difficult to believe how many stupid laws existed now in the different states of the USA. For example in the state Alabama residents should obey the following law: "It is illegal to sell peanuts in Lee County after sundown on Wednesday", in the state Hawaii the following:" Coins are not allowed to be placed in one's ears", in the Oklahoma: "Females are forbidden from doing their own hair without being licensed by the state." (http://www.dumblaws.com).

Some phenomenal abilities of fools were described by famous Abkhaz writer Fazil Iskander: "I have often observed that stupid and deceitful people often exhibit clever inventiveness. What's the secret? I think so: the habit of lying, and the need to constantly be turned out of the false situation trained their brains in the direction of the extraordinary mobility of the mental powers. Although the stupid man has a few sources, but being able to quickly assemble them into a single, he can find at this point benefits. …He is a good general of his small mental power."

In ordinary life, fools either follow the written and unwritten rules of society, or use smart tips from friends. This allows them to find solutions to most problems which they encounter in everyday life. Failure occurs only in unusual situations, which are quite rare.

At present, in addition to the power of the state, party-ideological authority (for example, Republicans or Democrats), and financial power is appeared an additional force - mass media. The last includes television, movies, radio, newspapers, magazines, books, records, video games and the internet. Thanks mass media the majority of people in today's society depend on information and communications to remain connected with the world and partake in daily activities like work, entertainment, health care, education, socialization, travelling, and anything else that we have to do. Mass media has a huge impact on society, because can form or modify the public opinion in different ways depending of what is the objective.

From early childhood we were sure that all parents, teachers, professors and scientists are clever, that all politician, businessmen and businesswomen are smart and that especially clever people manage a state with democracy. Unfortunately, the history of humankind demonstrates the opposite.

The ancient Greek philosopher Plato rejected Athenian democracy on the basis that such democracies typically run by fools.

From Sebastian Brant medieval satire in the poem "Das Narrenschiff (1494; The Ship of Fools)" the idea that all human are sailing on a ship of fools has intrigued many people.

Heinrich Heine wrote: "There are more fools in the world than there are people". Oskar Wilde wrote: "The world has been made by fools that wise men should live in it!" The famous German philosopher Arthur Schopenhauer also wrote about the number of fools: "The wise have always said the same things, and fools, who are the majority, have always done just the opposite". And at last, George Bernard Shaw warned that fools are dangerous: "Power does not corrupt men; fools, however, if they get into a position of power, corrupt power". As the world population grows, fools increase in numbers. Today because of unwritten laws of democratic societies it is very impolite to inform people about their stupidity.

The indirect demonstration of the abundance of fools: it is the principle which all modern productions are using: Fool proof. If fools were not so common, this principle wouldn't be necessary. There are two exceptions: the companies, in which a lot of fools work (usually it is visible on how the manual of production done), or companies that produce products which fools don't buy.

In essay (Iserlis, 1996) was assumed, that fools govern the world. Of course, it was only my observations, because I didn't have sufficient scientific statistics and proofs. Recently I found CBS News an article by Dave Logan "Why geniuses don't have jobs" which concluded with the concept "there are lots of fools in management, so they are often the power structure in companies".

"Always and inevitably everyone underestimates the number of stupid individuals in circulation….., the fraction of stupid people is a constant σ which is not affected by time, space, race, class or any other sociocultural or historical variable." –wrote Italian economic historian Carlo Cipolla in his essay "The basic laws of human stupidity" (Cipolla, 2011).

Jonathon Gatehouse in the article "America dumbs down" wrote:" Everywhere you look these days, America is in a rush to embrace the stupid" (Gatehouse, 2014).

There is one of the unwritten laws of management (though not absolute): each manager chooses his subordinates dumber than himself. If he will not do this, the more intelligent subordinates can be put in its place. Thanks to this rule, if a fool is hired at a high level job position, then the lower levels of the pyramid will be occupied by even dumber people. It is common for fools to be hired for high level job positions because their quality of mind is not important. There are other more important criteria, such as loyalty, devotion, relative, and party or love relationship.

Of course, donkeys and fools within the government and amid senior staff of corporations are very dangerous. However, it is not any less dangerous when there are a lot of fools around us. Because of this and other reasons, smart people scatter into small minorities. Fools like to be united in groups; the more, the better. They feel comfortable only in the company of their own kind because their collective intellect makes up for what they lack of.

Mass media is a significant force in modern culture, particularly in America. Communities and individuals are bombarded constantly with messages from a multitude of sources including TV, radio, newspapers, to name a few. The power of mass media can be used for good, for example for education, propaganda of health, sport and so on. However, it can also be used for bad, for example, by exposure any type of porno, violence, junk food ads, etc.

Using for bad, mass media may transform whole or big parts of populations into "Ship of fools".

If the world had less fools, there would be no wars, no accidents at work, on the roads, and there would not be any more states with foolish laws.

Some experts speak that we now live in a VUCA world, i.e. under conditions of Volatility (dynamic contexts), Uncertainty (missing information), Complexity (multiple bases for categorization), and Ambiguity (unclear meaning of available cues). In such circumstances, the strength and flexibility of the intellect, as well as the ability to learn are fundamental for each individual and for the society as a whole.

In order to resist somewhere the waves of anti-science, anti-intellectual thinking in the USA and other countries appeared plenty of groups of people with high levels of intellect, including Colossus Society, Grail Society, Mensa International, Prometheus Society, to name just a few.

For example Mensa International, the high IQ society, provides a forum for intellectual exchange among its members. There are members in more than 100 countries around the world. Mensa activities include the exchange of ideas and knowledge by lectures, discussions, journals within special-interest groups by local, regional, national, and international gatherings.

2.7. Difficulties of overcoming the resistance to the changes

To overcome the described above categories of resistances to change, for any creative person there is not enough willpower and his/her persistence This person must be armed with the abilities necessary to achieve his goals: for example, the abilities to argue and persuade, attract the right people, to be patient in the dangerous complex situation and so on.

Many countries created social laws and special rules for regulation relationship between people and groups of people. For example, there are created systems and methods of conflict solving via justice institutes; realized systems of author rights defense; created systems of learning for any type needed on job market skills; supported different kind competition (tournament) systems (like the show "America's Got Talent"); used special mass media systems to regulate social conflicts, and so forth. Because many categories of resistances to change can be without laws violation, an owner of developed intellect to realize himself or his innovation must know many additional things kind of unwritten rules and laws of society where he lives, algorithms of success and many other things.

In the industry, many companies of different size used several rules of management. Inside of these companies lower and mid-level managers do not have high levels of intellect because it can disturb an existing order. Extra initiative and an ability to critically think can harm those who do it. Quite often successful business processes, once established, should not be changed without the serious, for example, economically substantiated needs. This is particularly true for a franchise, when one party (franchisor) transfers the right to a certain kind of business using the developed business model to another party (franchisee) for a fee. An example is the famous McDonald's. In the world of theater or cinema art there are other laws or rules.

In some corporations, it is authorized only to one group of people to think, for example from senior staff or founders, and all others must be only fast working executors. It can be reason why in some companies employers rarely pay attention to intellect. Human resource departments focus their attention mainly on past experience for any pretended on open job positions. If a person sometimes worked as a leader, and he/she worked more than year, HR inquires recommendations, and if they indicate that this person worked well, then this is sufficient for moving to a new same or higher job position. This blocks the perspective for gifted people who could not lift in top position in the first or next companies. The very frequently gifted people and especially geniuses have unsociable nature and do not love to work in the team. Obvious for them is big problem to receive a positive reference in process of job changing. Very often HR looks for new employees by checking team compatibility and a knowledge of tools or technology which is included in the job position's description.

The existing selection methods in the majority of companies allows for them to give an advantage not to the applicant with the highest power of intellect who can be trained quickly but an applicant who has experience in the necessary area.

Many companies afraid to hire overqualified people who need bigger salaries. They invite people with low qualifications that have the ability to solve the existed problems without guaranty to solve more complex problems in the future.

Very often, the manifestations of intellect for the surrounding people can appear strangely. For managers it is difficult to operate someone who is too clever, too skilled or too professional. Intellectuals always have their own opinion and are not afraid to state it. Very often, they like their work as equally as a hobby, sport or traveling. They would be happy to receive a lot of money, but they do not have an innate need for money. Here is a good example. A Russian mathematician Grigori Perelman In 2003 proved formulated and posed in 1904 by Henry Poincaré well-known amid mathematicians the Poincaré conjecture, which was one from seven The Millennium Prize Problems declared by the Clay Mathematics Institute in 2000. In May 2006, a committee of mathematicians voted to award Perelman a Fields Medal for his work. Perelman declined to accept the award or to appear at the International Congress of Mathematicians in Madrid, stating: "I'm not interested in money or fame; I don't want to be on display like an animal in a zoo."(BBC news, 22 August 2006). In 2010, Perelman was awarded a Millennium Prize for solving this problem. On June 8, 2010, he did not attend a ceremony in his honor at the Institute Océanographique, Paris to accept his $1 million prize. Unfortunately, such subordinates as Grigori Perelman aren't always wanted by some employers.

It is necessary to keep in mind, that the main difficulty in considered problem that sometimes people with high intellect cannot be distinguish from people with average intellect. Today there is not exist a scientific method for measuring intellect; usually IQ measurements give wrong assessments. Described in next chapters method allows to make a step for this problem solution.

In the real life for developing of a complex problem smart people with powerful intellect is not necessary all of the time. Therefore, companies divide their employees into experts who can find the solutions to key questions and other people for performing routines. The first of these people work on a contract (temporary workers), the second are permanent workers. There is a huge disadvantage to those who work on a contract because they do not have a stable income. If such an expert wants to find permanent job, he or she can encounter problems because in many cases he or she is considered overqualified.

To finish the description of very important problem concerning the fate of gifted and genius from birth people, it makes sense to recall the following old parable.

After death one general went to heaven and met an angel. When the angel was showing him paradise, the general asked him:

- Would you please show to me the greatest commander of all time and of all countries.
- It is good, answered Angel, let us go.

And he brought him to the man, who deals with the repair of footwear.

- Sorry, said the general, I wanted to see the greatest commander not the greatest shoemaker.
- Look closely, this is the greatest commander of all time and of all countries, only he did not found himself in life, replied the angel.

Because we cannot ask angel why shoemaker didn't make a carrier of great commander, we try to imagine some possible reasons for this. Probable he was younger than his brothers, who were killed in the war, and mother did not allow him participate in this war. Other reason probably was a necessity to save life his bribe or all members of family. Maybe his father was shoemaker, and grandfather was also shoemaker. Here must be thousands of other reason, but easy to united them in only one: he didn't find during his life a right place and right time to become a greatest commander.

2.8. Is it necessary to be ringing the alarm bells?

Mentioned at the beginning of this chapter, Dr. Loren has proved, that in our society can exist people who practically have no a brain and nobody beside of specialist cannot reveal it. It means that if a large number of brainless people gather somewhere, then there canl be a threat to society. It can be much more worse if people with low intellect seize power in this society or country.

It is possibly, because modern public relationship and institutions create opportunities to climb the career ladder for people who are having low intellect and they getting some power (authority) can create a risk of destroy (bankruptcy) to company, to society, or to state (nation) fast or by long time. Persistence of financial crises or recessions, military conflicts, and even revolutions demonstrated in the past and demonstrate now how these people can influence the situation in the country, or the world.

The development of technological progress using automatic systems and robots reduces the needs for the labor market in many specialties. For example, according to the forecast data from the Bureau of Labor Statistics from 2014 to 2024 farming, fishing, and forestry occupations must decrease on 5.9%.

On the other hand, the creation of special computerized tools and technologies reduces in many cases the need for certain types of intellectual work (for example, when performing mathematical calculations) and reduces the demands on the high level of intellect for workers (for example, when working in conveyor production). This can have bad consequences because lack of exercise for the brain can lead to the mental degradation of workers.

In this context Prof. Stephen Hawking, one of Britain's pre-eminent scientists warned that the invention of artificial intelligence could be the biggest disaster in humanity's history. He told the BBC: "The development of full artificial intelligence could spell the end of the human race" (Cellan-Jones, 2014).

Around the world only approximately 10 people per year are killed by sharks, 100 people are killed by lions, 1000 people are killed by crocodiles, 10 000 people by Tsetse flies, 50 000 people by venomous snakes, and about a half million are killed by other people. This means, that Latin proverb "homō hominī lupus est", translated in English as "man is a wolf to man", is acceptable for modern human life. However, this is nonsense for a man who by definition is King of nature.

Recently, the chairman of the Business Roundtable, chairman and CEO of JPMorgan Chase Jamie Dimon intuitively feeling this problem is sounding the alarm on a number of occasions to highlight problems in America. Here are the six problem areas, which he identified in interviews with Matt Turner from the Business Insider:

"Over the last 16 years, we have spent trillions of dollars on wars when we could have been investing that money productively".

"Since 2010, when the government took over student lending, direct government lending to students has gone from approximately $200 billion to more than $900 billion — creating dramatically increased student defaults and a population that is rightfully angry about how much money they owe, particularly since it reduces their ability to get other credit".

"Our nation's healthcare costs are essentially twice as much per person vs. most other developed nations".

"It is alarming that approximately 40% (this is an astounding 300,000 students each year) of those who receive advanced degrees in science, technology, engineering and math at American universities are foreign nationals with no legal way of staying here even when many would choose to do so. We are forcing great talent overseas by not allowing these young people to build their dreams here".

"Felony convictions for even minor offenses have led, in part, to 20 million American citizens having a criminal record — and this means they often have a hard time getting a job. (There are six times more felons in the United States than in Canada.)".

"The inability to reform mortgage markets has dramatically reduced mortgage availability. We estimate that mortgages alone would have been more than $1 trillion higher had we had healthier mortgage markets. Greater mortgage access would have led to more homebuilding and additional jobs and investments, which also would have driven additional growth."(Turner, 2017).

According to Adam Smith nation's intellectual wealth depends on the extent to which these and other problems will be solved.

According to last State of Mental Health in America issued by Mental Health America in 2018:

- Over 44 million American adults have a mental health condition. Since the release of the first State of Mental Health in America report (2015), the number of adults who have a mental health condition practically do not change many years (from 18.19% to 18.07%)
- Rate of youth experiencing a mental health condition continues to rise. The rate of youth with Major Depressive Episode (MDE) increased for last 4 years from 11.93% to 12.63%. There was only a 1.5% decrease in the rate of youth with MDE who did receive treatment. Data showed that 62% of youth with MDE received no treatment (http://www.mentalhealthamerica.net).

There is one more confirmation of the dangerous situation, that is now observed in the USA and poses a threat to intellectual wealth of nation, and explain why it is necessary to be ringing the alarm bells. This is a hard point to really absorb for the era of technological progress, that here is the fact of the decline in labor productivity observed recently in the USA according to the Bureau of Labor Statistics (Sprague, 2017).

One of the main ways to solve this problem is to pay attention to intellect for all nation starting from proper education of children or retraining of adults. However, this is not so simple.

An information explosion that began with the onset of the computer era can lead to brain overload, especially in children. Therefore, special methods and measures are needed to regulate and manage the information flows in the people brains.

Since the labor market is changing more dynamically than in previous years, it is often necessary for "blue collar" and "white collar" workers to retrain. Today the ability to retrain can be more important than the ability to learn the necessary tools.

The history of our planet shows that our civilization is not the first. Nobody knows why the great ancient civilizations have died, but it is easy to believe, that only collective intellect can save our civilization in the next force majeure situation, for example like Great Flood of Noah's Time or collision of our planet in cosmic space with an asteroid or something else.

It is obvious that the future of our civilization depends on how will develop a level of intellect for vanguards layers of human society in developed countries of the world. Otherwise, the 21 century can be last for our civilization.

Performed analysis, which certainly cannot claim to be complete, nevertheless allows us to formulate the following question, which of course interested in the majority of thinking people, and which can be formulated as follows: what level of intellect is needed for a successful life?

2.9. What level of intellect is needed for a successful life?

The American psychologist Dr. Lewis Terman at Stanford University tried to answer this question in project "The Genetic Studies of Genius", today known as "The Terman Study of the Gifted", is the oldest and one of the longest running longitudinal studies in the field of psychology. This project was begun in 1921 to examine the development and characteristics of gifted children during life span. Dr. Lewis Terman selected from California's schools about 1500 the best of the best children aged 8 to 12 years who had IQ at least 140 - the minimum required value, allowing a person to be considered a genius. The average IQ score of the group's participants was 150, 80 people had results above 170 points. Participants of the project were called the "Termites". After the death of Dr. Terman in 1956, several psychologists continued the original study, observing the same participants. This project is not finished until now, in 2003, there were over 200 "Termites" still alive. The study will continue until the final "Termite" either withdraws or dies (Terman, 1959, 2013).

The study has the disadvantages of any longitudinal study: it is possible that the characteristics and behaviors of the project participants are a partial result of the era they lived in. Indeed, many members of the sample could not attend college, due to the Great Depression and World War II. Almost half of women in the sample were homemakers for most of their lives.

Did the participants in the study of Professor Terman succeed in life? I was not able to review all the publications relating to this remarkable experiment, but even familiarity with the Volume V, devoted to the results of 35 years of the research (Terman, 1959, 2013) and several articles, allows coming to some conclusions.

In 1955, the average annual income was $ 5,000, the average income of the subjects of Terman's study was impressive - $ 9,640. Two thirds of the Termites received higher education and a large number of them received higher degrees. Many members of the group became doctors, lawyers, business leaders, teachers and scientists. However, rest part pursued more "humble" professions, which did not require any high levels of intellect. It very interesting that amid Termites levels of divorce, alcoholism and suicide were about the same as the national average.

Early, by the 4th volume of Genetic Studies of Genius, Terman concluded: "At any rate, we have seen that intellect and achievement are far from perfectly correlated." Some of Termites reached great eminence in their fields. Among the most notable were writer Jess Oppenheimer, American Psychological Association president psychologist Lee Cronbach, American physiologist Ancel Keys, etc. Maybe still early to speak, researchers could not reveal the conventional geniuses from approximately 1500 gifted children with highest IQ. The latter proves once again that IQ cannot be a criterion of genius.

Authors of the volume V of "The Genetic Studies of Genius" understood that life success depends not only from level of innate intellect but from how every participant understood goal (success) of his/her life and how he/she strive to achieve this goal. They revealed by questioning the Termites "what constitutes success in life". The five most frequently mentioned definitions of life success were:

- Realization of goals, vocational satisfaction, a sense of achievement;
- A happy marriage and home life, bringing up a family satisfactorily;
- Adequate income for comfortable living (but this was mentioned by only 20 percent of women);
- Contributing to knowledge or welfare of mankind; helping others, leaving the world a better place ;
- Peace of mind, well-adjusted personality, adaptability, emotional maturity (Terman, 1959, 2013).

It is very interesting, that "rather than basking in their successes, many reported that they had been plagued by the sense that they had somehow failed to live up to their youthful expectations", for them "high intelligence has been a burden rather than a boon" (Robson, 2015).

From point of view of science of 21st century, project "The Genetic Studies of Genius" demonstrates, that to have even high level innate intellect is not sufficient to achieve an eminence and have success in real society.

Everything described above allows us to conclude that a person's success correlates not with the integral level of intellect measured by one parameter (for example g-factor), but with level of each his/her intellectual ability, personal traits and, finally, circumstances in which this person is (recall the parable of the ingenious shoemaker-see section 2.7). Today we can talk about the level of human intellect (genius, talent, mediocrity) in relation to the type of main activity. Likewise, depending on which intellectual abilities of a person are most developed, we can talk about the type of intellect (intelligence) in the Howard Gardner classification.

Professor Yale University Robert Sternberg coined a term "successful intelligence" (Sternberg, 1997). According to his triarchic theory (see section 1,3) successful intellect (intelligence) consists primarily of three intellectual components: analytical (which involves judging and evaluating ideas), creative (which involves inventiveness and imagination in problem solving), and practical (which involves using, utilizing, and applying strategies, ideas, and facts).

In my opinion it is very narrow definition of successful intellect. It can be more correct, if intellectual-psychological portrait meet the requirements of the chosen profession or lifestyle, than it ultimately determines success of any person in his performance and practice. It is obvious that everything depends on the initial position of a person in the society and his intentions. Most of the people receiving intellect from birth, but intellect either can be developed and received a faceting as a diamond or can be destroyed under life circumstances. To develop intellect, need special education and training, but not everyone has such possibilities.

Many people believe that successful intellect is needed only in intelligence agencies, in business intelligence services and so on. It is wrong, successful intellect is needed everywhere.

At present, there are certain professions, which require extensive education, and training, but not extremely high intellect. Let take the engineering profession as an example.

An engineer is a professionals who can invent, design, analyze, build, and test machines, systems, structures and materials to develop solutions for solution of technical, societal and commercial problems. To become licensed, engineers must complete a four-year college degree, sometimes work under a Professional Engineer for at least four years, pass some competency exams and get a license from a state's licensure board.

Here are many branches of engineering, each of which specializes in specific technologies and products. Typically, engineers must have deep knowledge in one area and basic knowledge in related areas. For example, mechanical engineer typically must know something from physics and chemistry, electrical engineering, computer science, materials science, metallurgy, mathematics, economics, technology, and so on.

Of course, among the engineers there are those who have a successful career, and those who make ends meet. It all depends on how far their knowledge, psychological and intellectual abilities meet the conditions in which they work and the goals of their lives.

Fyodor Dostoyevsky wrote: "The mystery of human existence lies not in just staying alive, but in finding something to live for." Expression "to live for" means finding a purpose and a meaning in life. The goal of life is different for different people. In each generation people exist who have many interests but is unable to decide on a single direction for their lives. These people, also known as scanners, love variety because their brain processes information rapidly, and they are more ready for new things than other people are. Scanners are frequently gifted in many fields.

In the human nature laid enormous opportunities, which in the overwhelming majority of people remain underdeveloped. Here is one of main reasons why around us are very few people who are capable of generating new brilliant ideas and injecting them into industry, farming or society. If the person does not know his capabilities, his life will be accompanied by a feeling of dissatisfaction and stress due to internal psychological conflicts.

Many geniuses found their profession after the second and more attempts. When young people choose their profession, in many cases, they do not know what they like and what they can do. As a result, some of them in a period of time change their professions. However, unlike scanners, they meet many difficulties when transitioning professions.

If a man has a purpose and knows the meaning of life, this still does not mean that he or she can apply himself. There can be different objective and subjective reasons. The main problem is how to sustain yourself and your family, how to make a living.

To summarize, the main problem for any person is possibility to find the right place and right time for realization his goal. Today on the internet, in the books and articles around the world can be found many secrets for being in the right place at the right time even many times. Because our item is powerful intellect, we will not consider these recipes and methods.

The main task of any person or his/her parents in the beginning of life is define what kind of innate intellect he/she has and what profession do best correspond this intellect. The famous Russian writer Leo Tolstoy wrote: "If you do not know your place in the world and the meaning of your life, you should know there is something to blame; and it is not the social system, or your intellect, but the way in which you have directed your intellect."

A man armed with a powerful intellect should have else certain personality traits: will power, persistence, self-confidence, ability to be calm in dangerous or strange situations, sometimes the ability to compromise, and so on depending on his/her type of activity.

Let consider briefly to whom and why is needed a developed and advanced intellect for people of different professions and life tasks.

People providing technical and scientific progress need advanced intellect not only to make an invention or a scientific discovery, but and to convince others sometimes through conflict resolution that it gives a positive effect, and also to make real results of their works. Artists, writers, composers need bright intellect to find their original style and the way in the corresponding art. Managers need developed intellect to organize the work in the full of risks complex situations and ensure fulfillment of tasks. Financial staff needs powerful intellect to know how to earn capital and where it can be invested. Commanders need advanced intellect not only to defeat the enemy, but for that to have been happen with minimal human and material losses. Experts need intellect to understand that he did not know before, for example, if it concerns a new invention. Scientists and researchers need developed intellect not only to acquire new knowledge, but also to combat dogmatism. A scientist should be able to always doubt and therefore be able to accept criticism.

Politicians must have developed intellect to find ways to control society to ensure its development, and at the State level, to find ways of ensuring peaceful coexistence and interaction with other States. The governments of many countries as well as large corporations must wide use a principle of foolproof as well as it used in the construction of complex machines and devices.

By the way, intellectual people of such professions as medical doctors and teachers cannot be soulless.

In the friendship, advanced intellect is needed to relate to poor friend just as to the rich, to manage relationship without the lie and fraud, to not have prejudices against the color of the skin, religious persuasions, and sexual orientation. Even in love intellect is needed to belief each other, and to help each other.

Each person must have a needed level and quality of intellect to recognize the symptoms of starting disease and take care about it on time, to fight with the stress and malaises without the medicines. Everyone needs a developed intellect to overwhelm any resistance to change and resolve multiple types of conflicts.

Many people think that for family happiness is enough to have common sense and life experience. Next example from real life demonstrates that common sense cannot be sufficient for some life situations.

Daily Mail on 30 November 2015 published an article by Lucy Elkins, "The mother who saved her daughter's life by lying to doctors after they failed to spot she had a brain tumor". This article describes a case in which Amanda Davies faked an emergency call and said that her kid had fallen and hit her head, then vomited. Amanda did this because she believed that the four different medical experts her three year old daughter, Lil, had seen, missed a serious disease. The CT, in the Accident & emergency department (A&E) of Cardiff hospital, diagnosed Lil's tumor in the brain, and after that she was successfully treated. Lil's life was saved by her mother's power of intellect.

Generally, power of intellect needs inside of a family to support state of love and mutual respect between the family members throughout life in a dignified manner to grow, to bring up and to train the children helping the old and the sick members.

What is the relationship between intellectual abilities of marriage partners and marriage satisfaction? A direct answer does not exist, because in each family must exist very important power – the love, that can be very strong and very weak. Any family is like a cell in a state. On the family level, the laws of state must be used, but there are also may be used unwritten internal laws. The intellect abilities of family members is necessary not only in order to create and to use these laws, taking into account the satisfaction of the necessary needs for each one in the food, love or sex (for husband and wife), or in the satisfaction of emotional hunger. Until now, the lawmakers in different countries are looking for recipes of family happiness. Unfortunately they usually forget about intellect of man, woman and especially children.

Today maybe the general problem is how to maximally use a mind of each potential worker so that necessary job positions would occupy people not only in accordance with its knowledge but also with intellect and destination. Here are two problems. The first - to determine before a graduation of the school capability of a student, and define where this student can use his talents at the best, for the mutual benefit of him and society; the second – to make accent not only on knowledge (here is impossible to compete with the internet), but more on ability of the intellect development.

This means that we need to develop and support our intellect in any forms. Ken Robinson writes: "First, it is essential to generate ideas for new product and services, and to maintain a competitive edge. Second, it is essential that education and training enable people to be flexible and adaptable, so that business can respond to changing markets. Third, everyone will need to adjust to a world where, for most people, secure lifelong employment in a single job is a thing of the past" (Robinson, 2011).

The solution of problem lays not in correct instructions, but in the development of the gifted and talented children. It also lies in the rules of the creation of a creative situation, the rules of the generation of ideas, the rules of their analysis, and the abilities of the introduction of these ideas.

An estimated 3 million gifted children sit in classrooms across the U.S. today. To help them in developing their intellect, the USA has now created a huge number of associations and centers for gifted and talented children:

- National Associations (for example, National Association For Gifted Children (NAGC)),
- State Associations and State Departments (for example, California Association for the Gifted, and so on in many states)
- University Resources (for example, Center for Gifted Education & Talent Development (UCONN) in the Connecticut state).

Without any noise, the industry of computer games made a very big contribution to the development of the intellect of the rising generation. This, in particular, led to the fact that the students of the youngest classes, were exposed to computers at a young age and did not want to learn or use manual calculation.

Modern life has many features that people have not met in previous centuries, first of all a wide range of possibilities that can be realized with the help of not only good luck but also developed intellect. Born today in the Middle West, a boy can be not only a cowboy, but also a physician, scientist and even an astronaut. The same is for a girl. Even if he/she chooses a profession for himself as a businessperson, he/she can, like Donald Trump, become not only a billionaire, but also a president of the country. If a person even chooses a profession in art, he/she will need not only talent (intellect), but also a lot of knowledge to win the inevitable competition.

To understand why each from us needs to have an advanced intellect, it makes sense to add to our analysis something else.

Intellect is this invisible double-edged weapon of the human being, which can be used into the good or into the evil. Being united from many people this weapon can have the enormous creative force or as its opposition- destructive force.

If we look from the position of sociology, people uniting their intellect in the so-called collective intellect for the good can develop society in the ideal form on the democratic principles. Even a slight increase in intellectual abilities can significantly accelerate the pace of economic growth in any country. The society does not controlling collective intellect can convert into the crowd, which lives according to the laws of the crowd (mob). Democracy aims to govern peacefully by the people, through civilized means. Mob rule is when "democracy" turns into "the masses brutalize everyone else that gets in their way".

At present, we can see numerous evidences of the existence of a spontaneous process of forming requirements for level intellect for workers in various professions necessary for success.

According to the wave theory of Alvin Toffler postulated him in the book (Toffler, 1980) in the very complex and contradictory human history can be seen three stages (waves) of our civilization development. There are:

- the First Wave, created Agricultural Society that started approximately in 2,000 B.C.
- the Second Wave laid the foundation for Industrial Society starting in about middle of 18 century
- the Third Wave created the preconditions for the emergence of post–industrial Information Society, which began to manifest itself in the middle of 20 century.

The Third Wave brought new demands to workers everywhere. Work is becoming increasingly diverse, less fragmented, each worker performing a larger, rather than shallower task. A flexible schedule and free pace replace the former need for mass synchronization of behavior. Workers have to cope with more frequent changes in their work, as well as product changes and reorganizations.

New technologies do not need millions of illiterate people doing an endlessly repetitive work but need those who familiar with the process of identifying problems and their causes, who familiar with developing and evaluating possible solutions, and who can be able implementing an action or strategy in order to achieve desired goals or outcome.

Thus, employers of the Third wave are experiencing an increasing need for men and women who take responsibility, understand how their work is related to the work of others who can cope with larger tasks and quickly adapt to changed circumstances and who feel the mood of people around them.

For the management of industrial enterprises and offices of the future, companies of the Third wave will need workers more capable of independent activity, more inventive than unquestioningly following instructions. To train such workers, schools will have to move far away from modern teaching methods.

"The illiterate of the 21st century will not be those who cannot read and write, but those who cannot learn, unlearn, and relearn. " wrote Alvin Toffler.

Obviously the answer to the question formulated in the title of this section cannot be a simple. It can be obtained strictly individually according to the analysis of Intellectual–psychological portrait of a person and also situation and his/her goals, which can have different level of complexity, from how to earn money via computer to how to get the planet Mars.

What has been said above leads us to the conclusion that for each person not only it is necessary to bring up the probably needed in later life character traits, but also necessary to tune and maintain during lifespan the appropriate level of intellect. The latter means tuning some intellectual abilities in the needed rate. How this can be done is shown in the following chapters.

CHAPTER 3

Methods of evaluation and measurement of intellectual abilities and characteristics

"You can discover more about a person in an hour of play than in a year of conversation" – Plato

3.1. Expert evaluations and psychometric measurements

Great physicist and astronomer Lord Kelvin stated, "If you cannot measure it, if you cannot express it in numbers, your knowledge is of a very meager and unsatisfactory kind."

At present, measurement of the characteristics of the intellect refers to the class of tasks, the solution of which cannot be obtained on the basis of direct measurements, because the nature of these characteristics is unclear and there was no tools for these measurements.

Today's practice uses two main methodologies to assess the intellectual abilities and psychological characteristics of people:

- Expert (heuristic) evaluations
- Psychometric measurements.

In general, an expert evaluation is a diagnostic method performed by experts, by means of which qualitative features of psychic phenomena are expressed in the form of quantitative estimates. The essence of expert evaluation consists in revealing the formalized objective information about the behavior and qualities of the individual in a number of subjective opinions and assessments. This method is widely used in psychological practice to assess the qualities of a person when composing psychological characteristics.

The method of expert assessments is usually applied in conditions of time deficit or extreme situations. Its use does not require long preparation and search or development of complex research tools. This method is flexible enough, it is easily modified when new tasks arise, and is also suitable for repeated use.

The methods of expert evaluation (assessment) is widely used by all people in the real life to select friends, recruit employees, assess students 'abilities, etc. This methods often used in medical practices, for example for evaluation of pain level in the scale from 1 to 10. For more serious use, for example in research work or to solve very important problems, additionally are used methods of mathematical (statistical) processing of evaluation results.

This makes it possible, for example, in the learning process, to make a more accurate forecast of learning ability, to select the means of pedagogical influence, taking into account the possible reaction of the individual to it, and opens the way to determining the prospects for its psychological and professional development.

Psychometrics as a science about psychometric measurements (Furr, 2007) originated from the classical test theory (CTT) and the more recent item-response theory (IRT), using advanced mathematical methods as a factor analysis (Gorsuch, 1983)), multidimensional scaling (Cox, 2001), cluster analysis (Everitt, 2011), and so forth.

At its core psychometrics involves two major parts, namely:

- the construction of instruments and procedures for measurement;
- the development and refinement of the theoretical approaches to measurement.

The first part is a set of methods that allow constructing verbal and mathematical models of mental processes in the form of special scales or questionnaires. The second part allows, according to these models, to estimate in digital form the amount of one or another mental characteristic using appropriate mathematical processing in order to increase the reliability and validity of the measurements. In fact, these models become tools for measuring (evaluating) psychological characteristics. Psychometric tests include personality profiles, reasoning tests, motivation questionnaires, and ability assessments. Many of these tests are completed using software programs, and some can even be completed online. Many psychological tests include a set of questions, each of which is directed on measurement of the certain aspect of the psychological characteristic. Usually the point for each right answer is put down, then points are summarized. The total score is considered the final value of the measured construct.

Psychologists have developed reliable methods of tracking down, minimization and the account of various types of regular mistakes in answers. They also have developed ways of reduction of influence of the experimenter on process of measurement.

Unfortunately here are too many problems. Dutch academic Denny Borsboom wrote that "after a century of theory and research on psychological test scores, for most test scores we still have no idea whether they really measure something. Or are no more than relatively arbitrary summations of item responses" (Borsboom,2009).

Since objectivity is main factor to using psychometric methods for measurement of skills, knowledge, abilities, attitudes, personality traits, educational achievement and so on, these methods must provide fair and accurate results each time it is given. To ensure this, the any test must meet the following three key criteria:

- Standardization – The test must be based on results from a sample population that is truly representative of the people who will be taking the test,

- Reliability - The test should produce repetitive results and not be significantly affected by random factors,
- Validity – The valid test has to measure what it is intended to measure.

Here we must indicate the main difference between the psychometric measurement methods and the physical methods of measurement.

If a physical measurement is a comparison with a fundamental unit, such as kilogram, meter, candela, etc., then the psychometric measurement should correspond, by the definition of American psychologist Stanley Smith Stevens, to "assign numbers to objects or events according to some rule". This definition usually refers to measurement scales. These scales contain a certain number of positions placed in some kind of correspondence with psychological elements. In accordance with the classification of Stevens in the use the following scales: nominal, ordinal, interval and relative scales (Stevens, 1946).

Because by definition any measurement is result of comparison of measured value with a standard unit of measurement, methodologies of the psychometric measurements must be closer to diagnostics than measurement methodologies.

3.2. IQ-tests and its accuracy

Most of the early researches in the field of psychometrics was based on the desire to measure an intellect. Frances Galton, known as the "father of psychometry", included mental measurements in anthropometric data. Later psychometric theory has been applied in the measurement of personality, attitudes, beliefs, and academic achievements. Psychometric testing, as a rule, is used by employers during selection of candidates on this or that a position. In this regard usually used psychometric tests of personal qualities concern specific interests of the employer, because results of testing must precisely enough show, how the specific person can lead itself in the various circumstances which have developed at work, and what will be its reactions and the decisions.

The manifestations of intellect are not always obvious. Everyone could be convicted in availability of high intellectual abilities of such intellectual giants like Archimedes, Isaac Newton or Albert Einstein. The real intellect of many people is not so obvious, especially if speak about the cognitive component of intellect. Archimedes as like as later Rights brothers, or Tomas Edison were innovators and everybody could see the results of their innovations and therefore could be convinced of their genius. But to understand some unusual ideas of many scientists especially which contradicted existed dogmas, it was not easy. This was the main reason why many new ideas and theories were not recognized during the lives of their authors, and consequently, their authors were denied recognition of the exclusivity of their minds.

Since the beginning of the 20th century, for the evaluation of human intellect have been developed and widely used a variety of standard psychological tests. These tests are based on the calculation of the so-called IQ. As an abbreviation of the German term "Intelligenz Quotient", the term IQ denotes for psychologists an attractive test, since it allowed them to measure the level of intellectual abilities of a person as a whole using one indicator. The concept of "IQ" has entered the German psychologist Wilhelm Stern in 1912, based on the work of Alfred Binet and Theodore Simon. The value of the quotient is a result of dividing the mental age by the value of the person's chronological age. Mental age expressed as the chronological age for which a given level of performance is average or typical.

$$IQ = \frac{Mental\ age}{Chronological\ age} \times 100$$

Stanford University psychologist Lewis Terman took Binet's original test and standardized it using a sample of American participants. This adapted test, first published in 1916, was called the Stanford-Binet Intelligence Scale and soon became the standard intelligence test used in the U.S.

The next development in the history of intellect testing was the creation of a new measurement instrument by American psychologist David Wechsler. Dissatisfied with the limitations of the Stanford-Binet, he published his new intelligence test known as the Wechsler Adult Intelligence Scale (WAIS) in 1955.

At present, the most commonly used individual IQ test series is The Wechsler Adult Intelligence Scale (WAIS), designed to measure intelligence and cognitive ability in adults and older adolescents, and The Wechsler Intelligence Scale for Children (WISC) designed to measure intelligence for children between the ages of 6 and 16. The fifth edition of this test WISC-V (Kaufman, 2016) is the most current version.

Rather than score the test based on chronological age and mental age, as was the case with the original Stanford-Binet, the WAIS is scored by comparing the test taker's score to the scores of others in the same age group. The average score is fixed at 100, with two-thirds of scores lying in the normal range between 85 and 115. This scoring method has become the standard technique in intelligence testing and is also used in the modern revision of the Stanford-Binet test.

Over the past 100 years has been invented a lot of tests to measure the "IQ": the current versions of the Stanford-Binet, Woodcock-Johnson Tests of Cognitive Abilities, the Kaufman Assessment Battery for Children, etc.

IQ classification varies from one publisher to another. As far as is known, at the present time, numerous attempts to find criteria that could replace IQ failed, and the main reason is that while we can consider an intellect as a complex function of our body.

Historically IQ and personality tests were used as a means to control immigration in the US in the 1920s. IQ tests have been very popular in the middle of the last century. Some schools have used IQ tests for identifying gifted children, and some companies have used them as a basis for hiring.

Even in the beginning of the 21 century TV companies in many countries (Netherlands, Belgium, Germany, France, the UK, Australia, the USA) broadcasted a television program "Test the Nation" when millions of people did the IQ test online. At present, there are many organizations uniting people with high IQ, such as Mensa International, Intertel, Prometheus Society and others.

Many researchers have determined in the largest online study on the intelligence quotient (IQ) that results from the test may not exactly show how smart someone is.

In 1971, for the purpose of minimizing employment practices that disparately impacted racial minorities, the U.S. Supreme Court banned from using cognitive ability tests as a controlling factor in selecting employees, where at first the use of the test would have a disparate impact on hiring by race and at second where the test is not shown to be directly relevant to the job or class of jobs at issue.

More recently, under the pressure of growing criticism, the popularity of all of the above tests has decreased significantly, because many scientists and experts have come to the conclusion that IQ tests have many limitations in their implementation, and not very reliable. Essential lack of IQ-tests consists that they are counted on well-defined culture of the developed society. These tests do not take into account the impact of heredity, some peculiarities of temperament, differences in abilities etc. and most importantly do not allow to estimate the impact of various factors on the development of intellect. Numerous studies in recent years have shown that the difference in IQ can be determined not only by social and economic factors, but even motivation. IQ-test have severe limitations because it does not test all situations that show intelligent behavior.

In the new study, published in the journal Neuron, scientists led by Adrian Owen from Canada's Western University in Ontario analyzed results of 12 cognitive tests more than 100,000 participants from around the globe, and found that a simple IQ score is misleading when assessing one's intellectual capacity (Hampshire, 2012). According to this study the differences in ability relate to at least three components of mind – short-term memory, reasoning and verbal aptitude.

Though in the normal population, offered by Spearmen the term g- factor and IQ are roughly 90% correlated and are often used interchangeably, this term (g- factor) has not received a wide circulation because of absence of a way of its direct measurement. Beside of this term was proposed others terms, for example, "Mind Power" (Atkinson, 1912), or "Mental force" (Schwartz, 2003).

Idea to use one or both suggested characteristics of intellect: "Mind Power" or "Mind Force" is very perspective. Theoretically both terms are suitable to estimate a power that drive human life, that compels a person to get up in the morning even, when he/she was deadly tired, to earn money and win battle of living. Unfortunately, these authors and their followers could not find methods of above mentioned criteria measurement.

Below we shall describe a new method of an integrated assessment and management of intellect by means of the g-factor (G-factor) as well as methods of assessments of intellectual functions by means of appropriating criteria.

3.3. The brain as an universal measurement device

Our brain, together with the organs of the senses may be an amazingly accurate, versatile, reliable and easy-to-use meter that people use in real practice.

Many people from birth or through large practice acquire an ability to determine the distance without using any appliances or devices. Since olden times good eyes was needed for such profession as hunters, land-surveyor, constructors, designers, artists, surgeons, etc. The human brain, as well as brains of many animals, can measure lengths or distance by processing the difference in the readings of the left and right eyes. Bats measure a distance using ultrasound. They emit ultrasound waves and receive back reflected waves. The time it takes to receive the waves back provides them with a very good estimate of the distance. A mother can using her hands or touching a child forehead with her lips to assess whether her kid has a fever. Almost everyone can estimate the weight of an apple or any item from household utensils with not big mistake. Many people can measure time, for example they possess capacity to wake up at night at time set on the eve.

The function of evaluation of the distance to any objects is used by each of us when we move along the street in a crowd, dance in the dancing room or drive a car in traffic, or when we play sports games, for example golf, soccer, volleyball or basketball. Let take, for example, the favorite American basketball game and consider what accuracy of distance measurement can be achieved in this game. Briefly, basketball is sports command game, in which players throw a ball in 'a basket ', consisting of a hoop with a grid below and locating at height equal 10 feet from a floor. The primary skill in basketball is shooting. A basketball basket (hoop) is circular in shape and has an inside diameter of exactly 18 inches from edge to edge. A standard NBA basketball is 9.43 to 9.51 inches in diameter. It means that basketball player to get the ball into the basket cannot make a mistake more than 8.5 inches from edge of the basket no matter from what distance he throws the ball.

The longest successful basketball shot was measured 34.29 m (112 ft 6 in) and was achieved by Elan Buller (USA) in Oak Park, California, USA, on 9 September 2014. It is obvious that for successful shot the brain of player had to execute five main operations:

- Measurement of distance from player to the basket,
- Measurement of weight of ball, or comparison with weights of training balls,
- Calculation the trajectory of flight of the ball from hand of player to basket,
- Calculation of force and direction of a ball throw,
- Formulation the task to muscles of the hand and the arm: what muscles in what sequence and with what force should are moved, and on what trajectory the hand with a ball should be.

For our analysis, it is interesting to evaluate accuracy only for the first operation. If we calculate the magnitude of the maximum shooting angular error of basketball shot Θ, then it will be approximately less than 4.32 degree

$$\text{Tan } \Theta = (18-9.5)/112.5 = 0.0755.$$

This is a possible accuracy of the human measurement device hidden in his brain. It is not bad for device maiden from human cells.

Let consider the other popular game – golf. The longest putt ever made in the history of golf – at least as verified by Guinness World Records – is 395 feet. The putt was made by the Australian YouTube trick-shot team How Ridiculous on the fifth hole at Point Walter Golf Club in Australia in February 2017. Putt is a gentle stroke that hits a golf ball across the green towards the hole. Under the rules of golf, a golf ball has a diameter not less than 1.680 in (42.67 mm). The size of the golf hole on every standard golf course in the world is 4 1/4 inches (4.25 inches) in diameter. It means, that if we calculate the magnitude of the maximum shooting angular error of the golfers putt Θ, it will be approximately less than 1.7 arcmin (angle minute)

$$\text{Tan } \Theta = (4.25-1.68)/395\text{x}12 = 0.0005.$$

It is accuracy of an industrial measurement device.

I may be objected that the ability to measure distances for a basketball player or golfer is a simple remembering the position of the hand for throwing the ball obtained after many trials and errors. If we consider another game, it becomes clear why this is wrong.

About 300 million people worldwide play ping-pong, or table tennis, according to the International Table Tennis Federation (ITTF). In this game the brain of player has for successful shot to execute one very important complex operation – based on evaluation of coordinates and speeds of flying ball, the calculation of coordinates of a place where a paddle of player must meet with ball sending by an opponent player, on the one hand; and on the another hand- direction and force of the his shot by paddle. It means that inside of a brain there are one or several programs of fast calculation of distances, directions and forces based on various algorithms. Formation or use such програм develops mental abilities of a brain. According to researches of professor Wendy Suzuki from New York University, ping-pong players after training in this game increased attention, enhanced memory and other mental functions (McLendon, 2016).

The difference between our brain and measuring devices is that for many cases, a brain can operate with various units of measure for different measured parameters, which are latent for consciousness, but it does not influence on the accuracy of internal calculations. We have to admit as a hypothesis that the brain can create itself these latent units.

However, this does not mean that the human brain cannot use common physical units for its measurements. Information in the next section proves that our brain with sufficient accuracy can measure lengh (distance), volume, vibration and other physical values using existed units for measurement, i.e. can do physical measurements.

Beside this, a brain using latent units of measurement can measure not only the physical values but also the mental and psychological traits, which no other tool can measure directly. For example, it can be the level of intellect, level of intuition or stress, and so on.

Measured information during the actions described above, such as for instance basketball or golf games, is formed in the depths of the brain and appears for internal use most likely in a language of vibrations incomprehensible to our conscious part of the mind.

Thus, if we shall find a way which will help to provide, first, access to this information using password or access code, secondly, to deduce this information in a form convenient for perception, and after that to decipher it, then we can use a brain as a measuring instrument in a wider range.

Currently, there are many methods of different quality to extract information getting and stored in subconscious part of the mind. Below we will consider several of them to find the best to have possibility for direct quantitative measurements of intellect functions. It is obvious that this type of measurement much better than the existed above-described expert evaluation methods and psychometric methods for IQ.

Because a form of extracted from subconscious part of mind information depends on method of extraction, access code for this information we will describe after description of extraction information methods.

3.4. Radiesthesia as subjective method of extraction of information from subconscious part of mind

Applied kinesiology (AK), innovated by a chiropractor George J. Goodheart in 1964, is a technique in alternative medicine that allows by testing muscles for strength and weakness to receive from subconsciousness answer about a condition of many organs and evaluate, for example, effectiveness of treatment (Horowitz, 2011). According to the American Chiropractic Association, in 2003 about 40 % American chiropractors employing AK methods in their practices. Experienced chiropractors can use this method for evaluation of functionality and diagnostics different organs of human body. The method of applied kinesiology was used by Dr. David Hawkins to assess the levels of human consciousness according to a scale developed by him as a result of numerous experiments (Hawkins, 2002). But this method has the limited possibilities.

The methods of dowsing do not have these drawbacks. A wide class of the dowsing methods that maybe have the same roots as AK or clairvoyance can be very useful for goals of intellect functions measurements. The term "Dowsing" came from the German word "dowsing," meaning, "to seek, to find." Dowsing (also called biolocation) is practices, historically declaring the possibility of finding hidden items, usually located underground, such as cavities, water sources, mineral deposits, buried treasure and even missing people with the help of bare hands or keeping in hands special tools like vines, angle rods, Y-shaped rods, wands, pendulums and so on. It has also been used to guide people to make daily decisions or to determine if someone is telling the truth or not.

Many dowsers as well as AK professionals can detect a reason of allergies and other ailments, and even accurately determine the gender and birth date of unborn babies.

Dowsing has a very ancient history. Herodotus writes of its use by the Persians, Scythians, and Medes. There are records of its use by the Etruscans, Hindus, Egyptians, Greeks, and Romans several thousand years ago.

In our era one of the first wide known dowsers was a French Catholic priest Abbe Alexis Bouly. At the end of World War I, he became well known as a water dowser after finding commercially important supplies for French manufacturers and contracting to do likewise by other industrialists in Belgium, Portugal, Poland and Romania. Bouly eventually founded the Society of Friends of Radiesthesia, a new word he coined for dowsing. This word comes from the Latin word "Radius" (ray) and from the Greek word "Aisthesis" (sensation). In 1953 UNESCO study of radiesthesia conducted by leading European scientists came to conclusion that there was "no doubt that it (dowsing) is a fact" (Webster, 2010). In the United States of America were created the "American Society of Dowsers" and regional dowsing group in the many states, which acts for qualified therapists, energy workers and those involved in alternative or complementary medicine, who are either actively using dowsing in their work or would like to learn how to use dowsing in their work. There are the British Society of Dowsers and the International Association of Health Dowsers (I.A.H.D.) in England. The similar societies exist in Canada, Italy, Japan, Portugal, Scotland, Latvia, Russia, etc.

In the USA, about 30 % chiropractors use dowsing with a pendulum in their practice. In Europe, especially in France and Russia, physicians have used the pendulum to assist them in making diagnoses and healing.

At first time I saw a pendulum for this goal in the San Jose, CA in the hands the licensed Acupuncturist, President of Alliance of World Traditional Medicine Dr. Andrew Q. Wu.

According to Chantal Cash (Cash, 2013), today a dowsing practice divided in Medical Dowsing, Psychic Dowsing, Dowsing for Spirits, Dowsing for water & other resources, Dowsing for Information, Archeological Dowsing, Missing Items and Map Dowsing.

Experience with the use of classical dowsing shows that professional dowsers can measure, for example, the depth of the source of usable water (or oil, metal, etc.) in meters or feet.

According to Richard Webster, an experienced dowsers can find also for example the possible yield of the vein in gallons of water per minute (Webster, 2012). In addition, they can measure almost any physical state (temperature, pressure, weight, speed, etc.). Pendulum even can be used as a lie detector (http://www.diviningmind.com/dowsingcharts.html).

Russian studies shows that dowsing in arms of experienced dowser allow to measure a frequency of patients pulse (Puchko, 2015).

In 1930, a French physicist and archaeologist A. Bovis innovated the Bovis Scale (Biometer) as a way of measuring the radiation of objects and expressing these values as wave lengths. A Bovis Biometer is a drawn scale coupled with typically a pendulum held by the operator, that is used to indicate the measurement. Bovis invented this for the purposes to understand the mystery of the phenomena of the Great Egyptian pyramids and evaluation of cosmo-telluric radiation. After Bovis's death, engineer Simoneton developed 4 scales for measurement of subtle human bodies - Etheric, Astral, Mental and Causal. The next and last person who worked on this scale was clairvoyant Blanch Merz. She added another 2 scales (Spiritual and Pure Spiritual) to the existing (Haxeltine, 2008). Unfortunately, Bovis and his followers did not reveal principle of design of Biometer. It is unknown also why he used angstrom as unit of his scale[3]. At present Biometer is very popular amid radiesthesists and dowsers[4]. The units of Bovis scale, to his memory, are called "Bovis units".

The Bovis Scale used to measure the Life Force (or some other) energy level of any substance, food, medicine, living beings, objects or geographic places.

Italian dowser Frank Zapa created another chart, which used to help measuring a level of vital energy in humans (in %) before and after treatment.

Unfortunately, no one single theory has been found to satisfactorily explain what goes on when dowsing takes place. Many people believe that dowsing is a natural intuitive skill. Unlike general intuition, which gives information on one or other ways in random 'hits', dowsing is focused intuition. It can be more understandable, if remember a definition of intuition of the Carl Jung: "Intuition is perception via the unconscious that brings forth ideas, images, new possibilities and ways out of blocked situations".

Some hypothesizes have been offered to explain dowsing from point of view Quantum physics and non- locality, but here is how write Haxeltine (2008) "the problem is that the more one tries to explain dowsing the more complicated it gets".

When inventor Thomas A. Edison was once asked, "What is electricity?" He replied: "I don't know - but it's there - so let's use it". The same can be said about radiesthesia because this method really works in the hands of sensitives and trained skilled people. (http://www.biolifestyle.org).

---

[3] The angstrom is obsolete unit. It has been largely superseded by the nanometer (nm), 1 nm = 10 angstroms.
[4] It can be found in the Google in the different editions.

In my opinion, dowsing and applied kinesiology allow to receive and decode information from subconscious or superconscious. Below we will use only one synonym of dowsing - "radiesthesia".

In essence, radiesthesia includes a method of receiving answer on a question to subconscious mind or a method of subjective (by operator) measurement of all necessary parameters using pendulum (or other devices) through mechanisms of operators subconscious mind with the help of appropriate charts or scales. Commonly used dowsing tools include not only the pendulum, but and L shaped swivel rods, forked sticks, and so on.

From the point of view of biophysics, the dowsing and AK effect can be explained as a way of transforming the vibrational information being the internal language of communication between the organs of the body and the brain by way understandable for our sense organs. Using a pendulum or other dowsing tools in connection with human mental faculties allows obtaining some wave information, based on ideomotor activity effect (ideo-dynamic response or Carpenter effect).

The existence of an internal vibrational language of information exchange is confirmed by the following response of a person who could read other people's thoughts, Wolf Messing to correspondent of journal Oreshkin P.: "... It's not mind-reading, it's, like the "reading of muscles"... When human thinks hard about something, the brain cells transmit impulses to all muscles of the body. Their movements, invisible to the eye, I can easily feel.... Often I'm performing mental tasks without direct contact with the inductor. The pointer to me here is the breathing frequency of inductor, the beating of his heart, voice timbre, his walking nature etc."(Oreshkin, 1961).

A pendulum is a weight suspended from a string held between the thumb and index finger; and its movements can be deciphered on answers "yes", "no", "don't know", and also on results of measurement with the special charts or scales. Most professional pendulum dowsers use a gem or crystal suspended on a light chain or special strings. Some of dowsers believe that using crystals can help absorb excess energy in order to be able to concentrate on the task at hand which is to dowse properly.

In some literature, method of decoding information using questions "yes" or "no" received name as Bar Kokhba method (maybe more known as Bar Kokhba game). Simon bar Kokhba was leader of a rebellion of the Jews in the Roman province of Judea circa 132–136 CE, establishing an independent Jewish state, which he ruled for three years. According to legend, once was brought to him a disfigured warrior, who had his tongue pulled out and his hands cut off. Unable to speak or write, the victim could not tell who had mutilated him. To find out the truth, Bar Kokhba decided to ask simple questions to which the dying warrior could nod or shake his head. This was enough to get the right information, and the attackers were captured based on the received by this way data.

A technology of pendulum use is very simple. A person asks a question when, for example, a pendulum hangs over something. The pendulum responds by swinging or spinning. Although there are many factors attributed towards how the pendulum swings, it is most likely connected to our subconscious via our micro muscle movement. For skilled persons (professionals) the method gives right results in the condition of a clear mind, and a quiet area to dowse (Longley, 2015).

Now, a guide "How to use a pendulum "can be found in many books, articles and blogs in the Internet, for example in the book (Webster, 2012). Let consider one else very important question: who can be a dowser?

It is very interesting question. The first systematic studies of sensory thresholds were conducted by physiologist Ernst Weber at the University of Leipzig in Germany. Weber's experiments were designed to determine two types of sensory thresholds: absolute and relative (like a minimal difference in intensity between two stimuluses). Researches show that different people has a different threshold of sensitivity for this information acceptance (Harris, 2007).

Based on available information, all human beings, depending on the magnitude of the sensitivity and ability to concentrate their attention can be divided into three categories. The first category is sensitives (2-4% of population), they can feel different kind radiation in the space surrounding us like as a sensation of heat, cold, tingling, or see it with the help of the third eye. The second category - the dowsing and radiesthesia operators (20-30%), who can be easily trained to spontaneously form ideomotor act in their setting (focusing) on a certain object and receiving radiation from it. The third category is the majority of people (70-80%), which either cannot concentrate their attention for the acceptance unknown for them signals, or their threshold for radiation is higher than the level of radiation from surrounding objects. Most of them can form ideomotor act as dowsing and radiesthesia operators after the special training (Mermet, 1991). Lyudmila Puchko, who spent her life developing the methods of radiesthesia for diagnosing and treating people, believed that almost every adult person can awaken the radiesthetic effect (Puchko, 2008).

Author in his studies used as tool for radiesteshia the different types of pendulum, charts and scales, demonstrated below, and leave to readers unlimited opportunities for selecting other tools.

3.5. Regimes and some rules for assessments and measurements

In technology of radiesthesia can be consider the following informative regimes:

1. receiving answer on a direct question to subconscious part of operators mind. Questions should be compiled to get unambiguous answers "yes", "no", or "I do not know"
2. receiving by operator results of measurement of needed parameters with use pendulum located over scales
3. receiving by operator an answer on a direct question or results of measurement of needed parameters for any other person, who needs to touch a scale or paper with written questions; in this case subconscious part of operators mind collect information from subconscious part of other persons mind
4. for sensitive operators receiving answer on a direct question to Universe, maybe Gardian Angel, or Universe Knowledge Base.

The technology of using the pendulum is described in many books on radiesthesia, for example (Mermet, 1991; Eason, 2005).

To ensure the necessary quality of answers to the questions posed and to obtain stable measurement results, it is necessary to fulfill certain requirements for the operator's thought processes, which must contain the following stages:

• Concentration of attention on the task and orientation of the mind to obtain a possible answer

- Mental or verbal formulation of the question
- Creating a special state of mental condition for efficient and selective perception with defense from as any undesirable form of suggestion.
- Collecting a result in the form of neuro-muscular reflexes reaching the operator's fingers, amplified by the pendulum,
- Analysis of the result of credibility. If the result seems implausible, then you must either repeat or modify the questions.

Experienced radiesthesists execute all of this stages automatically, sometimes with only bare hands.

3.6. Basic scales for radiesthesia (dowsing)

Experience proved that for the intellectual functions measurements very appropriate to use a pendulum in conjunction with the radiesthesia scales represented in Fig.3.1, or special charts with scales described in the next section. In this figure R-scale can be constructed in the form of a sectored semicircle or a circle for different diapasons, for example O-10. 0-100, 0-1000. 100-200, and so on. In this scale, and in scales below any sector has a different tone to increase an accuracy of measurement.

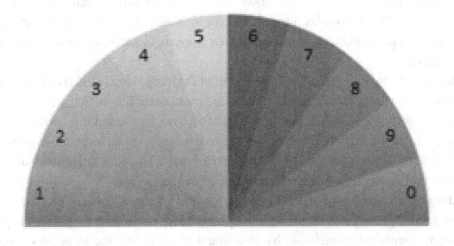

Fig. 3.1. R-scale (0-10)

If we return to the regimes of using the pendulum, in the first and second regimes (section 3.5) suspended pendulum is placed above the scale on the list of paper or above computer screen (what can be easily do for a tablet PC, or smart phone), dowser asks a question, after that pendulum starts oscillations. In the second regime the plane of the pendulum's oscillation turns, and when it stabilized, we define line of intersection of a plane of fluctuation of a pendulum and a plane on which the scale of measurement is located. This line shows a result of measurement.

In the third regime is needed a phantom of the patient (for example, photo or lock of hair) laid on the chart (scale), or that the patient touched by his/her finger to a paper with the chart (scale) or to a computer screen with this chart (scale).

Objective methods of measurement obtained by biofeedback and represents the measured value by one or another way allow to make optimization of characteristics of intellect or intellectual functions.

The executed experiments demonstrated that a pendulum allows to measure some psychometric parameters as g-factor, stress level, intuition level, creativity level, etc, i,e, the parameters, that cannot be measured by other methods today.

3.7. Access codes for subconscious part of mind

Today is known many ways to access subconscious and unconscious parts of mind, for example:

- Neuro-Linguistic Programming (NLP)
- Hypnosis
- Meditation
- Visualization
- Being Consistent And Repetitive
- Etc.

In its essence, the idea of an access code was used by Sigmund Freud when he suggested using special words ("switch-words") for direct access to the subconscious mind and the impact on it. Switch-words (or switchwords) can be written, spoken, chanted, and even used as part of a meditation practice.

Later James T Mangan in his book "The Secret of Perfect Living" (1963) devised the first list of switch words. Now thanks to Liz Dean there are more than 200 switch words in circulation (Dean, 2015). Unfortunately switch words don't allow to read a needed information completely.

For this goal, among different methods of the radiesthesia seems to be more perspective the Method of vibration series innovated by prominent Russian scientist, Academician of the Russian and International Engineering Academies Dr. L. G. Puchko.

According to her testimony, the idea of her method was connected with discovery of the little-known French pharmacist G. Lessour (maybe Lessur), according to which it is possible to eliminate a pathogenic microorganism by selecting a plant having the same wavelength as the pathogen and taking this plant inside with homeopathic doses. The discovery of G. Lessour is difficult to overestimate. He first developed a method for determining the presence of infections by their radiation, and instead of the empirical method of selecting herbs, he used an individual search using a resonating pendulum.

Practical use of the methodology developed by Puchko L.G. allow revealing pathological abnormalities of the organism that are invisible and cannot be registered by any other means. Here are many evidences that this methodology conducts early diagnostics of oncological, infectious, psychic, psychosomatic and other diseases, to determine the root causes of these diseases and to find the most effective ways of healing for a particular person. The experience of the application of this method in alternative medicine practice demonstrates also success in the treatment of some diseases.

Vibration series can be various types: therapeutic, cleansing, restoring, protective, etc. The new class of measuring vibration series, developed by author and described below, is very promising.

In my opinion, the Method of vibration series allows, at first, creating agreement between consciousness and subconscious part of the operators mind about reception and transmission of information using natural language, and, at second, to find an access code for input and output information for the subconscious and deeper structures of the operator brain. Access code looks like as magic words "Open, Sesame" used Ali Baba to open door for full treasures cave from ancient Arabian tale "Ali Baba and forty Thieves". Access code is more like as an individual password, because it is created only for operators use, it is like a personal key for door in the operator's brain between conscious and subconscious parts, or between conscious and superconscious parts of the mind. Famous Ukrainian cardiac surgeon, a medical scientist, and cybernetician Nikolai Amosov wrote about coding information between brain structures in the book "Modeling of thinking and psyche" (1965). At the same time, he noted that the best ancient code is pictures and verbal descriptions.

The hypothesis of using symbols as an access code can be confirmed by a similar use of symbols in Japanese practice Reiki, an alternative form of healing developed nearly 100 years ago in Japan by a Buddhist monk named Mikao Usui. Using in the Reiki, for example, Power symbol named Cho-ku-rei means "I have the key", The Mental/Emotional Symbol' Sei Hei Ki means "Key to the Universe" (www.reiki-for-holistic-health.com/reikisymbols.html).

Let's remember the saying "a picture is worth a thousand words". It can be assumed that the internal language of the subconscious makes extensive use of symbols, including those that are essential elements of the collective unconscious (Brenner, 2013).

The source for vibrating series design is a set of geometric, alphabetic, or numeric characters, cosmic and zodiac symbols, esoteric signs, to name just a few. All of them can be a reflection of reality, and carry some information and energy. The symbols and characters were known to humankind even before the appearance of ancient writing, they were passed from generation to generation. Their storage is a part of the collective unconsciousness (according to Carl Jung' theory).

Lyudmila Puchko believed that this code works as amplifier of vibrating signals from consciousness to subconsciousness. It is a reason why she coined this method "vibration series" (Puchko, 2010).

Creating access code to subconscious part from conscious part of the mind for an measurement of needed criteria is carried out by means of special radiesthesia charts and a pendulum with setting the certain commands (Puchko, 2010; Green, 1994). Obviously, these access codes must be different for the different operators and for different intellectual components.

A developer of vibrating series determines the number, type, and location of each symbol of the series and the shape of the mark in which these symbols must be enclosed. In this process, a developer defines tasks and functions of that. Then a series is charged by operator and, after verifying its effectiveness, starts to work.

Without delving too deeply into this site, creating an access code begins with the formulation of the concept, which should be reflected in the main purpose of the series. To understand how it is doing, consider the example of constructing the access code for measuring the level of intuition.

Here is installation concept which operator formulate for its subconscious part of mind: "To create access code for measurement of intuition". After that he/she formulate the request to the subconscious mind: inside of which figure lies code for the measurement of intuition?" To answer a question he use radiesthesia charts (R-chart), represented on Fig. 3.2, and pendulum. To do is, he brings a pendulum to the center of this R-chart, and asks the question stated above. The plane of the pendulum swinging after initiation should stop within a particular sector of the R-chart. For example, in our case for intuition the answer is a rectangle.

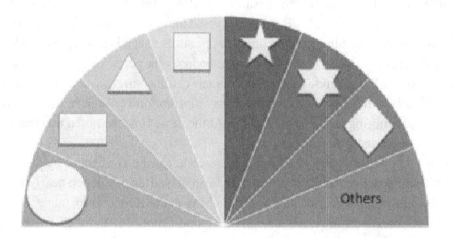

Fig. 3.2. R-chart for selecting a figure in which an access code is enclosed

Next question: how many rows in access code for measurement of intuition? -Answer by using the scale Fig. 3.1 is "1".

Asked a question: how many characters per line, get the answer using the same scale – "1".

Asked a question: what kind of symbol must be included in the access code, using a pendulum and R-chart Fig. 3.3, get an answer – "Number".

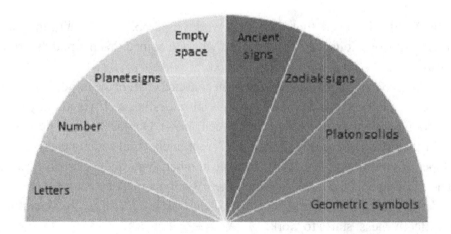

Fig. 3.3. R-chart for choosing a type of symbol

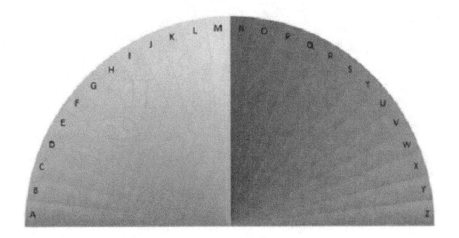

Fig. 3.4. R-chart for choosing a letter

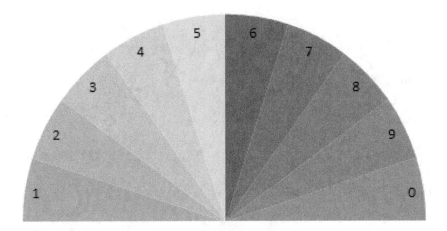

Fig. 3.5. R-chart for choosing a number

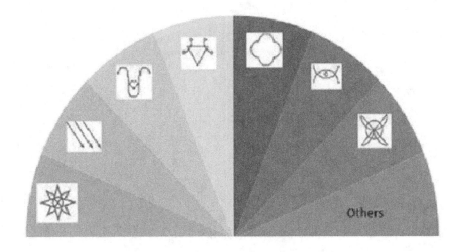

Fig. 3.6. R-chart for choosing an ancient signs

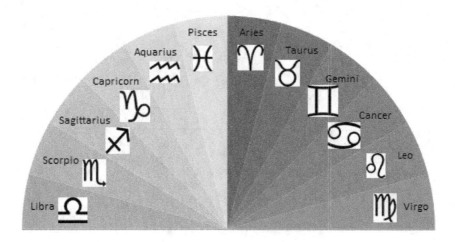

Fig.3.7. R-chart for choosing the zodiac signs

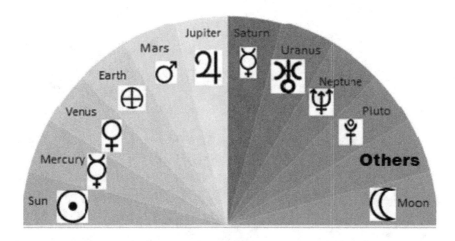

Fig.3.8. R-chart for choosing the planet signs

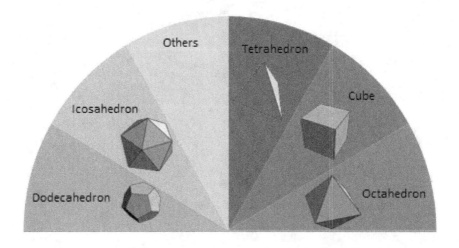

Fig.3.9. R-chart for choosing the Platonic Solids

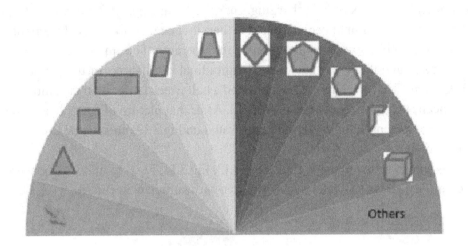

Fig. 3.10. R-chart for choosing geometric symbols

To find a number, an operator using scale Fig. 3.5 get an answer – "1". In common, if by chart Fig.3.5 he got answer Letter, he would need to use R-charts for Letter (Fig. 3.4.), if he got answer "Ancient signs", he needs to use R-chart "Ancient signs" (Fig.3.6) and in other cases to use accordingly the following R-charts: Zodiac signs (Fig. 3.7), Planet signs (Fig. 3.8), Platonic Solids (Fig. 3.9), Geometric symbols (Fig. 3.10).

If at a choice of a symbol the sector "Others" drops out, it means that it is necessary to pick up some element of an access code from any other sources which can be found in the Internet or in the books (for example Koch, 1965).

The access code must be defended by an even quantity of circles (2, 4, 6...), defined by pendulum help from Fig.3.1.

After that a needed access code is almost ready (see Fig. 3.11.). It can be cut out from the paper.

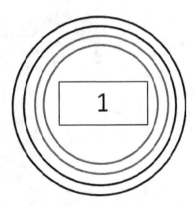

Fig. 3-11. The access code for intuition level measurement

Next, according to the Method of "vibration series" it is needed to charge the access code with internal energy of the operator. It is desirable that operator make this procedure as follows. Above the surface of the figure of the access code is placed a pendulum and the operator mentally or aloud commands: "Access code is charging, charging, charging." After a while, the pendulum will begin to make longitudinal oscillations. Further, it is needed to check the amplitude of the natural pendulum oscillations over the palm of the operator. The image of the access code is placed on the palm and the amplitude of the longitudinal oscillations of the pendulum is again checked. If this amplitude becomes at least one and a half times larger than the amplitude without a pattern, it means that the access code is charged. After that picture of access code unites with a scale for its measurement. It looks so as demonstrated on Fig.3.12, and can be used on ipad, ipod or telephone screen.

The scales shown below in the Fig.3.12, Fig.3.13, Fig.3.14, Fig.3.15 allow to measure levels of intuition, creativity, stress, G-factor and willpower with sufficient accuracy.

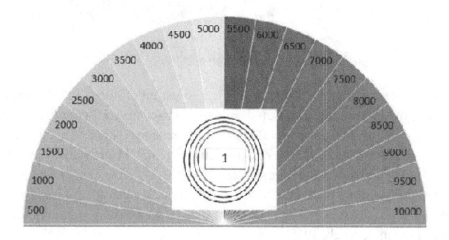

Fig. 3.12. R-scale with access code for intuition level measurement

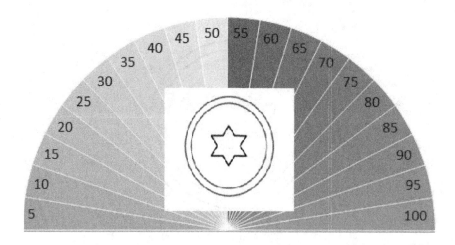

Fig. 3.13. R-scale with access code for creativity level measurements

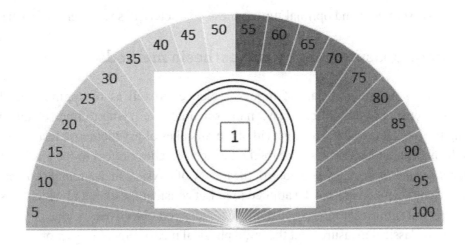

Fig. 3.14. R-scale with access code for stress level measurements

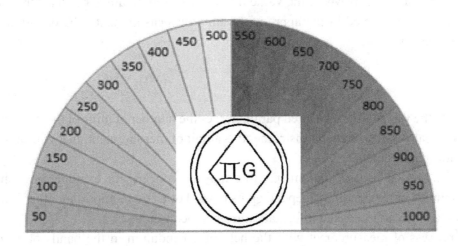

Fig. 3-15. R-scale with access code for G-factor measurement

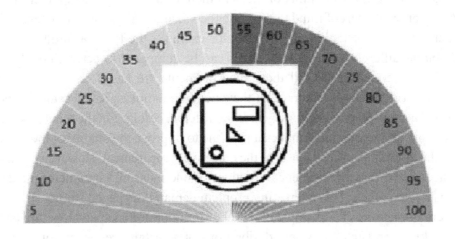

Fig. 3-16. R-Scale for willpower measurement

Results of measurement and optimal tuning these characteristics represented in Chapter 9.

3.8. Quality of measurement by radiesthesia methods

Measurement is an integral part of modern science as well as of engineering, biology, chemistry, commerce, and daily life. Today, many scientists are trying to find reliable methods of measuring the intellect, which would get rid of the mistakes of psychometric tests that strongly depend on the accepted testing rules. The method described above refers to the class of physical measurements and therefore many positions of the theory of physical measurements, in particular, concerning measurement errors, can be applied to it. Let consider the extent to which these errors of measurement are significant in order to achieve the desired objectives.

According to classical measurement theory, a physical measurement is a set of operations for determining a relationship of measured value for another value adopted by all participants as a base standard unit.

A direct measurement is where the value of the measured parameter is read from the scale of the instrument, graduated in the appropriate units of measurement. The equation of direct measurement has the form

$$F = c \cdot f, \qquad (3\text{-}1)$$

where F is the value of the measured parameter; c- the number of units of the measured value from the unit scale of measurement instrument; f- a unit of measurement, that is usually used as a standard for measurement.

In physics an object of pendulum research may be concern for example how should the rocking amplitude of the pendulum depend on its mass. For us more important other information of pendulum as indicator of asked information from subconscious.

In the process of measurement with the help of a pendulum in the hands of an operator, the value of unit of measurement is not known, and in principle for each operator (person), it is different, but as shown, this value is constant. As was demonstrated above in section 3.3, consciousness does not know the units of measurement of a particular parameter and does not prevent the subconscious part of mind to measure the value of a parameter in latent units.

If an operator uses a pendulum to measure, for example, the level of intuition for different people and obtains different values, these values can be compared to determine which of these people intuition is better, because all these measurements are based on the same latent unit of measurement. If different operators with their pendulums measure for example the level of intuition for one person, they will naturally receive different results. On the other hand, the use of different operators for measurement makes it possible to improve the quality of the measurement by averaging the results and bring it closer to the objective measurement. However, a small number of studies have shown that the difference of values of one measured parameter maiden by two operators was not too much. This interesting problem demands a special study.

If the operator measures for example the level of intuition for one person before (state 1) and after affecting his brain (state 2) by sound vibrations of a certain frequency for a certain time, then in this case the relative change of the level of intuition will not depend on value of the latent unit of measurement. This can be seen from the following formula

$$\delta F = (F2-F1)/F1 = (c_2 \cdot f - c_1 \cdot f)/f = c_2 - c_1, \qquad (3\text{-}2)$$

where δF is the relative change of the measured parameter in the second experiment with measure equal c_2 in comparison with the first one with measure equal c_1.

Every experiment executed with errors. Usually here are systematic and random errors.

The systematic error is the component of the measurement uncertainty, which remains constant or changes regularly with repeated measurements. A random error is a component manifested in the form of unpredictable deviations from the true value of the physical quantity, changing from one measurement to another.

Random errors can be evaluated through statistical analysis and can be reduced by averaging over a large number of observations. Systematic errors in experimental observations usually come from measurement methods and the measuring instruments. Systematic errors are difficult to detect and they can be found in the comparison theoretic and practice results, or in comparison of different measurement methods.

Depending on the causes of the occurrence systematic errors are divided into the following types:

1. The methodical errors of the method
2. Errors of external factors: external thermal, radiation, electric and magnetic fields
3. Errors arising from inaccurate actions or personal qualities of the operator, called personal errors.
4. Instrumental errors due to technological imperfections of measuring instruments, their state during operation. For example, wrong pendulum or R-charts.
5. Stress, fatigue, illness, bad mood, even inconvenient position of the body of operator.

Obviously, a radiesthesists body and his mind conditions define an accuracy of measurement. Linda L. Green defined different reasons that can influent on accuracy of measurements by pendulum or others pointers. Here are: magnetics storms, over limit radiation, extraordinary storms on the sun, earthquakes, EMF, works of art, negative energy, and so on (Green, 1994).

As shown by the results of comparative tests of pendulums of various shapes made of different materials, the differences in their readings are within the accuracy of the experiment with one pendulum.

The practice of comparative measurements for optimization of tuning the intellect and its functions makes it possible to minimize many systematic errors of measurements. Analysis of random errors of measurements we will show in chapter 9.

CHAPTER 4

Where are intellectual abilities hidden

Our present lives are subordinated to the god, name to whom - intellect. At the same time this is a very our greatest and most deplorable illusion. Because of the intellect, we assure ourselves, we conquered nature. But this is only slogan-not more. Since "the conquest of nature" turns around for us by the overpopulation of planet, adding troubles in the political plan, since people continue to quarrel and to seek superiority over each other just as always. About what "conquest of nature" then it is possible to speak?

Carl Yung

Most cells in the body are invisible to the brain and largely take their own decisions.

Brian J. Ford.

4.1. Tasks formulation

A question "Where are intellectual abilities hidden?" today has not a simple answer. The human body is a super complicated system for the study, and for its description sometimes is not enough the available knowledge of physical, chemical and biological laws.

Usually in studies of complex systems, researchers are trying to build a functional model of a system which allows to predict its behavior in any given situation. Models can be different: math, simulation, purely descriptive (phenomenological), etc. Instead of this in the scientific methodology usually used cause-effect analysis. In our case the latter can build links (cause-effect relationship) between what and how it affects on intellect. It aims at discovering possible or probable causal factors and their outcomes and may lead to the creation of a cause and effect diagram.

Collected and presented below information about brain, brain functions and intellect is not full and does not allow explaining the all, but using cause-effect analysis we can receive answers on the some very important questions. For this goal and for our super task of understanding an intellect and its functions (or intellectual abilities) we will consider many themes from the viewpoints of the various disciplines – from physiology to physics, from genetics to psychology, etc.

Let starts from the genetics.

4.2. Cell Intellect

Smart cells. Let start our study of the nature of intellect from the point of view of molecular biology, genetic linguistics, biophysics, biochemistry, etc.

The human body is multicellular organism, which has for adults about 50 -100 trillion cells. These cells vary enormously in sizes, forms, structures and functions (Alberts, 2014).

It is known, that the basic components of most cells are the nucleus that contains genetic material, the energy-producing mitochondria, the protective membrane at the outside rim, the cytoplasm in between, and more. By analogy with technical systems, a cell may have power stations, transportation stations, communication system, with great variety additional functions. It means that cells is a tiny, but very complex world, which must be under the autonomic internal control subsystem, allowing to defend and maintain itself in the different situations, to create new cells, to replace dying cells, to repair itself, and even moving during processes such as wound healing, the immune response, etc.

Each cell has the atomic and molecular structure. Each atom consists of a core with rotating around electrons, which leads to the creation the electromagnetic and spin (torsion) fields, on one side, and a different vibration frequency on the other. A cell is a system consisting of a plurality of synchronously working subsystems supporting an information exchange.

Now we have some evidences that human cells can communicate each other not only via light and chemical messages –words, composed of amino acid or nucleated alphabet, but and via electromagnetic, torsion, or acoustic radiation. All cells can communicate with each other almost instantly. From all this, now we can meet in not only esoteric literature paradoxical conclusion that cells of the body can think and talk on the special language. Moreover, according to yoga, a trained person is able to understand what cells say. On the other hand, thoughts, produced by people, can translate to the cell level because body cells are able to understand the language of the people. Now many scientists, including author, counts that human cells play their roles in our life and health as musicians in the orchestra.

Here is an excerpt from the book "A Guide to Trance Land: A Practical Handbook of Ericksonian and Solution-Oriented Hypnosis (2009) written by Bill O'Hanlon: "One time when I was doing a demonstration of hypnosis in Finland. I asked for four volunteers who understood and spoke English very well to come up to be part of the demonstration. During the trance, I invited people to come out of trance from the neck up and stay in trance from the neck down. After that, I invited them all to experience hand levitation. Later, one of the demonstration subjects told me that she realized during the trance that she could no longer understand English but that her body still could. She said that when I had invited the hand levitation, she couldn't understand what I was saying and had to wait for the Finnish translation that was being provided. But her hand began to twitch and move even before she heard the translation, so she knew that her body could still understand English even though her mind couldn't. There are parts of you that can understand things that other parts don't." This citation shows that cells or organs of our body can understand what sometimes our brain cannot.

The term "Cell intellect" was coined by Nels Quevli in the year 1916 in his book entitled "Cell Intelligence: The Cause Of Growth, Heredity, And Instinctive Actions…." illustrating that the cell is a conscious, intelligent being, and, by reason thereof, plans and builds all plants and animals in the same manner that man constructs houses, railroads and other structures (Quevli, 1916).

Bruce H. Lipton, in his book (Lipton, 2005) writes: "As a cell biologist I can tell you that you in truth a cooperative community of approximately 50 trillion single- celled citizens. Almost all of the cells that make up your body are amoeba-like individual organisms that have evolved a cooperative strategy for their mutual survival."

"My research for the past 30 years or so was devoted to examine whether cells have such signal integration and control centers. The results suggest that mammalian cells, indeed, possess intellect" –writing Professor Emeritus, Feinberg School of Medicine, Northwestern Univ., Chicago, Guenter Albrecht-Buehler in his blog (Albrecht-Buehler, 2013).

Smart genetic apparatus of cells. In 1957, researcher Dr. Dzang Kangeng from China began supergenetics experiments that echoed previsions Russian scientists A.G. Gurvich and A.A. Lubishchev. In the 20's — 30's, they predicted that the genetic system of organisms on the Earth works not only on real (material) level, but also on the field level, transmitting genetic information by means of electromagnetic and acoustic waves. Dr. Dzang Kangeng showed that light passed through a duck into 500 chicken eggs resulted in chicks in which 80% had flat bills, 90% had eyes moved to a position similar to ducks and 25% had webbing between their toes. Their offspring remained partial duck, partial chicken.

According to the theory of linguistic wave genetic created by famous Russian scientist and Director of Wave Genetics Institute in the Moscow Garyaev Peter, because everything in the human body depends from genes, human intellect depends from genes structure, information inside them and influence from outside (Gariaev,1994).

Deoxyribonucleic acid (DNA) is the molecule that carries the genetic information in all cellular forms of life and some viruses. Any gene is a part of DNA that carries the information needed to make a protein – a main building block of the human body. The latter has about 20,000 to 25,000 genes located on chromosomes. According to genetic science, the set of all genes forms the human genome. Generally, the human genome provides heredity and variability of human organism.

Heredity is characterized through the transmission (inheritance) of genetic material from parents to offspring, determining the similarities between the individuals of the same species. Variability represents the differences between the parents and offspring determined by molecular, biochemical, physiological and environmental factors.

The information in DNA is stored as a code made up of four chemical bases: adenine (A), guanine (G), cytosine (C), and thymine (T). Human DNA consists of about 3 billion bases, and more than 99 percent of those bases are the same in all people. The order, or sequence, of these bases determines the information available for building and maintaining an organism, similar to the way in which letters of the alphabet appear in a certain order to form words and sentences.

Usually all nuclear human cells except sperm and egg cells contain 46 chromosomes, 23 pairs to be exact. Pairs 1 to 22 are identical or nearly identical; the 23rd pair consists of the sex chromosomes, which are either X or Y. Each egg and sperm contains a different combination of genes. This is because when an egg and sperm cells form, chromosomes join together and randomly exchange genes between each other before the cell divides. This means that, with the exception of identical twins each person has unique characteristics.

With the fairly recent decoding of DNA in our genes via the Human Genomic Project (2003 year), scientists thought that now we will have a "full" understanding of life as dictated by our genes. Unfortunately, in this area of science there is still a lot of unclear, for example, there is no consensus among geneticists today how many genes contain the human genome (Saey, 2018). On the other side, the number of genes that we possess is not so different than most living organisms have – i.e., we are not so "special". Comparative genomics has revealed high levels of similarity not only between closely related organisms as humans and chimpanzees, but and similarity between such organisms as humans and for example rats.

Now scientists begin to understand that the "junk" DNA in the human genome appear to act as regulators that control the information that affects the operation of our body and hence its health.

In recent years, epigenetics has exploded into one of the most exciting and rapidly expanding fields in biology. This discipline allows to explain any process that alters gene activity without changing the DNA sequence, and leads to modifications that can be transmitted to daughter cells.

Russian linguists studying a branch of semiotics known as "Genetic Linguistics" discovered that the genetic code in DNA follows uniform grammar and usage rules virtually identical to those of human languages.

Generally 46 chromosomes can be viewed as a database or knowledge base consisting of 46 packages of information. In essence, genetic apparatus operates as follows. The texts, written in the "DNA language", are first translated by the organism into the "RNA language" and then into the "Protein language". And proteins are the stuff that we are mostly made of (not counting water).

The Muscovite group of Dr. Peter Garjajev, including highly-qualified molecular biologists, biophysicists, geneticists, embryologists after many years of hard work at the end of 20 Century came to revolutionary realizations, which let our understanding of the human genetics appear in a completely new light.

The Gariaev group has discovered a wave-based genome and DNA phantom effect, which strongly supports the holographic concept of reality (Miller, Webb, Dickson, 1975). They (Gariaev P.,1994) has proposed a theory of the Wave-based Genome where the DNA-wave functions represent organic device like a biocomputer.

It makes sense to mention here some of the results of their research. They discovered that the genetic apparatus works on two levels: material and wave. Material level can be concerned about 2% genes, responsible for the synthesis of proteins. Wave level – about 98% genes, working on the principles of bio-computing and holography. Genetic apparatus can be viewed as the trinity of its structure-functional organization consisting of holographic, soliton and fractal structures. There are genetic "texts", similar to natural context-dependent texts in human language; the genome reads and understand genetic texts in a manner similar to human thinking, but at its own genomic level of "reasoning". This cellular bio computer can form holographic pre-images of bio structures. The chromosome apparatus acts simultaneously both as a source and receiver of these genetic texts, respectively decoding and encoding them; DNA probably works as an antenna; chromosomes are able to distant translation gene-wave information. The chromosome continuum acts like a dynamical holographic grating, which displays or transduces weak laser light and solitonic electro-acoustic fields. The distribution of the character frequency in genetic texts is fractal, so the nucleotides of DNA molecules are able to form holographic pre-images of bio structures. This process of "reading and writing" is carried out in conjunction with its quantum nonlocality. Rapid transmission of genetic information and gene-expression unite the organism as a holistic entity. The experimental work of the Gariaev group shows how quantum nonlocality is directly related to laser radiation from chromosomes (coherent light), which jitterbugs its polarization plane to radiate or occlude photons.

To create an organism, two genetic programs are required. The first one is like a draft which describes how to design the body. It is entire anatomy. The second program is in the form of a meaningful text which contains instructions and explanations how to use the first program, how to understand and build the organism. That is like entire human physiology. These programs exist in the form of "DNA video tapes", which are used by the genetic apparatus, acting like a bio computer. When the bio computer reads these video tapes, sound and light images appear that constitute the movie program of the development of the organism. (Gariaev, 2011). When the creation of a grown-up organism is completed, the movie ends. Then the second movie starts, which contains the instructions for maintenance of the organism for an indefinitely long time. Unfortunately, the videotapes containing information about a perfectly healthy organism, get corrupted with time, errors accumulate (DNA mutations). It is very likely that these DNA video tapes can be renewed and corrected. With this new understanding of how our genetic apparatus works, completely new technologies for healing a person and extending a person's life become feasible. This is the essence of Wave Genetics and its practical applications to come.

In the course of experiments in the Wave Genetics Institute in the Moscow were observed very interesting unexpected effects. One attempt was made to record the sound of the laser beam reflected from the plate with the preparation of DNA. People in the lab could not believe their ears when via radio receiver heard fantastic trills resembling a bird. Other experiments demonstrate very beautiful music of DNA.

In addition, they found that anybody can interact with the DNA using voice modulated laser light. They were actually able to change the information patterns in the DNA with laser light and were able to convert a frog embryo into a salamander embryo.

Genes of intellect.

Some of scientists count that intellect (measured for example by IQ test) is 75 to 85 percent heritable; others, for example, a prominent cognitive psychologist Richard E. Nisbett thinks, it is less than 50 percent (Nisbett, 2015).

The founder of eugenics, Francis Galton in 1869, analyzed the pedigrees of many talented people came to two conclusions: 1) ingenious abilities are inherited; 2) every person carries a load of hereditary traits, which is expressed in the character, intellect, diseases.

In subsequent years, studies of separated in childhood monozygotic twins have shown that intellectual capacity by 40-80% inherited.

In trying to find the genes responsible for intellect, was identified only a few genes having a significant impact. In particular, researchers have discovered a gene, binding to the thickness "gray matter" in the brain. (Schumann, G, 2014). In their view, this discovery will help to understand why some people have difficulties in learning.

Dr. Michael Johnson, scientists from Imperial College London have identified two clusters ("gene networks") of genes that are linked to human intellect. Called M1 and M3, these gene networks appear to influence cognitive function, which includes memory, attention, processing speed and reasoning. Importantly, the scientists have discovered that these two networks are likely to be under the control of master regulator switches. Johnson says the genes they have found so far are likely to share a common regulation, which means it may be possible to manipulate a whole set of genes linked to human intellect (Knapton, 2015; Wighton, 2015).

Epigenetic factors. The behavior of a person's genes does not just depend on only the genes' DNA sequence - it's also affected by so-called epigenetic factors. Changes in these factors can play a critical role in intellect and diseases. Epigenetic changes can switch genes on or off and determine which proteins are transcribed (Deary IJ, 2006).

An American developmental biologist, Doctor Bruce Lipton in the article "Nature, Nurture and Human Development" wrote: "In addition to protein-coding genes, the cell also contains regulatory genes that "control" the expression of other genes. Regulatory genes presumably orchestrate the activity of a large number of structural genes whose actions collectively contribute the complex physical patterns providing each species with its specific anatomy. The research of Dr. Lipton showed that genes and DNA do not control whole our biology; but instead, DNA is controlled by signals from outside the cell, including the energetic messages emanating from our positive and negative thoughts (Lipton, 2012).

It is known that the genomes of humans and chimpanzees are identical 98.7%. Humans have one less chromosome than chimpanzees, gorillas, and orangutans. It is not that we have lost a chromosome. Rather, at some point in time, two mid-sized ape chromosomes fused to make what is now human chromosome 2, the second largest chromosome in our genome. Famous German geneticist Svante Pääbo (director of Max Planck Institute for Evolutionary Anthropology) finally figured out why we have significant differences from monkeys. Pääbo and his colleagues proved that the genetic difference between the blood and liver humans and primates was almost imperceptible, but in the brain differences were stunning. The distinction was in the characteristic, which is called "gene expression". In other words, the difference was found in the activity rate at which the through them produces new proteins. The genetics found that in the human brain, the genes expression of up to five times higher than in monkeys.

Phantom DNA and networked intellect. In the early 1990s, Dr. Peter Gariaev and quantum physicist Dr. Vladimir Poponin discovered the DNA Phantom effect. The Russian scientists irradiated DNA samples with laser light. On screen a typical wave in genes pattern was formed. When they removed the DNA sample, the wave pattern did not disappear, it remained. Many control experiments showed that the pattern still came from the removed sample, whose energy field apparently remained by itself. This effect is now called phantom DNA effect. And this phantom can be visible about forty days! Then vanishes. But not quite. Info, included in the genetic apparatus of human, never leaves without a trace. It was a sensation. Because no one else from scientists, biochemists and geneticists could not be so deep look into the inner sanctum of human nature. See: http://wavegenetics.org/en/issledovania/volnovyie-repliki-dnk/.

The German writers Grazyna Fosar and Franz Bludorf have written a book about entitled Networked intellect (Vernetzte Intelligenz). They argue that there is a networked intellect at the DNA level that enables hyper communication of information amongst all sentient beings. We are all interconnected in the DNA, and this accounts for a whole host of paranormal phenomena (Fosar, 2001).

According to their hypotheses, all DNA works as just like the bio Internet, because the DNA may:

- inject your own data in this network,
- retrieve data from this network,
- surf in the network,
- chat with other participants.

Grazyna Fosar and Franz Bludorf wrote: "The latest Russian scientific research directly or indirectly explains phenomena such as clairvoyance, intuition, spontaneous and remote acts of healing, self-healing, affirmation techniques, unusual light/auras around people (namely spiritual masters), mind's influence on weather patterns and much more. In addition, there is evidence for a whole new type of medicine in which DNA can be influenced and reprogrammed by words and frequencies WITHOUT cutting out and replacing single genes." (http://www.soulsofdistortion. nl/dna1.html).

Cells and water. In the body of a man is about 70% water by weight, approximately 2/3 is inside the cells, and about 1/3 is present outside cells in blood plasma and other body fluids. Japanese author and entrepreneur Masaru Emoto claims that human consciousness has an effect on the molecular structure of water. He proves it by water crystal experiments, which were consisted of exposing water in glasses to different words, pictures or music, and then freezing and examining the aesthetics of the resulting crystals with microscopic photography.

Water is the most abundant molecule in cells. Consequently, the interactions between water and the other constituents of cells are of central importance in biological chemistry.

Peter Gariaev' team and later Laureate of the Nobel prize in medicine in 2008, Luc Montagnier, is claiming that DNA can send 'electromagnetic imprints' of itself into distant cells and fluids which can then be used by enzymes to create copies of the original DNA. The claim of Montagnier's team is that the radiation generated by DNA affects water in such a manner that it behaves as if it contained the actual DNA (Montagnier, 2011). It means that water inside and outside of cells can be an important component of cellular and human intellect.

One hundred years ago famous American writer and philosopher William Walter Atkinson, author more than 100 books wrote about cells so if as he knows results of genetics theories appeared in end of 20 century: "The cells in the human body have sufficient mind to enable them to seek, select, and absorb their own food, and to move from one place to another in search for it when necessary"... "But the mind in the cells is more than merely the particular manifestation of mind in each particular cell. There is also found a sympathetic and coordinated mental activity existing between all the individual cells of the body; a "group mind" of certain groups of cells; and an "organ mind" of the various groups composing an organ of the body. The latter can group together in what we have called the Corporeal Mind, which is the great group mind of the cells of the entire body. Just how these cells coordinate and cooperate in this way is unknown to science, but there seems to exist a high form of telepathic communication between them, of which we get a hint in the psychology of human crowds, in which there is found a "contagion of thought" between the various members thereof." (Atkinson, 1918).

4.3. A little bit about human brain from anatomy and physiology

Brain size. "Your brain is the most complex and capable entity in the universe, far more chemically intricate and variegated than any star, and vastly more capable of fact storage and random information access than the world's most sophisticated computer. In fact, the brain is so complex that mankind will, in all probability, never fully understand it"- wrote President & Medical Director of Alzheimer's Research & Prevention Foundation Dr.Khalsa (Khalsa, 1997).

Let try to find out some of the functional features of the human brain needed to understand more human intellect and intellectual abilities.

Descriptions of some functions of the brain and the mind (in the division 1.5) are insufficient for apprehension of this problem because we have not determined the relationship between the brain and the mind at the level of their structure and cellular level.

During a long time scientists have tried to link an intellect' level of development and size of the brain. The intuitive assumption that 'big brain may mean more intellect,' was first destroyed some time ago. For years, scientists studied the brains of exceptionally smart people after their death, trying to locate the source of their intellect. After V.I. Lenin died in 1924, for example, the Russians invited the great German neuroanatomist Oskar Vogt to try to locate the "source of genius" in the leader of the Russian revolution. Vogt cut Lenin's brain into more than 1,100 slices, but he found nothing exceptional except unusually large pyramidal cells.

The brain of physicist Albert Einstein has been a subject of much research and speculation. Einstein's brain was removed within seven and a half hours of his death. The brain has attracted attention because of Einstein's reputation as one of the foremost geniuses of the 20th century, and apparent regularities or irregularities in the brain have been used to support various ideas about correlations in neuroanatomy with general or mathematical intellect.

To disappointment of scientists, Einstein's brain actually weighed 10 percent less than the average brain. Scientific studies have suggested that regions involved in speech and language are smaller, while regions involved with numerical and spatial processing are larger. But a study of Einstein's brain organization and synaptic molecule configuration still remains to be completed.

The last brain that the Russians studied in this way was that of Andrei Sakharov, the nuclear physicist and human rights activist who died in 1989. From the dozens of brains they studied, the researchers made many observations about brain size, the density of neurons and the number of convolutions of the cortex, but their findings revealed next to nothing about human intellect dependency from brain size.

Comparison of human brain with animal brains allowed to discover, that the brain abilities exactly depends not how much on sizes and weight of the brain, but from the relationship of its weight with the total weight of entire body. And here man have nobody equal.

For example: for people the average weight ratio of body to the weight of the brain is about 50, for cow it is about 1000, for dog - 500, for chimpanzee -120, whale -3000. On the whole, our only nearest "on the mind" relatives – are the dolphins which the weight ratio is about 80.

According to a meta-analysis summarizing themes from the existent literature, a team of Cambridge University found the evidence that brain size and structure are different in males and females (Ruigrok, 2014). This meta-analysis proved that males on average have larger total brain volumes than women (by 8-13%), and have differences in special fields of brain. This and other studies show that

- by structure - females often have a larger hippocampus and have a higher density of neural connections into the hippocampus. As a result, women tend to input or absorb more sensorial and emotive information than males do;
- in processing - male brains utilize nearly seven times more gray matter for activity while female brains utilize nearly ten times more white matter;
- in chemistry - male and female brains process the same neurochemicals but to different degrees and through gender-specific body-brain connections (Jantz, 2014).

The real gender difference of brain functions depends not only on difference in their physiology, but on difference in culture people grew up, and well described in the book (Moir, 2015).

About a structure of human brain. For many years people thought that man thinks with his heart. On the other hand, Indian yogists from ancient times was considered a solar plexus as an essential part of the nervous system and the brain kind. They assumed that in the "abdominal brain" is "warehouse" of Prana - life energy.

The followers of the Chinese teachings of the Tao are sure that a brain is our first mind – watching mind. Second brain - consciousness, located in the heart. But the third mind is located in the lower abdomen, and called "abdominal brain." All three of the mind meld together in the abdominal cavity and form a whole mind. In China it is called "I".

The founder of the West-European medicine, Hippocrates believed that the brain is a gland, designed to remove from the body extra fluid, which is released through the nose, eyes, ears, throat, as well as the spinal cord and the lower back. Around the 4th century BC, Hippocrates wrote: "People should know that from the brain and from the brain only, our pleasures, joys, laughter and jokes arise in the same way as our sorrows, pain, sadness and tears".

Over the next more than two millennia, thousands of scientists from various countries have tried to understand how the human brain works.

Beginning with the 20th century, the study of the brain was performed using modern methods and instruments such as electroencephalography, magnetic resonance imaging devices and many others. These devices allowed to find inside the brain the causes of various physical and mental diseases, and which brain regions are activated when we try to concentrate, laugh, sing, cry, look, imagine something or perform other functions (Amen D., 2010). Performed under the supervision of many researchers work, including research by Jeff Hawkins (Hawkins,2004) showed that the basic thought processes occur in the cerebral cortex.

Therefore, the main organ of man as "Homo sapiens" is the brain. The human brain is capable of processing huge amounts of information received through various communication channels with the outside world. It also receives signals from internal organs and controls them. Through him people are able: to think about problems, to remember what happened, to talk, to dream, to be pleased, sad, angry, love, jealousy, etc. The brain is responsible for any decisions, for creating wonderful paintings, or music and poetry. This organ gives us a sense of identity and makes us feel like a person.

At present, researchers have assumed that a fully functioning brain can generate as much as 10 watts of power (Hermann, 1997). This is much less than used famous Deer Blue chess computer, which famously won against the human world champion Garry Kasparov. The human brain, which possibilities capacities still surpass the advanced computers, can remember in 5 times more information, than the British encyclopedia has. On some estimates, its memory may contain a fantastic volume – about from 3 up to 1000 terabyte. For comparison, a question answering supercomputing system Watson that IBM built to apply advanced natural language processing, information retrieval, knowledge representation, automated reasoning, and machine learning technologies, has a 15-terabyte data bank.

From anatomy is known, that the human brain can be divided into several major sections. The largest part of the human brain is the cerebrum, which is divided into two hemispheres. Underneath lies the brainstem, and behind that located the cerebellum. The outermost layer of the cerebrum is the cerebral cortex, which consists of four lobes: the frontal lobe, the parietal lobe, the temporal lobe and the occipital lobe.

Each part of the large brain receives signals from sensory nerves originated in the certain parts of the body. The frontal lobe is responsible for speech, thinking, feeling, and synchronized movements. Parietal part of the brain is responsible for a variety of tactile sensations, such as throughout human touch, pressure, pain, temperature. Temporal lobe of the brain is involved in recognition of sounds, takes part of the function of memory. Occipital part of the brain processes information of visual images.

Structures located deep inside the big brain known as the limbic system, control emotions and memory. These include:

- The thalamus, which acts as a gatekeeper for the messages from the spinal cord to the cerebral hemispheres,
- The hypothalamus, which controls emotions, such things as eating or sleeping and regulates body temperature
- The hippocampus, which transmits information from memory.

According to research of Suzanne Gerkulano-Housel in the brain is about 120 billion neurons (Herculano-Houzel, 2009).

At the end of the twentieth century, a new concept of brain function appeared. This conception includes the following provisions (it is a partial list):

- The brain determines our mental abilities and intellect
- In the process of development (growth or aging), the brain can be trained, and the intellect to change
- The main function of the brain is reserved by various methods, for example, using 2 hemispheres.
- The human brain is not homogeneous, but consists of several functional components that affect our thinking and emotions

- Illnesses, injuries, chemicals, energy, electromagnetic and other fields, as well as the aging process can affect the functioning of the brain
- The brain can operate in different modes, such as Mover mode, Perceive mode, Stimulator mode, Adaptor mode (Kosslyn S., 2013) by determining type of personality.

Recently, Associate Professor, Departments of Psychology and Bioengineering of Beckman Institute for Advanced Science and Technology Aron Barbey and his team carried out studies that allow identifying some areas of the brain responsible for intellect (Barbey, 2012)

On the other hand, a group of neuroscientists at the University of California, Berkeley, reported that they are close to having a possibility to read people's minds with a special device (the Neurosky MindSet brainwave sensor), which is worn on the head (Berekeley School information, 2013)

Recently the US and Japanese team found that the brain "doubles up" by simultaneously making two memories of events. They found, that one memory is for the here-and-now and the other memory for a lifetime (Kitamura, 2017).

Let continue our research, focusing on the main part of the brain - the cerebral hemispheres.

Cerebral Hemispheric asymmetry. The cerebral cortex is responsible for all the processes that a person does consciously, for cognitive processes (sensation, perception, thinking, memory). The activity of the cerebral cortex is consciousness, the subcortex responsible for the subconscious. The cerebellum processes the signals that come from the cerebral cortex and are related to physical activity. The cerebral cortex of the brain is a layer of gray matter, which consists of 10-14 billion vertically oriented neurons of various types, ranging in 6-7 cell layers. The main part of the cerebral cortex of the human brain (about 96%) is the so-called neocortex. The latter plays a very important role in the implementation of higher nervous (mental) activities involved in the regulation and coordination of all functions of the body. In humans, the cortex is about 44% of the volume of the hemisphere, its surface average 1468-1670 cm2.

According to research by Jeff Hawkins, everything we think, perception, language, imagination, mathematical abilities, drawing, music, planning - occurs in the neocortex practically, and it generally determines a person's intellect (Hawkins, 2004).

Of course, the neocortex cannot operate without the other structures of the brain, such as the hypothalamus, brain stem, cerebellum, etc., but these structures have less influence on the intellect.

During the last decades we observe almost in the geometric progression an increase of the number of articles and books about the hemispheric asymmetry (Hellige, 2001).

Conducted studies unveiled some features of hemispheric asymmetry, which refers to anatomical, physiological or behavioral differences between the two cerebral hemispheres. The hemisphere that is larger, more active, or superior in performance is dominant. This asymmetry of mental processes - functional specialization of the cerebral hemispheres: the left hemisphere is leading in the process of the implementation of some mental functions, the right hemisphere is leading to others. The right brain-left brain theory originated in the work of Roger W. Sperry, who was awarded the Nobel Prize in 1981 (Sperry,1961).

It is now considered that the left hemisphere in right-handers plays the predominant role in the expressive and impressive speech, reading, writing, verbal memory and verbal reasoning.

The right hemisphere is leading to the non-verbal reasoning, visual-spatial orientation, nonverbal memory, criticality and so on.

In the left hemisphere are concentrated mechanisms of abstract thinking, and in the right - the particular imaginative thinking. At the same time, within certain limits, there is interchangeability of the cerebral hemispheres

It is important to note that a particular type of hemispheric response is not formed at birth. In the early stages of ontogenesis for most children revealed imagined, right-brain type of response, and only at a certain age (usually from 10 to 14 years) fixed a particular phenotype, mainly specific for this population (В. Аршавский, 2007).

Areas of specialization of the left and right hemispheres of the brain represented (in average and not in full volume) in the table 3-1.

Table 3-1

Function	Left hemisphere	Right hemisphere
Monitoring and control of the body	For the right half of body	For the left half of body
Type of thinking method	Linear and analytic: Ideas appear one directly following another, often leading to a convergent conclusion	Holistic and synthetic: Whole things appear at once, perceiving the overall patterns and structure
Way to conclusion	Logical: drawing conclusions based on reasons or facts	Illogical, intuitive or miscellaneous, not requiring a basis of reason or facts
Method of the thought description	Verbal: using words and numbers to name, describe, define	Nonverbal: using the symbols, pictures, colors
Ability to abstract thinking	Taking out a small bit of information and using it to represent the whole item or object	Using analogies, seeing relationship between things or events
Time connection	Keeping track of time, sequencing one thing after another	Without sense of time

Function	Left hemisphere	Right hemisphere
The information processing method	Consequentially	Parallelly and synthetically (as a whole) the right hemisphere is responsible for the holistic (whole, complete) perception of the world, synthesis, creative thinking and sensory processing of information, interaction with the universal information field, and as a result of an intuitive understanding of the world.
Subconscious	Control muscles of the body, Management functions of the internal organs and hormonal system, Mind control through emotions, Place for intuition function	
Way of thinking	Analytic way based on verbal information. Responsible for logic and analysis	Imagination way, expressed in the use of non-verbal information such as characters, symbols and images
Memory	Memorizes facts, names, dates	Remember the feelings, smells, emotions
Music and art	Professionals	Ordinary people
Comprehension	Understanding the essence	Understanding metaphors and the results of another's imagination, the ability to form and understand associations
Creativity	On the basis of logical reasoning	The ability to quickly sort options. A place for insight. The ability to dream and fantasize
Spatial orientation		Responsible for the perception of the location and spatial orientation
Emotions, sex and dreams are more closely related		More closely related

Function	Left hemisphere	Right hemisphere
Dominant hemisphere The absence of a pronounced dominance of one hemisphere (number of people with equal hemispheres - 48-50%. It's either right-handers with evidence of left-handed or left-handed with signs of right-handedness (on the leading eye fellow southpaw, and in his hand - the right-hander, and others.).	Number of absolute right-handers - 42%. They have leading: right arm, foot, eye, ear, Such people have in 95% cases the speech center in the left hemisphere). They have verbal-logical character of cognitive processes, the tendency to abstraction and generalization;	Number of absolute left-handers - 8-10%. They have leading: left arm, leg, eye, ear, Speech center located in the right hemisphere. They are characterized by concrete creative thinking, development of imagination.
Time orientation. Operates in the present and future functioning at present and the past	Operates in the present and future	Operates at present and the past

At birth, each of the two human hemispheres has the potential to control the majority, and perhaps all kinds of bodily functions. However hemisphere soon begin to "specialize": the majority of people - those who write with his right hand - the left hemisphere become responsible for speech, logical processing of information and more, including the movement of the right side of the body. The right hemisphere starts to "control" creative thinking, intuition, creativity, imagination, and it is the link with the feelings (sensations). At the same time, the person can advantageously process the information using the left or right hemisphere or be representative of a "mixed brain." Some scientists suggest that there is a connection between the right hemisphere and a sense of humor.

People, processing information mainly through the left hemisphere like to deal with the problem to be solved logically, usually they are active and talkative people. Rather conformists, they are looking for accurate facts and love design specifications (task). They would rather draw conclusions than produce new ideas, and in fact, they are likely to improve the existing process or product than invents something new. They like to work in a problem-oriented organizations, with a clear structure, streamlined control and clearly defined circle of responsibility.

People, processing information with the right hemisphere, "strong" matters of the intuitive way, and very good imaginative thought. They like the invention, finding the main idea, the opening through a problematic situation. They often are not conformists (adaptive,). People with a "mixed" strategy used by the brain left or right hemisphere, in accordance with the situation. But not certain, that the mixed information processing has the advantage, for example, in the field of personnel management.

People with a "mixed brain" used strategies of the left or right hemispheres, in accordance with the situation.

An interesting detail noted by scientists - the ability to change with age asymmetry. Experimentally proved that the dominant hemisphere works more economical and slower aging.

How to determine the dominant hemisphere? There are several simple tests. Fold hands in "lock", and if the top would be the right thumb, most likely, you are right handed. The same can be done with the "posture of Napoleon" - with his arms crossed. If the right hand is on the left, you are exactly right-handed. In the case of the dominance of the right hemisphere of the above should be left handed.

A researcher in mind-body dynamics David Shannahoff-Khalsa studied the electroencephalogram (EEG) of simultaneous activity of the left and right hemispheres of the brain and found that the dominant hemispheres activity shifted from one hemisphere to the other in a wave-like rhythm (Shannahoff-Khalsa, 1993).

Right and left hemispheres operate on different frequencies. Twice a day, at the time of falling asleep and waking frequency synchronized. At this point, a person has incomparably greater opportunities.

Functional systems of the brain. According to theory of complex systems, the human functional system is a dynamic, sometimes self-regulating system, selectively uniting structures and processes based on the nervous, humoral regulation mechanisms and wave genetics mechanisms to achieve necessary for this system (and usually the organism as a whole) values of main functional parameters.

In connection with the intellect above (section 1.5) we described shortly 5 functional units of the brain: Input/output Interface, Operating system, Memory, Body Control, and Psychological Control. All these parts and units are functional systems. Let us take a closer look at the description of some characteristics of these systems.

It is obvious that intellect as an operating system is the main functional system of the brain.

The next by its importance is the Body Control system (BCS), working according to programs recorded in the cells of the brain and body (see section 1.5). BCS is information-energetic functional system, which provides support for all functional systems of the human body, providing interaction with the outside world, with reception and transfer of the energy. There are structures within the brain that control body temperature, heart rate, respiration and hormonal control of thyroid, renal glands, kidney and reproductive organs and much more. In terms of cybernetics and theory of control systems, it is complex multilevel automatic control system given to any person from birth.

The mind is the third main functional system of brain. By analogy with computer the brain is a computerized machine that runs a complex of programs called "mind". In humans, consciousness is a manifestation of the activity of the mind. This is the behavior and cognition management system, which task is to assess the current state of the body on the basis of sensory information, creation and development of intentions, by implementing actions to realize these intentions, and evaluate the results of that action in particular through emotions. It is controlled or managed by Psychological Control Unit (PCU) (see section 1.5). On the other side, work of mind is impossible without Memory unit.

In psychology, there are many theories of mind, in which thinking was viewed as an association of representations, as an action, as the functioning of intellectual operations, as a behavior, and so on (Armstrong, 1993).

We will not bore the reader with a description of these theories, especially since not all of these theories allow the inclusion of intellect in the process of thinking. For further analysis it is now important to consider only the following.

An Austrian neurologist and the founder of psychoanalysis Sigmund Freud identified three different parts of the mind, based on our level of awareness: conscious mind, preconscious mind (many refer to as the subconscious), unconscious mind.

The conscious part of the mind contains all of the thoughts, memories, feelings, and wishes of which we are aware at any given moment. The subconscious part of the mind consists of anything that could potentially be brought into the conscious mind. According to Freud the unconscious part is least accessible part of mind, which is maybe the primary source of human behavior. The unconscious part is a complex of feelings, emotions, thoughts, urges, etc. and also is the storehouse of all memories and past experiences. This is also hides the primitive, instinctual wishes.

The concept of three levels of mind of Sigmund Freud did not cover the entire spectrum of unconscious phenomena of the human psyche, but only a part, which today is usually called individual unconscious. This drawback tried to correct his student Carl Jung. Jung's research opened a new field of the human psyche - the collective unconscious. In Jung's interpretation, the collective unconscious is the level of the human psyche, which stores the experience of all humankind, presented to us in the form of archetypes, i.e. certain images that have symbolic meaning.

Subconscious mind is always alert and awake, does not have verbal language, unable to process negative words, takes everything literally, can work much faster than conscious part of mind. Usually, most of the contents of the subconscious and unconscious parts of the mind are unacceptable without special tricks.

Subconscious mind is responsible for storage and processing of the entire incoming information. This part of mind is the store room with an unlimited capacity; all memories, past experiences, deepest beliefs, everything that has ever happened with its owner is permanently stored. Subconscious mind may create automatic programs to make functioning easier in life, without having to pay attention on everything you are doing. Mundane routines such as brushing teeth, combing hair, gum chewing, washing dishes, etc, are examples of such programs. In addition, our deepest core beliefs, values and imprints are "anchored" and programmed into our subconscious mind.

The subconscious mind is in charge of our emotions. Intuition is the voice of subconscious mind also. This part of mind by the volume is much more (maybe tenfold) than consciousness. This is colossal information source, it knows all about his owner and very much other.

Our subconscious prefers to talk to conscious part without words. Words are the domain of your conscious and logical mind – subconscious prefers images, music, vibration and sounds. Subconscious mind is available to conscious part of mind when steady interaction between these parts of thinking is established.

For many people, the level of intellect depends on a brains ability to access and exchange information between above mentioned parts of mind.

The mind is the manifestations of thought, perception, emotion, determination, memory and imagination that takes place within the brain. The mind is the awareness of consciousness about what we know, the ability to control what we do, and knowledge about what we are doing and why. It is the ability to understand everything inside and around us. From other point of view, consciousness is the arena, in which interact simple and complex thoughts, feelings and emotions, premonition and what that the intuition gives.

It is possible that some of what might be perceived to be unconscious becomes subconscious, and then conscious (e.g. a long-forgotten childhood memory suddenly emerges after decades). We can assume that some unconscious memories need a strong, specific trigger to bring them to consciousness; whereas, a subconscious memory can be brought to consciousness more easily.

In the state of sleep or hypnotism, only the conscious and the subconscious layers of mind are set into total peace. This helps natural elimination of mental fatigue.

By research of scientists, consciousness of our mind can only handle about 12 to 20 bits of information per second, our subconscious mind can process around 11 million bits per second.

The critical faculty of the human mind is an invisible, protective barrier located somewhere between the conscious and subconscious minds. Its function is to constantly filter flow of information from one part of mind to other part and vice versa.

The unconscious part of mind constantly communicates with the conscious part of mind via subconscious part, and that help us with the meaning to all our interactions with the world, using our beliefs and habits. Communication is provided through feelings, emotions, imagination, sensations, and dreams. Unlike the conscious part of mind, or even the subconscious part of mind, the unconscious part may hold information that a person either did not earn or cannot acknowledge. This includes the two basic types of information, one that pertains to this person specifically and one that pertains to humanity as a whole. Biological instinct lives within the unconscious mind, and these instincts are more primal. Because human emotions begin there, the potency of powerful or disturbing information is tucked there for self-preserving purposes.

Today many scientists and esoterics speak about superconscious mind. What is it? While the conscious and subconscious parts of our mind are closely aligned with our physical body, the superconscious mind is out of our body. In other words, it exists at a level extending beyond our space time continuum.

If our brain is example of very complex computer, this computer wireless linked to Internet gives us access to every other computer (every other mind). By analogy with this, our brain can wireless connect to Universe knowledge base known as the Akashic records, Universal Mind, The Infinite Mind, The Universal Consciousness, The Source, Divine Mind, God Mind, or simply God. According to theosophists believes the Akashic records is memorized and saved thoughts of all people, events, and emotions in a non-physical plane of existence known as the astral plane.

Spiritual literature uses the term "soul" to refer to subconscious mind. Spiritual literature also refers to the subconscious mind in expressions "as you think in your heart", or "as you believe in your heart".

Super consciousness is the level of awareness that we experience when our mind is in uplifted state. It has unknown mechanism at work behind intuition, spiritual and physical feeling. Very probably, that well known Leonardo da Vinci, Edgar Cayce, Nicola Tesla, Jeane Dixon, and Baba Vanga had this type of superconscious mind.

The hypothesis of morphic resonance developed by Rupert Sheldrake (see also in the section 3.7) leads to a radically new interpretation of memory storage in the brain and of biological inheritance. Nevertheless is close to hypothesis of superconsciousness.

A.K. Maneev based on the idea of Heraclitus that "The Power of thinking is outside the body," stated that consciousness and thinking is not based on protein structures of the human body, but in the structure of the bio field. Going further, he argues that the brain is just a "reader unit" (Манеев А.К. 1974)

Human memory (more correctly memories) do not reside in single brain cells, but in vast chains of brain cells called memory traces (Khalsa, 1997). There is a hypothesis, that our memory is located not in one particular place in the brain, but in several different areas of the brain acting in conjunction with one another.

The functional system Input/output Interface provides interaction with the outside world for import-export of information (see sections 1.11D, 1.5).

Brain-computer interface. Brain-computer interfaces (BCI) are systems allowing communication between the brain and various machines or devices, used for different goals. These interfaces allow to collect brain signals, interpret them and output commands to a connected machine or device. At present, there are many different techniques to measure brain signals: non-invasive, semi-invasive and invasive. This interface with non-invasive technique is used in many devices described below.

4.4. Neuroplasticity of the brain

Phenomenon of the neuroplasticity is one of the most extraordinary discoveries of the twentieth century. In 2016, three neuroscientists Michael Merzenich of the University of California, San Francisco, Carla Shatz of Stanford University, and Eve Marder of Brandeis University were being honored for discovering "mechanisms, that allow experience and neural activity to remodel brain function, and awarded a Kavli Prize in Neuroscience.

According to the theory of neuroplasticity, thinking, learning, and acting actually change the brain's anatomy and physiology. Scientific studies and some medical practices, for example, fantastic results of the American neuroscientist Paul Bach-y-Rita, proved that the brain can change itself during practically whole life.

According to Dr. Fred Travis, the brain is river, not the rock. The central argument of his book that our brain is, literally, in an unceasing flow. Even as grown-ups, our brain retains amazing plasticity as 70% of our brain connections change every day (Travis, 2012).

Environmental stimuli, thoughts, emotions, and even behavior may cause neuroplasticity changes, which has significant implications for learning, memory enhancing, and recovery from some unhealthy conditions. When our brain encountered with unknown situation it can reorganize and restructure itself to respond to that situation. The more often our brain is exposed to that new challenge – the more it reorganizes and makes that path more established.

The brain is an extremely complex network of neurons that are in contact with each other through special formations - synapses. We can distinguish two levels of neuroplasticity-macro and micro. Macro is associated with the change in network structures of the brain, providing communication between hemispheres and between areas of each hemisphere. At the micro level occurring molecular changes in neurons themselves, and in the synapses. On the both levels, brain plasticity can occur quickly or slowly (Kharchenko, Klimenko, 2004).

On the other side, there are two types of brain neuroplasticity: functional and structural. Functional neuroplasticity is the ability of the brain to redistribute the functions of damaged areas of the brain between intact, structural neuroplasticity is the ability of the brain to change its physical structure during cognitive activity.

Studies showed that the brain is capable of changing its own structure and functioning, stimulated by thoughts and actions of his owner. The researchers even found that thinking, learning and active actions can "turn on" or "off" some of our genes.

The attributes of plasticity are inherent in all brain tissues. The last years funding proved that plasticity exists in the hippocampus and in those regions, which control respiration, process primitive emotions, and pain. Plasticity is present even in the spinal cord (Doidge, 2007).

Thanks to neuroplasticity, the human brain is capable of joining many of our feelings and sensations with a system of pleasure or pain system.

There are different ways how neuroplasticity works. To be more efficient our brain can produce new cells or losing cells from pathways not currently in use as well as form new connections or change existing connections to be more efficient. When we learn something new, neurons are activated at the same time and communicate with each other within a process called "long-term potentiation» (LTP), that enhances the connection between neurons. When the brain "forgets" the association, another chemical process called "long-term depression » (LTD) occurs.

Jordan Grafman described four types of plasticity:

- Expansion that occurs primarily at boundaries between brain areas (brain maps) and results from daily activities
- Sensory reassignment that occurs when one sense is impaired and its sensory input are relocated to another sense
- Compensatory ability of the human brain to achieve certain functions in several different ways
- Mirror region that occurs when part of one hemisphere fails, the mirror region in the opposite hemisphere changes to compensate (Doidge, 2007)

The brain neuroplasticity is not a permanent property of the brain, this property may vary depending on age, feeding, environment, emotions, and most importantly, depending on the activity of the brain. Usually to develop this are recommended widely the following simple strategies:

- Intellectual activities
- Physical activities
- Brainwave entrainment (see chapters 6-7)
- Meditations
- Brain workout regimes – learning new things
- Memory and attention training, Lumosity.com is the largest provider of memory and attention training games.

Described above the concept of neuroplasticity of the brain and the effects of meditation give hope to countless victims of traumatic brain injury. These people can have a chance of restoring the connections in their brains, rediscovering memories, and re-learning the skills they had forgotten.

4.5. Some specific features of consciousness and mind

Relationship between consciousness and matter. Consciousness is a functional system of mind which includes set of mental processes. According to many medical studies, there are many different states of consciousness, called by different authors differently, for example, as modes, levels or states. From a medical point of view, there is a framework for the normal mode and disorders of consciousness: waking state, quick dream, hypnotic state, anesthesia (mild), minimally conscious state, stupor, seizures, dementia, delirium, etc. (Demertzi, 2008).

At present, consciousness presents a complex of problem for science in explaining its structure, functions and dynamics (Henriques, 2018). A large number of scientists of various specialties in many countries are involved in solving these problems, for example, in such organizations as Center for Research on Consciousness and Anomalous Psychology, The Esalen Institute, International Consciousness Research Laboratories, and so on (see: https://www.scientificexploration.org/links).

In 1979, an American plasma physicis Robert Jahn organized the Princeton Engineering Center for the Study of anomalous phenomena (Jahn, 1986). Since then, the researchers of the Center not only provided convincing evidence of the existence of relationship between consciousness and matter, but also collected many data on it.

In one series of experiments, Robert Jahn and his colleague, clinical psychologist Brenda Dunne, used a random number generator (RNG) - a device for multiple automatic flipping a coin with instant recording of the results. In an experiment, the operators, seated a few feet from machine, initiated each run by pressing a remote switch and then attempted to influence the process to produce a higher number of counts or lower number of counts, or to generate a baseline, in accordance with pre-recorded intention or randomized instructions.

Extensive data from a variety of man/machine experiments indicate that human operators can influence normally random output distributions in accordance with pre-stated intentions. The proposed model represents consciousness as a quantum mechanical wave function confined in an environmental potential well (Greenberger, 1989).

Experiments also revealed that different volunteers produced different, but clearly identifiable results; the results were so individual that Jahn and Dunne began to call them "signatures"

Jahn and Dunne believe that their findings may explain the tendency of some people to cause damage to the devices and machines. One such person was the physicist Wolfgang Pauli, whose abilities in this regard have become so legendary that physics phenomenon called "Pauli effect". His colleagues discovered that the mere presence of the Pauli in the laboratory could cause an explosion or damage to the glass unit of the measuring device.

This work was continued under the Princeton Engineering Anomalies Research (PEAR) program (Jahn, 2011) and after 2007 in International Consciousness Research Laboratory.

Approximately at the same time in Russia, first at Sankt-Petersburg Optic-Mechanical Institute and later in the Center of Energo-information technology united group of scientists, managed by Dr. Dulnev, studied phenomena of telekinesis. Researchers came to conclusions, that it is impossible to explain the result of researches based on the existing physics paradigms. And after that was written (citation) the following: "Perhaps the closest you can come to an explanation of these phenomena on the basis of the theory of physical vacuum, i.e. involving fundamental torsional interaction" (Dulnev, 2000).

There is a quiet revolution underway in theoretical physics. For as long as the discipline has existed, physicists have been reluctant to discuss consciousness, considering it a topic for quacks and charlatans. Indeed, the mere mention of this word in contexts of paranormal phenomena could ruin careers in some countries.

Today a new information about consciousness is spreading through the theoretical physics community. A world-renowned quantum physicist Dr. John Hagelin, Director of the Institute of Science, Technology and Public Policy at Maharishi University of Management in Fairfield, Iowa, is developing a Unified Field of Consciousness theory. He has pioneered the use of Unified Field-based technologies proven to reduce crime, violence, terrorism, and war and to promote peace throughout society—technologies derived from the ancient Vedic science of consciousness. He has published groundbreaking research establishing the existence of long-range "field effects" of consciousness generated through collective meditation, and has shown that large meditating groups can effectively defuse acute societal stress—thereby preventing violence and social conflict, and providing a practical foundation for permanent world peace (http://istpp.org)

In 2012 year, an American psychologist and parapsychologist known for his psychological work on the nature of consciousness (particularly altered states of consciousness), as one of the founders of the field of transpersonal psychology Dr. Charles Tart published the book "The end of materialism. How evidence of the paranormal is bringing science and spirit together", that helps to understand many other questions about relationship between our consciousness and matter.

Altered states of consciousness. An important feature of the functioning of consciousness is an ability to change its state in various ways, which opens up the possibility of immersion into the deep layers of the psyche. At present, theory of consciousness is on the stage of development. The normal and abnormal (altered) states of consciousness have not yet been fully classified. This is, first of all, because the fact that what for some nations is the norm, for others is considered as a deviation. This concern, for example, the states of trance and meditation. Therefore, we will try in the first approximation, to identify and enumerate the states of consciousness that are not normal for the average person of Western culture. In this case we rely on the latest brain research, carried out in different laboratories around the world using the arsenal of advanced methods and techniques. In addition, we will not consider common condition that relates to mental illness and birth defects.

Range of consciousness states is quite wide, from full absence (coma) to the state of maximal activity or productivity (e.g., insight).

Wakefulness - normal state of consciousness, corresponding activation of the whole organism. The level of activity of consciousness includes right perception of the external world and the events in it. Sleep with or without dreams are other normal state of consciousness.

According to modern ideas, sleep state- not just the recovery period for the body, but is a state of paramount importance to maintain the normal state of consciousness. Brain during sleep continues to function. At night, we typically go through several sleep cycles NREM (non-rapid eye movement) and REM (rapid eye movement), normally in that order and usually four or five of them per night. REM sleep is associated with the capability of dreaming.

Studies carried out in recent years, allowed to excrete to the class of specific states of consciousness the Altered states of consciousness (ASC). At first this expression was used in 1966 by Arnold M. Ludwig (Ludwig, 1966, 1969) and brought into common usage since 1969 by prominent psychologist and parapsychologist Charles Tart (Tart, 2001, 1990). At present are created many classifications of ASC proposed by Tart, C.Maxwell Cade, John F. Kihlstrom, etc. In all kinds of ASC we see the capacity for reflection, expressed in most or less (as, for example, in a trance and suggestive states of consciousness) power than in normal state of consciousness.

In common an altered state of consciousness is a condition of awareness in terms of the level of awareness, quality or intensity of sensations, perceptions, thoughts, feelings and memory activity (Tart, 1974).

Charles Tart proposed that consciousness, as a system, is capable of assuming a variety of steady states each of which can be referred to as a discrete state of consciousness (d-SoC). According to Tart, examples of d-SoCs are our ordinary waking state, lucid dreaming, hypnosis, alcohol intoxication, marijuana intoxication, and some meditative states.

It makes sense to distinguish between ASC, which are innate or the result of post traumatic or diseases induced changes in the body and brain, and ASC, which are a reaction on some artificial practices or drugs. The first group encompasses the following ASC providing abilities to: telepathy, clairvoyance, somnambulism and so on.

The second group comprises the following ASC which may be created intentionally, for example:

- induced by psychoactive substances (such as psychedelics) or procedures (e.g., holotropic breathing);
- caused by hypnosis, various meditations, religious rites;
- arisen spontaneously in normal human conditions (with considerable tension, listening to music, sports game, orgasm), or in unusual but natural circumstances (e.g., in process of birth delivery), or in unusual and extreme conditions (e.g., peak experiences in sport, near-death experience of various etiologies).

There are also many common experiences that can create altered states of consciousness, such as daydreaming, sexual euphoria or even panic.

Altered states of the second group can be organized and disorganized. Organized state that is familiar to many creative people, scientists call a flow state (Csikszentmihalyi, M. 1997) or state of inspiration (Reese, 2005). Often, people intentionally try to alter their conscious state. There are many reasons, for example, people try to attain an altered state of consciousness by religious and spiritual reasons, for relaxation or to increase health.

A psychedelic experience is a temporary altered state of consciousness induced by the consumption of psychedelic drugs (such as LSD, mescaline, psilocybin mushrooms, salvia divinorum, cannabis, etc.). Psychoactive drugs can create hallucinations and delusions, making people see and hear things that are not there. The legal status of these drugs varies worldwide.

Hypnosis is a state of intense relaxation and concentration, in which the mind becomes remote and detached from every day cares and concerns. Psychologists believe that hypnosis is an altered state of consciousness that allows a person to be more open to suggestion. Hypnotic suggestion can increase a person's ability to concentrate, have a stimulating effect on the mental and creative abilities. If under a hypnosis a person received is suggestion that he is famous chess player or an artist, his ability to play chess or painting have to increase significantly.

Undoubtedly amid ASC enlightenment (or inspiration) state is the brightest representative of the second group. This state sometimes called as a higher state of consciousness (insight). Usually, in state of inspiration a person totally absorbed in what he doing, enjoying every moment, and everything seems him easy. Productivity of work in this state can be maximal.

Clairvoyants and healers come programmed with an expanded range of perception, or can alter either the state or frequency of their consciousness in order to perceive energy vibrations outside of the normal human range.

Trans state distinguishes from the usual state of consciousness by orientation of attention - when a focus of attention directed inside (i.e, attention is focused on images, memories, feelings, dreams, fantasies, etc.) rather than outside, as in the ordinary state of consciousness.

The term trance may be associated with hypnosis, meditation, magic, prayer, and so on. When a person enters a trance-like state intentionally, then this is called by many spiritual schools of east as meditation.

In a trance state, some people may be very long, even many months. However, they have a completely normal life without needing constant supervision. When they exit from this state there is some amnesia for the time spent in a trance.

Holotropic brain condition. The famous psychiatrist, one of the founders of the field of transpersonal psychology, Stanislav Grof discovered a holotropic brain condition and innovated the method of Holotropic Breathwork to transfer brain from normal state to how he called it to non-ordinary state of consciousness for goals of treatment (Grof, S. 1992).

Stanislav Grof distinguish two modes of consciousness: hylotropic and holotropic (Grof, 1988). The hilotropic mode of consciousness is a normal everyday experience of a common reality. The holotropic mode has to do with states which aim towards wholeness and the totality of existence. The holotropic is characteristic of non-ordinary states of consciousness such as meditative, mystical, or psychedelic experiences.

According to Grof, these non-ordinary states are often categorized by contemporary psychiatry as psychotic. Grof associates the hylotropic mode to the Hindu conception of namarupa - the separate, individual, illusory self. On the other side, he associates the holotropic mode to the Hindu conception of Atman-Brahman as the divine, true nature of the self (Grof, 1988).

The holotropic states refer to states which aim towards wholeness and the totality of existence. Patients encounter transpersonal experiences which go beyond the usual boundaries of human body and ego. Transpersonal experiences extend feeling unusually personal identity, including in it the elements of the external world and other dimensions of reality.

Transpersonal experiences can lead to the episodes from the life of relatives and ancestors. Patients can experience the full conscious identification with other people, groups of people, animals, plants and even inorganic objects and processes. A holotropic experience is often accompanied by extraordinary changes in sensory perception with profound changes in color, shapes, sounds, smells and tastes as well as profound perceptions that have no counterpart in this realm. A person with eyes closed is often flooded with visions drawn from personal history and with involving various aspects of the cosmos and mythological realms.

A key for physical explanation of nature of the holotropic condition can be find in Hameroff –Penrose's hypotheses. Director of the Center for the study of consciousness at the University of Arizona Stuart Hameroff and the famous British physicist and mathematician from Oxford Sir Roger Penrose developed and is now defending the so-called quantum theory of consciousness (Hameroff, 1996, 2014)

The Penrose-Hameroff theory of "orchestrated objective reduction" ("Orch OR") proposes that consciousness depends on quantum computations in structures called microtubules inside brain neurons, occurring concomitantly with and supporting neuronal-level synaptic computation (Hameroff, 1998).

Our soul also has a quantum nature and is an example of a nonlocal holographic fractal space-time geometry. In this quantum information cannot disappear after death. Once the body dies, the information is merged with the universe and can exist indefinitely (Hameroff et al, 2011).

In January 2014 Stuart Hameroff and Roger Penrose announced that a discovery of quantum vibrations in microtubules as the hypothesis of Orch-OR theory was confirms by Anirban Bandyopadhyay of the National Institute for Materials Science in Japan. A reviewed and updated version of the Orch-OR theory was published along with critical commentary and debate in the issue of Physics Of Life Reviews (Hammeroff, 2014).

Awakened mind and brain waves. Billions neurons of brain use electricity to communicate with each other. This produces an electrical activity in the brain, which can be detected using sensitive medical equipment (such as an EEG), measuring electricity levels over areas of the scalp. Electrical activity emanating from the brain is displayed in the form of brainwaves.

Thanks to many years research fulfilled by British biophysicist and consciousness researcher Maxwell Cade and his colleague Anna Wise with help of designed by Geoffrey Blundell multi-channel electroencephalograph (EEG) called them the Mind Mirror, the mind science have received instrument allowing look at the relationship of different rhythms of each brain hemisphere and display these simultaneously (Cade, 1989; Wise, 1997,2002). The Mind Mirror was specifically designed to measure states of consciousness, as opposed to the medical EEGs that were used to identify pathology. Cade presented his remarkable breakthrough discovery of the Awakened Mind brain wave pattern in his book (Cade, 1989).

Using the Mind Mirror Cade measured wave activity in brains of many sensitives: the famous yogis, healers and people skilled in meditation, - and discovered consciousness' conditions, which he called an active intellect or the Awakened Mind. Later the term Awakened mind has been widely used to describe the phenomenon of a spontaneous change of consciousness. Maxwell Cade demonstrated by many experiments that any brain activity and condition of consciousness depends on a combinations (by frequencies and amplitudes) of 4 conditional categories of brain waves (alpha, beta, delta, and theta). Anna Wise, high level specialist in Neurotherapy and Humanistic Psychology, a member of the Academy of Certified Neurotherapists, after collective work with Cade in England returned to the United States in 1981 year to develop her own work with the awakened mind expanding the research into areas other than spirituality and healing. Instead of special sensitives researched by Cade, she measured brain activity for artists, composers, dancers, inventors, mathematicians, and scientists. In addition, she measured brain activity for high-level managers of corporations, and found that the brain wave patterns in states of high performance, creativity, the bursts of peak experience, were the same patterns that the yogis and healers lived in. Anna measured the "ah-ha" or the "eureka" experience and found that the brain waves flared into an awakened mind at the exact moment of insight.

As a bottom line, she came to conclusion that when people are in an awakened mind state of mastery, their mind became clearer, sharper, quicker, and more flexible; emotions are more available, understandable, and easier to transform; intuition, insight, and self-healing abilities increased (Wise A., 2002).

She worked with both the state and the content of consciousness, learning how they interrelate, using content to develop and train the state, and vice versa using the state to access and transform content. The state of consciousness is displayed using the actual brain wave pattern as measured on the Mind Mirror, while the content of consciousness includes the thoughts, feelings, emotions, images and all of the material of the mind. To understand the achievements of Anna Wise we need to make a short description of the brain waves.

There are several main categories of these brainwaves with blurred boundaries: alpha (8 Hz to 12 Hz, beta (12 Hz to 40 Hz), delta (0.1 Hz to 4 Hz), theta (4 Hz to 8 Hz), and gamma (40 Hz to 100 Hz).

Alpha waves along with beta waves were discovered by German neurologist and EEG inventor Hans Berger. Alpha waves are also called Berger's waves. Alpha waves are present during daydreaming, fantasizing, and visualization. Person who has completed a task and sits down to rest is often in an alpha state. This state is associated with lighter meditative states and with super learning. Alpha is the gateway to meditation and provides a bridge between the conscious and the subconscious mind. Alpha brain waves are also associated with the ability to learn, process, store and recall information. When our brain works in the alpha rhythm, may appears connection not only with the subconscious or unconscious parts of the mind, but and powerful connection with the entire world. As scientists established, the alpha rhythm of brain can enter into the resonance with the basic rhythm of the fluctuations of the Earth's atmosphere, is known as Schumann waves.

When the brain is aroused and actively engaged in mental work, it generates beta waves. A person in active conversation, playing sports or making a presentation would be in a Beta state. Generally, beta is the human normal thinking state, active external awareness and thought process. When functioning at their best, beta waves are associated with logical thinking, concrete problem solving, and active external attention.

Frequency of theta brainwaves is the frequency the subconscious mind. Theta is present in dreaming sleep and provides the experience of deep meditation. Theta also contains the storehouse of creative inspiration and is where people often have their spiritual connection. Theta brainwave states have been used in meditation for centuries. Research has proven that thirty minutes a day of Theta meditation can dramatically improve a person's overall health and well-being. Researchers have noted that the more theta waves you make the more creative you are (Harris, 2007, 2015). Theta is also associated with some types of superlearning, as well as with increased memory abilities.

Delta brainwaves are human unconscious mind, the sleep state. When they present in combination with other waves in a waking state, delta acts as a form of radar – seeking out information. Delta provides intuition, empathetic attunement, and instinctual insight.

Gamma waves are associated with a remarkable state: compassion and loving kindness (Harris, 2007, 2015).

Studies of brain waves showed that Beta waves provide concentration; Alpha waves are ideal for learning new information, data or facts, creativity also; Theta waves, provide access to information from the subconscious and appears to awaken intuition; and Delta brain waves, associated with deep sleep, may help the brain reorganize all the experiences of the previous day (Oster, 1973).

At bottom line, brain waves have different responsibility: beta-conscious, alpha -bridge between conscious and subconscious, theta -subconscious, delta – unconscious (Wise, 1997, 2002).

Starting with an American linguist, philosopher and cognitive scientist Noam Chomsky, many researchers raise the issue of the existence of an internal brain language allowing the brain and mind components communicate effectively. Most likely, the media in this language are brain waves.

Our brain does not operate only with one brainwave type, but instead use all these waves, with one of the waves being dominant at given time. The dominant state indicates level of consciousness. Because different areas of the brain may have different activity at any given time, a man may have activity in one brainwave state in one area of the brain while at the same time a different brainwave state may be more active in another area of the brain.

Max Cade and Anna Wise found empirically optimal combinations of above mentioned brainwaves categories of frequencies and amplitudes, and coined corresponding brain state as Awakened mind.

Based on her 6000 hours researches Anna Wise created a specific protocol to teach people to develop an awakened mind using some instruments for biofeedback. It was the ability of the state to allow the flow of information (content) from the conscious mind to the unconscious mind and vice versa.

To achieve the condition of the Awakened Mind is needed something like meditation to reduce the rhythms of the brain to the alpha and theta level and remain at this level for a long time without falling asleep.

Not so long ago in Bath (Pe) in USA was founded the Institute for the Awakened Mind (IAM) - an international consortium of certified practitioners trained in EEG-led biofeedback meditation and awakened mind development.

Awakened mind practitioners, neurofeedback therapists, and consciousness researchers use last version the Mind Mirror (The Vilistus Mind Mirror 6) to guide brainwaves into optimal meditation states that connect people with their deep, subconscious wisdom to clear issues, release suppressed resources, and expand into the state of clarity, creativity, insight, intuition, and spiritual awareness. When desired, the awakened mind pattern amplifies into the evolved mind of illumination, bliss, absolute understanding, and pure mind, or pure spirit. Methods developed in the Institute allow relaxing and reducing stress, improve health, sharpen thinking, enhance creativity.

Autopilot for brain. Marcus E. Raichle, the American neurologist at the Washington University School of Medicine in Saint Louis, Missouri, discovered a default mode of brain function and the brain's default mode network. The default mode network (DMN) is a network of brain regions that are active when the individual is not focused on the outside world and the brain is at wakeful rest (Raichle, 2000, 2007)

Andrew Smart in his book published in 2013 (Smart, 2013) substantiated the hypothesis of the existence of a default mode of brain function like an autopilot.

4.6. Abdominal brain and heart intellect

As we wrote in section 1.5 any operating system of developed alive organism must include survive function needed first of all in the super dangerous for live situations. According to many studies in the human organism must exist additionally to the brain other systems which can be named the systems of emergency control. These systems may not work continuously in parallel to the brain, but be activated only in the period of danger. It is wide known, that in body of many people exist some mechanism known as the "flight or fight" response. It engraved in us to recognize when to run from a perceived danger after injection adrenalin into our blood systems by adrenal glands.

Dangerous situation can be recognized very rapidly without brain estimation via five main sense organs. An example of such situations is the attack of predator like a tiger or snake in the jungle or automobile emergency like a collision on the road. This system makes it possible to warn these attacks or collision for time less than a second does and more prior to the beginning of the real events.

Probably each of us has experienced the strange sensations in different places. Either people hear an inner voice, or feel something wrong in stomach or gut (gut feeling), some women in such situation have uterus feeling.

Some scientists associate the sensitivity like as "gut feelings" with the existence of something of type of brain - the "second brain"- in our gastrointestinal systems (Sonnenburg, 2015; Enders, 2018).

In the early twentieth century British physiologist Newport Langley calculated the number of nerve cells in the stomach and intestines - 100 million, it is more than in the spinal cord.

There is a vast network of nerves surrounding the esophagus, stomach and intestines of such complexity what it allows to compare that with the complexity of the brain or part of brain, and name it like as "second brain." Its real name is the enteric nervous system (ENS) and it has more neurotransmitters than anywhere else in the entire nervous system does.

Professor at Columbia University in New York, Michael Gershon after thirty years of research has found that the nerve cells in the gastrointestinal tract do the work of the second brain. Brain and second brain are autonomous units, and are in constant contact.

In his book "The Second Brain" Michael Gershon confirms the assumption that the ENS is not a stupid cluster nodes and tissues performing tasks of the central nervous system, but an unique biological network that can independently carry out some complex mental processes including responsibility for some emotions, intuition and so on" (Gershon,1998).

For centuries, people have known in varying degrees about the unusual properties of the heart (especially for female), which allows to feel the future events and prompted some people to the right decisions.

The heart is no longer viewed by scientists as just an organ with its sole purpose to pump a blood; rather today in the scientific world are appeared more and more supporters to consider that heart plays also an important role in the body controlling and communicating with the brain.

In the 60-70 years of the last century, John and Beatrice Lacey found a significant effect of the heart on the brain in the perception of the world (Lacey, 1978). In 1991, after extensive research, Dr. J. Andrew Armour introduced the concept of brain in heart. He proved that the heart has significantly developed nervous system (about 40,000 neurons), which can be qualified as a small brain, because this little heart brain allows yourself to sensate, feel, and even remember. In this case, cardiac nervous system, solving their tasks under the control and management of heart is working closely with the brain sending information that affects the perception, decision making and cognitive process (Axmour, 1991).

According to study in the HeartMath Research Center, the heart through its extensive communication with the brain and body, is intimately involved in how we think, feel, and respond to the world. Scientists at the HeartMath institute discovered that the heart has its own "little brain" with elaborate circuitry and neurons that enable it to learn, remember, feel and act independently of the cranial brain (McCraty, 2006;Childre, 2017).

The heart communicates with the brain and body in the following communication ways:

- Neurological via nervous system
- Biophysical via pulse waves
- Biochemical via hormones
- Energetic via warm and electromagnetic radiations and waves.

The heart's electromagnetic field much greater in amplitude and in strength than the same activity generated by the brain and can be detected a number of feet away from the body. Researches made in the Institute of HeartMath showed some real interaction between hearts of people and their brains when these people in touch or proximate. It means that when two people are at a conversation distance, the electromagnetic signal generated by one person's heart can affect the other person's brain rhythms (Emmons, 2008).

The study of McCraty, Atkinson, & Bradley presents compelling evidence that the body's systems may continuously scanning the future, and the heart plays a significant role in the body's sensing and processing (McCraty, 2004-2005).

For example, experiments demonstrated that hearts of normal people, not sensitive, receive intuitive information even before the brain (~4.8 seconds before an event versus ~3.5 seconds before an event, respectively) (Bradley, 2006).

4.7. Super complexity of the human cerebral system. Expanded definition of the human intellect.

In addition to cases that had been collected and described Dr. John Lorber (see section 2.1) below you can get acquainted with absolutely unbelievable facts which for today have no full explanation. Why, for example, people with brain stroke when hematoma or blood clot formed, lose many functions (for example, stroke condition) or die, while some, who has for one reason or another destroyed a large part of the brain, live and do practically everything not even notice this. In 1978, in Moscow suburb after accident on one of the most powerful accelerator in the world, physicist Anatoly Bugorsky received impact a flow of protons of 70 billion electron volts. Radiation dose at the entrance was more than 200,000 roengten. He had to be completely burned brain, and by the laws of medicine could not survive. However, he survived. Anatoly continued to work, ride a bike and play football.

In the summer of 2007, numerous magazines and newspapers reprinted excerpts from an article in the British medical journal "The Lancet» (Feuillet, 2007) about the unusual case of dropsy of the brain (hydrocephalus) in 44-year-old French clerk. It all started with the fact that the employee appealed to doctors Marseille (Hôpital de la Timone and Faculté de Médecine, Université de la Méditerranée, Marseille, France) with complaints of weakness of the left leg. Among other things, he undergoes a survey scan of the brain with which revealed that almost all of the cranial cavity is filled with fluid, leaving on its inner surface a thin layer of the medulla. The treated his doctor, Lionel Feuillet said. "These images of the brain were most unusual and shows that virtually here is no brain." Despite it all, "brainless" clerk lived, worked as a civil servant, had a wife and two children. Neuropsychological testing showed that the man had an IQ =75. If no problems with his foot, so that no one would have learned that his mind does not occupy the bulk of the brain cells, and cerebrospinal fluid. The Wechsler Intellect Scale states that an IQ Score of 75 is borderline IQ, which is also called borderline mental retardation.

The famous American doctor, a psychologist and neuroscientist Karl Pribram in the 60s of the last century explained the reason why some people, having a tiny amount of brain substance, do not suffer from dementia or any disorder at all. Numerous experiments of this outstanding scientist led him to create the hypothesis that all brain functions (including memory, intellect, reflexes, etc.) are not stored in a separate area of the brain, as it were distributed throughout the volume of the medulla. Dr. Pribram called this phenomenon neural holography. But why it happens only a few people, and everyone else killed or maimed? The reason for this probably lies not only in the brain, but also in the circulatory and lymphatic systems of the brain that, unfortunately, are not duplicating exactly the same, and therefore affects the damage of nerve cells fatal.

Pribram published his first article on the alleged holographic nature of the brain in 1966 and over the next few years continued to develop and refine his theory. As it met other researchers, it became increasingly clear that the distributed nature of memory and sight - is not the only neurophysiological puzzle that you can solve using the holographic model.

Together with Karl Pribram at Stanford University the physicist David Bohm from University of London, came to conclusion about the holographic nature of the universe, after several years of failed attempts to explain all processes and phenomena of quantum physics with traditional theories. For both scientists holographic model suddenly filled with meaning and served as the answer to many unanswerable questions before.

Their work found a ready response in the scientific community. Cambridge scientist, Laureate of the Nobel Prize in Physics 1973 Brian Josephson called Bohm and Pribram theory "a breakthrough in understanding the nature of reality." With this view agrees Dr. David Peet, physicist Royal Canadian University and author of "The bridge between matter and brain", stating that "our thought processes are much more closely related to the physical world than many assume."

Described above cell and cerebral structures makes possible for us to arrive at the conclusion that human organism has distributed brain structures as members (parts) of united control systems. Human body is made of trillions of cells, which fulfill three major functions: growth, development, and repair. It is difficult to imagine such a control system that can reliably and safely manage trillions units, for example, in the processes of man's growth and aging. Most likely, this system has kind of a self-management by the specified algorithms, recorded as programs in DNA.

In the scientific literature, it is possible to find the opinion that the activities of our cells, so closely correlated and combined in their actions that they form a unity, and may be regarded as one mind working in harmony and unity. This mind can be sufficient for solving the simplest problems of existence and was called by some scientists "The Corporeal Mind". There is no doubt that beside this any man has the one more or several coordinated control subsystems, located not only in the brain under skull and having about 86 billion cells (according to studies of Brasilian neuroscientist Suzana Herculano-Houzel), but and in abdominal part (about 100 million cells) and in heart (about 40000 neurons). Certainly, the main functions of human existence are carried out with a brain, and other parts of the united brain structure - auxiliary, maybe alternate or reserve functions.

The human being is an open system because it receives energy, information, and sometimes control from the outside (e.g., from people, nature and space).

Therefore additional complexity of human intellect is connected with the fact that the intellect is also an open system having the possibility of joint functioning with intellects of other people simultaneity receiving information and influences from the outside by extrasensory perception. Therefore, we can specify definition of intellect given in the section 2.1. **In the terms of theory of complex systems, the human intellect is real-time distributed operating system, ensuring the work of human super complex cerebral system, including cells (corporeal mind), abdominal brain and hearts part of brain. We can assume also that this is a multi-level system, that includes biochemical, physical, and maybe other control levels, and can communicate with outside world.**

4.8. Extrasensory perception from contemporary science viewpoint

Extrasensory perception (ESP) is the ability to receive information beyond the ordinary five senses sight, hearing, smell, taste, and touch. It can provide the individual with information of the present, past, and future from unknown source.

As a rule, this category includes paranormal and psychic phenomena (psi-phenomena) which include telepathy, precognition, clairvoyance, etc. The Parapsychological Association divides psi into two main categories: psi-gamma for extrasensory perception and psi-kappa for psychokinesis.

More than 100 years ago, clairvoyance was described in English in the several books of famous theosophist Charles Webster Leadbeater "Clairvoyance", "Outline of Theosophy", "Astral Plane", etc. Perhaps the most publicized early experiments in the field of extrasensory perception were published by Dr. Joseph Banks Rhine in 1934 in a monograph entitled "Extra-Sensory Perception". At this time first post about "feeling of being stared at" was written by one of the first specialists in the science of psychology at Cornell University (NY) Titchener E.B (Titchener, 1898). Professional doctor - psychiatrist Shafica Karagulla devoted her life for study of supernormal manifestations of human possibilities and proved their objectivity. She investigated numerous instances of clairvoyance and clairsentience, telepathy and foresight (Karagulla, 1967).

Many historical figures, such as Nostradamus, Marie Lenormand, Wolf Messing, Vangelia Pandeva Dimitrova (baba Vanga), Edgar Cayce possessed these amazing abilities. In my opinion, the most amazing of them was Wolf Messing. In his youth, he went to Berlin markets and read thoughts peasant who sold vegetables, potatoes and meat. Much later, he drove through the streets of Riga by car, sitting in the driver place blindfolded. How to drive him mentally dictated sitting beside a professional driver. He could read not only the thoughts of others, but also to suggest his thoughts to other people. It is known that due to this capacity, he managed to escape from a German prison and enter the Kremlin without a permit, in order to prove their ability to Dictator Stalin. Thanks to another the ability of foresight, he saved the life of Stalin's son Vasyli Stalin, who, thanks to the intervention W. Messing, did not fly on a plane that crashed.

To some extent, these abilities also had one of the greatest engineers and innovator at all times Nicola Tesla (Cheney, 2011).

Many surveys indicate that 70% people believe in the existence of premonitions, 25% report of having experienced premonitions and around one third of the general population considers premonitions in dreams as real. The acquisition of precognitive information has often been linked with sleep or, more specifically, to dreams and dream-like altered states of consciousness (Dossey, 2009).

An English author and researcher in the field of parapsychology, Rupert Sheldrake had spent more than 25 years researching many minds phenomena, such as the ability to telepathy, distant vision, foresight, and so on. The huge material of researches (more than 5,000 case histories, 4,000 questionnaire responses, and the results of experiments carried out with more than 20,000 people, as well as reports and data from dozens of independent research teams), all of this allowed to Sheldrake to prove that most unexplained by official science human abilities are not paranormal but normal, part of our biological nature.

He showed that our intellectual abilities extend beyond our brains into surroundings us space with invisible connections that link us to each other, to the world around us, and even to the future (Sheldrake, 2013). His results correspond with studies of the chief scientist at the Institute of Noetic Sciences (IONS) professor of Stanford University Dean Radin, who in the his remarkable books (Radin, 2010, 2013, 2018) based on the results of thousands of lab experiments and some physic theories of conscience demonstrated and explained the evidences of extrasensory perception phenomena like psychokinesis, remote viewing, prayer, jinxes, and more. Amid these no wide published experiments were results of a US government supported program provided first at Stanford Research Institute (SRI) in Menlo Park, California, from 1973 to 1988, and then at Science Application International Corporation (SAIC), a large defense research and development company, from 1988 to 1995.

Dean Radin has convincingly shown that the proposed by such outstanding scientists as physicist David Bohm, physiologist Karl Pribram, Laureate of the Nobel Prize Brian Josephson, mathematician Sir Roger Penrose from Oxford University, and neurologist Benjamin Libet, theories and hypotheses argue that all manifestations of extrasensory perception are statistically significant and has a strict physical and biological basis.

Is extrasensory perception some manifestation of intellect? It depends on. For example, if a hunter can hear or see his victim very far from him, this is not intellect; this a good sense. But the ability of skilled photographer Edgar Cayce in curing incurable deceases fascinate. Decades ago, he was emphasizing the importance of diet, attitudes, emotions, exercise, and the patient's role - physically, mentally, and spiritually - in the treatment of different kind of illness (http://www.edgarcayce.org/edgar-cayce1.html). Without a doubt, that is high-class intellectual abilities. Is ability to telepathy an intellect? It also depends on. Many spouses in many years after marriage can read thoughts each other not literally with words but with intentions. It is not intellect. But if in dispute one person can predict what his opponent will say in next time, it is a high level of intellect. The same is for people who can predict future, because the information which they receive must be rightly interpreted.

Subsequent researches showed that the people who were trained in martial arts - such as Aikido, develop the skills to fill space around them. Some of the experiments described in the Rupert Sheldrake' book show doubtless significance of the results, confirming this phenomenon for people who are hypersensitive (Sheldrake, 2011).

Modern psychologists classify the manifestations of the supernatural potentials of human mind in the following categories:

- Clairvoyance – the faculty of perceiving things or events beyond normal sensory contact or in the future
- Psychometry-the ability to "read" information from objects
- Premonition- sense of anxiety or alert connected with real or future extraordinary events
- Precognition – knowledge of the future
- Retrocognition –knowledge of the past
- Biocation- the alleged ability to be in two places at the same time
- Telepathy – distant communication through mental signals
- Hypersensitivity - sensation of the glance of another person, the sensation of objects in the surrounding space (in the dark room), orienting in space, etc.
- Astral Projection: the ability to spiritually separate from body and travel in space with the mind alone.

The latest achievements of contemporary science make it possible to open slightly curtain above these phenomena, model them and use in medical practice. A group of neuroscientists at the University of California, Berkeley, reported they might have come up with a scientific way to read people's minds. Led by post-doctoral researcher Brian Pasley, the scientists have developed a method for deciphering the electrical signals in a person's brain as they listen to words or conversation. Upon figuring out these signals, they were then able to use them to recreate the imagined speech of the same person (Grush, 2012).

More than decade in the Research Centre for the development of abilities and consciousness in the Moscow (Russia) have passed research with people with and without extrasensory perception abilities. Studies of patients' brains were performed using encephalograph, which receives data from 16 sensors attached to the head. And then the resulting information processed by a special computer program that registered the local energy sources inside of the brain (Gnezditskiy, 2004)

After computer processing examples of encephalograms usually were seen as pictures with white dots on a black background. For ordinary people, these white dots scattered throughout the brain like stars in the black sky. For extrasensory perception people these "stars" are arranged in the form of a cloud and even a beam directed to the crown or forehead (Коёкина О.И, 2000, 2003). Moreover, the researchers found that in the process of inductor –recipient contact occurs extrasensory synchronization of biorhythms for the inductor and the recipient.

4.9. Automatic writing, channeling, mediumship, etc. from psychology and esoteric knowledge viewpoints

Automatic writing (psychography), channeling, or mediumship - all of this cases have a common roots of retrieving information from external source thanks to a superconscious.

Term automatic writing designates a person's ability, in a normal state, as well as in a state of hypnosis or trance, writes meaningful texts outside conscious control of the process. A person there may be quite busy in other activities and do not realize that he did write. For example, the writer can carry on a conversation, while his hand automatically writes a text of poetry, or a text on the some unfamiliar for writer language.

The famous Scottish scientist James Clerk Maxwell, who developed a theory combining electricity and magnetism, admitted on his deathbed, saying that not himself invented the famous electro-magnetic equations, but "something inside of him". In the same way, the famous Johann Wolfgang von Goethe claimed that he wrote his short story "The Sorrows of Young Werther" in a strange way, as if he was holding a pen that was moving by itself.

Channeling or mediumship is phenomena of receiving information from external sources of unknown origin, referred usually as the Supreme Mind, spirits, angels, mentors, representatives of other civilizations, and so on. The person who receives and transmits information is usually called a medium. To receive information some mediums come in full trance, others use another meditative state.

Amid the best known mediums were Edgar Cayce (1877-1945), Aleister Crowley (1875-1947), Jane Roberts (1929-1984) and Allan Kardec (1804-1869), who fathered the term "spiritism". In 1882, a New York dentist John Ballou Newbrow wrote a book "Oahspe: A New Bible" He reported, that this book was a result of automatic writing allegedly as "dictation of spirits." Sometime in 1912, Elsa Barker, an accomplished American Poetess, was visiting in Paris, when one night she found herself automatic writing, meaning that is someone other than her subconscious was writing using her hand. The entity inspiring the writing, claimed to be Judge David Patterson Hatch. The judge explained that he had recently passed away and that he wanted to document his experiences on the other side in the form of letters that he would write through Elsa's hand. In 1914, Elsa Barker published work "Letters of the deceased" (in 2004, the book was republished under the title «Letters From The Afterlife: A Guide to the Other Side»).

Rosemary Brown (1916 – 2001) was a spirit medium who claimed that dead composers dictated new musical works to her. She created a small media sensation in the 1970s by claiming to produce works dictated to her by Franz Liszt, Johannes Brahms, Johann Sebastian Bach, Sergei Rachmaninoff, Franz Schubert, Edvard Grieg, Claude Debussy, Frédéric Chopin, Robert Schumann and Ludwig van Beethoven.

Brazilian popular medium and philanthropist Chico Xavier (1910 - 2002) wrote 468 books using an automatic writing.

Nowadays, Andrey Trofimov for many years almost daily opens a special notebook in which he write a few pages in an unknown ancient language. What language he does not know. But he has this need. Experts from the University of Jerusalem recognized that he wrote on a mixture of ancient languages of Aramaic, Hebrew and more.

With the appearance and propagation of the internet the number of people owning automatic writing increased significantly. On the internet is appeared concept of channelers, or contactee, the latter means a person who perceives information from some creatures or entities.

After graduating with a business and economics degree from California Western University in California, Lee Carroll started a technical audio business in San Diego. During the four years he telepathically felt and heard a voice in his head, but was too scared at first this phenomenon. The voice identified as Kryon, incorporeal creature that is present in the world "from the very beginning." Finally, making the goodness of the "voice", Lee publicly stated (1989) about it and started channeling communications with earthlings.

Now Lee Carroll describes himself as a combination between an engineer and a "spooky guy", a melding of roles that sees him working as a `channel' or mouthpiece through which a spiritual entity called Kryon speaks. Lee Carroll is the author of twelve Kryon books, and co-author of The Indigo Children, An Indigo Celebration, The Indigo Children Ten Years Later (15 books total in print in twenty four languages worldwide. He has created also on the internet the channel Kryon http://www.kryon.com/menuKryon/menuKryon.html.

Globally, Lee's work channeling Kryon is well received. Many times he has held channelling sessions at a United Nation's group called Society for Enlightenment and Transformation.

In Russia, General –Major Retired Liventsov E.I during many years (1997 to 2002) received information from the Highest Universal Mind (HUM). In 2001, Liventsov E.I. published texts adopted them from the HUM, in his book (Ливенцов Е.И, 2011)

Can we trust the information received because of channeling? The ancients Latin proverb said: "Trust, but beware whom". Nearly every contact information contained mental refinement contactee, because he or she is trying to understand the resulting information. Here we can talk about the purity of the channel in terms of ability to distortion of information. If the information is simple enough in its content, the purity of the channel is determined by how contactee succeeds at reception completely stopping the flow of his thoughts or analysis and created an a state of meditation or "awakening mind." If the information contains complex knowledge or unknown before, the purity of the channel depends on the level of intellect of the contactee and his knowledge. Experience shows that the purity of the channel is not the worst. More importantly, who pass information. In ancient times, said, "Not everything that is from above - from God." Transmitters can be, for example, so-called low-level entity not holding true information. Often these contacts occur in beginners mediums causing spirits. It may be misinformation transmitted for some unknown reasons.

Freud and other psychoanalysts used a variety of techniques including automatic writing for access to the subconscious thoughts of their patients. For example, it was used by Pierre Janet in France, and later by Anita Muhl in the United States.

Research psychotherapist Dr. Anita Muhl shown (Muhl, 1963), that most mentally healthy individuals under the appropriate conditions can be taught to write automatically. Dr. Anita Muhl used automatic writing to get her sick patients possibility to look inside themselves and revive the repressed memories, ideas and associations. One of the patients was capable of automatically recording music. Other reproduce something mysterious and enigmatic or drawings with gorgeous colors, symbolic content which then stood for the process of automatic writing. All this was the perfect material for psychoanalysis.

Some old-school psychologists counted existing works written or recorded by automatic writing as clearest declaration of dissociated personality, which is buried in the depths of the subconscious.

CHAPTER 5

Intellect and multi-dimensional biofield

"The human biofield - the energy field that surrounds and permeates our bodies…"

Eileen Day McKusick

5.1. The energies in the human body

From an architectural perspective man is a complex biological system consisting of functional organs with the cellular principle of construction.

If consider primitively, the man is composed of cells as building constructed of bricks. Because the cell has a complicated structure also it is more accurate to say that the human body is a cellular structure like as a multination state, comprising about 10-100 trillion areas. That means one more time that the human is a super complex system.

The man is heterogeneous system, because includes coexisting organisms, such as bacteria, viruses, and parasites living in the different parts of body in symbiosis. There are 10 times more microbial cells on and in our bodies than there are human cells. There is also an estimated 100 times more microbial genes than the genes in our human genome.

More than 400 types of bacteria live in the human digestive system. The human body encounters both good and bad bacteria. Bacteria that we call "good" helps us digest our food and protects us from getting inside us bad organisms that can make us sick or even kill us.

As we mentioned above, the human being is an open system. Any person cannot live normally without Earth's magnetic field and Earth gravitation. Energy and all necessary components for his existence, man receives from food, air, water and in the form of radiation from our planet and space.

The human being is a dynamic system, because his cellular composition is changing constantly: some cells die, others are born again, and those and those - not in equal amounts, and secondly, because a person must compensate any changing and influence of his environment to support a condition of homeostasis. In fact, all our internal organs like as: a heart, respiratory system, intestines systems, or brain work rhythmically with different frequencies.

The man is an adaptive system that has a possibility to adapt with external systems of the environment.

The human body includes visible and invisible parts, the first are the material systems and second are the energetic systems. All material components of the body have their corresponding energetic components (Liu, 2018).

Any living organism cannot exist without energy and its transformations. The yogis in India call this energy Prana, in China it is known as Chi, in the Hawaiian culture it is known as mana, in Tibetan Buddhism as lüng, in Jewish culture as ruah, and, finally, as the life (Vital) energy in Western philosophy. Other expressions commonly used to describe this type of energy include such terms as "immanent energy", "subtle energy", "electromagnetic energy" and "universal energy".

According to Hinduism, Prana is physical, mental, and spiritual energy, which is the fundamental energy and the source of all knowledge. This energy continuously flows inside us, creating vitality in our bodies. Too little prana in the body can be reason to have a lack of motivation, lack drive, be depressed, and of course to be ill. The Hinduism philosophy believes that state of mind is directly linked to the amount of prana within us. The more balanced and peaceful a person is, the more prana is inside his body.

Chemical reactions occurring in a living cell in the body with a high speed accompanied by the release of heat, sound, and electricity. Back in the 1940-45 years, Professor Harold Burr of Yale University found that all living organisms have electricity, and therefore the magnetic field. According to the theory of physical vacuum of Shipov G.I. (Шипов Г.И. 2002), the electromagnetic field is the source of spin or torsion fields. Thermal, electromagnetic and torsion fields of all cells join together to form flows energy that permeates and surrounds the physical body. Of school courses in physics, chemistry, etc. is known that energy can exist in various forms, such as electrical, mechanical, chemical, thermal, nuclear, but also can be transformed from one form to another.

Energy carriers of these fields are all body fluids (protoplasm of cells, intracellular fluid, lymph and blood), which are electrostatic colloids, since their particles have a negative charge. Free electrons and ions produce an electric current flowing through the blood vessels, the nerves, on the skin surface. Fluid, tissue and bone perceive and convey various types of waves: electromagnetic, acoustic, light, torsion and gravity.

According to ancient Chinese (oriental) theories, the structure of the biofield is made up of a network of meridians, which are conduits for the flow of subtle energy, and a number of chakras, the subtle energy centers whose job it is to regulate the physical functions of the body (Rich,2004).

Studies carried out by Professor Kim Bong-Han of North Korea revealed that in humans and animals, there is a system of canals, apparently corresponding to the meridians of acupuncture. Total was found four systems of channels: 1) channels, free-floating within the blood and lymph vessels, 2) channels, passing over the surface of the internal organs, 3) channels running along the outer surface of the blood and lymphatic vessels (it seems that it is familiar with these channels acupuncture), and 4) channels distributed in the central and peripheral nervous system. These channels can transmit some types of energy. Voltage and current in these channels are so low that they require hypersensitive devices for measurements. Almost 50 years later, Kim Bong-Han 's discoveries have been confirmed by a variety of studies with animals and humans (Milbradt, 2009). Stereo-microscope photographs and images from transmission electron microscopy in the research papers show assemblies of tubular structures 30 to 100 μm wide (red blood cells are 6-8 μm in diameter).

In the 60-ies of the last century, American physiologist and sleep researcher Professor Nathaniel Kleitman discovered the existence of a basic the 90-120-minute cycle of rest-activity (BRAC) during both sleep and wakefulness. The BRAC repeats itself every 90-120 minutes. During the first half of this cycle, brain is in a fast brainwave state. A person feels focused and alert. During the last half, his brainwaves start to slow down. In addition, for the last 20 minutes of the BRAC, he had to feel day dreamy, and perhaps a bit tired, because it is a sign of his brain was restoring the sodium/potassium balance. Psychobiologist and hypnotherapist Ernest Lawrence Rossi calls this approximately 20-minute transition period the Ultradian Healing Response (Rossi, 1991). According to many studies, BRAC is the key factor in our mental activity - our ability to cognize, think, create, remember; and our physical activity - our energy, speed of reaction, strength, endurance, etc.(Hutchinson, 2014).

A lot of research in this field was carried out also by the Director of The Research Group for Mind-Body Dynamics at UCSD's Bio Circuits Institute David Shannahoff-Khalsa (2007).

The BRAC has something in common with Chinese Body Clock. According to ancient healing traditions in Asia, in the human body vital energy flows through the twelve organs and completes one cycle every twenty-four hours. Each organ has maximum energy for two hours.. The 24-hour cycle begins in the lung meridian at 3am and continues for 2 hours then moves to the colon meridian from 5am to 7am and so on energizing each of the 12 major meridians and returns to the lung meridian at 3am the next day to begin the cycle over again.

The rhythmic changes to the endogenous biorhythms include metabolic and energy variations and can exist on different levels of human body- from cellular to organs levels, and whole body levels.

5.2 The human organism is like as an orchestra

From classical physics, we know that the atoms and ions, which are bonded with each other with considerable interatomic forces, are not motionless. Due to the consistent vibrating movements, they are permanently deviating from their equilibrium position. Elastic waves of different lengths, frequencies, and amplitudes run through each matter of human body at all times.

In nature, there is nothing in the state of absolute rest. Everything vibrates and radiates. The planet on which we live, vibrates at a frequency of 8-14 Hz. The air we breathe is a carrier of low-frequency vibrations (sound), we undergo the influence of high frequency (light and electromagnetic radiation) vibrations, as well as ultra-high-frequency (X-ray and gamma rays) vibrations.

Study of the processes occurring in the human body shows that the presence of the vibrations is true for humans as an integral part of nature and the universe. Try closing your eyes and stretch your arms forward, after a while your hands will start to oscillate. The internal vibrations clearly manifested in extreme situations, for example, when a person is very cold, he is in a fever, or he is in a state of excessive fear.

Let start with genes waves described above in section 4.2. The technology, called "cell quake elastography" allow to scientists from at the University of Montreal Hospital Research Centre

(CRCHUM) to observe mechanical vibrations in the cell lasting less than a thousandth of a second (Grasland-Mongrain, 2018). Thanks to new technologies scientist proved that molecules of the proteins in human body vibrate constantly.

In common, were observed and registered in the issues and bones of body three types of infrasonic vibrations: waves of the first type are connected with the heartbeat; the second with the human respiratory rhythm; the third with states of emotional tension.

An American anthropologist and ethnographer Frances Densmore wrote "The human organism is like an orchestra and man, to some extent, is its conductor. There are large rhythms in which the organism acts as a whole, and small rhythms in which individual organs go through their parts. Health is the normal balance of all these rhythms, like the playing of an orchestra in which each instrument does its part and keep its proper relation to the others" (Densmore, 1927).

From the point of view of physics, it is absolutely right.

According to studies of A. Gurvich, F.A. Popp, F. Morell, etc., human body and its functioning systems are sources of mechanical and electromagnetic oscillations in a wideband spectrum of frequencies. For example, source of vibration is not only heart and lungs, but and diaphragm, muscles in processes of chewing, walking and so on. These oscillations are called "physiological" or "harmonious" oscillations. When a human body is disposed to pathogenic factors or is ill, some new sources of oscillations are being formed there, i.e. pathological or disharmonious oscillations. They distort physiological balance breaking self-regulated system of the human body.

All organs, systems and cells of the human body are controlled by exact cyclic patterns of rest and activity, which we can call the "universal laws of life." Cyclical processes in the human body called biorhythms. Biorhythms are periodic changes in the intensity and character of biological processes that are self-sustaining and self-replicate under different conditions. Biorhythms are characterized by a consistent alternation of phases of tension and relaxation, when a given parameter is consistently reaches the maximum or minimum values.

In biology, rhythms are divided into exogenous and endogenous. The endogenous rhythms are the physiological rhythms of the body. They have a very wide range of fluctuations in time (ranging from hundreds and tens of fluctuations per second to several fluctuations per minute, hour, day, week, and month).

Exogenous rhythms are influenced by an external stimulus for example such as the movements of the sun. Exogenous rhythms clearly reflect periodic environmental processes, both in the surrounding and in the social environment. For example, the rhythm of photosynthesis in plants is determined by the ratio and intensity of the light and dark time of day. Light is the main cue influencing such exogenous rhythms as circadian rhythms, turning on or turning off genes that control an organism's internal clock. While all human have circadian rhythms, there are some diversity in the length of the cycles, which help to explain, why some of us are "night owls" and others are "early birds". About 25% of people have a circadian period, which is slightly less than the 24-hour day, and 75% have a circadian period slightly more than 24 hours.

According to the criterion- the frequency of oscillations, all biological rhythms can be divided into three groups:

- high frequency rhythms with a period not exceeding the half-hour interval. These are the rhythms of contraction of the heart muscle, respiration system, the brain, biochemical reactions, intestinal motility, and so on;
- medium frequency rhythms with a period of half an hour to seven days. These include changes in sleep and wakefulness, activity and rest, daily changes in metabolism, fluctuations in temperature, blood pressure, frequency of cell divisions, fluctuations in blood composition;
- low-frequency rhythms with a period from a quarter of a month to several years: weekly, lunar, seasonal changes 11 year periods rhythms. The biological processes of this frequency include endocrine changes, sexual cycles, etc (Endres, 2002;Шноль, 2009).

In a living organism, oscillations of various types are closely intertwined, for example, mechanical and electrical, and the excitation of one type of oscillations can cause excitation of others (for example, mechanical movements are due to the process of propagation of a nerve impulse). It is reasonable to assume that an external resonant effect of one type (for example, mechanical) can lead to the buildup of oscillations of another type (electrical).

This means the existence of links between the rhythms of different processes; the presence of multiplicity and synchronicity between rhythms, and subordination of some rhythms to others.

The human body is non-linear oscillating system, which has a system of resonators, well-established "devices" for energy replenishment, and some a nonlinear limiting devices for increasing oscillations and feedback between the resonators and the energy sources.

Amplitude-frequency characteristics - the most objective characteristics of the oscillating system.

It is possible to select in the human body a complex electromagnetic vibration that interacts with the environment in different frequency bands. They can be divided into:

- Basic frequency
- Supporting frequencies
- The frequency of the energy-exchange cell.

The basic frequencies in the body are in the range from 7.8 to 14.1 Hz. This frequency alpha and beta rhythm of the brain. They are close or identical to the frequencies of the magnetic field of the Earth. Thus, human bio-rhythms resonate with the electromagnetic field of the Earth and synchronized.

Supporting frequencies lay within the 750-850 Hz. Almost every organ of the human body having its own frequency range. For example, for the heart, it is 700-800 Hz with an increase in angina up to 1500 Hz, for the kidneys - 600-700 Hz with an increase in inflammation of up to 900 Hz, for the liver - 300-400 Hz with an increase in inflammation of up to 600 Hz. Since each human organ has its own frequency, the appearance of additional electromagnetic fields differing from this frequency can enhance or retard the metabolic processes in the body.

Furthermore, changing the frequency of electromagnetic radiation can affect an activity of bacteria and viruses. At low frequencies up to 450 Hz can be activated viruses, and at 350 Hz - microbes.

The value of the internal energy of each person cannot be measured by means of modern measuring equipment, and it is practically impossible to calculate. However, it is possible to measure some of the kinetic components. For example, was found that the human body radiates heat from the surface layer of the skin in an infrared wavelength range of about 10 milliwatts from 1 cm^2 in average, with the entire surface of the body is emitted over a hundred watts. All these components of the thermal energy, sometimes called the animal heat is usually taken into account when calculating the heating of buildings, concert halls, and theaters. On the other side, it is possible to select in the human body a complex electromagnetic vibration that interacts with the environment in different frequency bands.

Among the many varieties of biological rhythms, the most well-known are those relating to sleep and wakefulness, which are part of the circadian rhythm. Three US scientists Jeffrey C. Hall, Michael Rosbash and Michael W. Young have won the 2017 Nobel Prize in Physiology or Medicine for their discoveries of molecular mechanisms controlling the circadian rhythm, otherwise known as our biological clock.

Ancient Chinese medicine, unlike the traditional one, takes into account the influence of internal rhythms of a human body, connecting them with fluctuations in the vital energy (chi), which consecutively passes 12 basic paired meridians in 24 hours, being found in each of them 2 hours.

In the 60-ies of the last century, American physiologist and sleep researcher Professor Nathaniel Kleitman discovered the existence of a basic the 90-120-minute cycle of rest-activity (BRAC) during both sleep and wakefulness. The BRAC repeats itself every 90-120 minutes. During the first half of this cycle, brain is in a fast brainwave state. A person feels focused and alert. During the last half, his brainwaves start to slow down. In addition, for the last 20 minutes of the BRAC, he had to feel day dreamy, and perhaps a bit tired, because it is a sign of his brain was restoring the sodium/potassium balance. Psychobiologist and hypnotherapist Ernest Lawrence Rossi calls this approximately 20-minute transition period the Ultradian Healing Response (Rossi, 1991). According to many studies, BRAC is the key factor in our mental activity - our ability to cognize, think, create, remember; and our physical activity - our energy, speed of reaction, strength, endurance, etc.(Hutchinson, 2014).

The BRAC has something in common with Chinese Body Clock. According to ancient healing traditions in Asia, in the human body vital energy flows through the twelve organs and completes one cycle every twenty-four hours. Each organ has maximum energy for two hours.. The 24-hour cycle begins in the lung meridian at 3am and continues for 2 hours then moves to the colon meridian from 5am to 7am and so on energizing each of the 12 major meridians and returns to the lung meridian at 3am the next day to begin the cycle over again.

The rhythmic changes to the endogenous biorhythms include metabolic and energy variations and can exist on different levels of human body- from cellular to organs levels, and whole body levels.

The physiological parameters (hormonal levels, quality of metabolic processes, body temperature, etc.) can be regulated by clock genes.

Can biological rhythms affect human intellectual abilities? To answer on this question, let consider what is a chronotype of people.

Individuals who are characterized by a more activity during morningness are usually known as "early larks," whereas those who show a more activity during eveningness are known as "night owls", the third chronotype whom we sometimes call hummingbirds is between them (Smolensky, 2015). "Larks" are early risers, perform mentally and physically at their best in the morning hours, and go to bed early in the evening. "Owls" stay up late at night, rise later in the morning, and perform best mentally and physically in the late afternoon or evening. "Hummingbirds" may be ready for action both early and late.

The number of "larks" usually reaches about 40 % from population. They go to bed on the average on 1,5 hour and rise on 2 hours before "owls", the dream of "larks" from night by the night is more stable, they are more satisfied by their dreams. At "owls" which number equally about 30 %, working capacity rather higher in the evening and at night.

If any "lark" or "owl" works during not optimal for him working hours, his productivity can be worse than usual. Currently, 30% of the U.S. workforce has unusual work schedules, such as alternating or extended shifts, and on-call duty. These unusual schedules are linked with health and safety risks. Unfortunately, no one can calculate the amount of economic losses associated with the non-optimal use of intellectual abilities. However, there are already abundant evidences that chronotype-tailored schedules might help minimize those risks (Alterman, 2013).

5.3. The human biofield and its components

According to the modern understanding, because there are processes in the body related to the transmission of electrical signals, gravity, magnetism, vibration, absorption and release of heat, etc. we can speak about an existence of a multicomponent field system, usually named biofield.

Historically, the term biofield was coined by Russian doctor A.G.Gurvich, who in 1923 year discovered a mitogenic glow around a cell. However, he could not explain this phenomenon. Dr. Beverly Rubik In her 2002 article "The Biofield Hypothesis: Its Biophysical Basis and Role in Medicine" wrote that the "biologic field is seen as a holistic or global organizing field of the organism" (Rubik, 2002).

A number of well-known the West scientists and researchers as Albert Szent-Györgyi, who laid out theory bioenergetics, Valery Hunt - the Executive Director of the BioEnergy Fields Foundation, Richard Gerber, M.D., - author of Vibrational Medicine, Dr. Karl H. Maret –president of the The Dove Health Alliance, and many others believe that the biofield is the bearer of the all information in the body.

Because, as we discovered above, any human cells created and support some fields, measured by the different apparatus as electromagnetic, torsion, infrared, acoustic and even like laser emanation, the whole human body must possess these types of fields.

These fields is carried out environmental effects, various types of radiation, man-made and natural. Many fields components of human as well as of experimental animals and plants (electrostatic, electromagnetic, acoustic, thermal, etc.) were either visualized by various devices like Kirlian photography apparatuses or measured by special measuring devices in the various laboratories (Brennan, 1988; Godik, 2010). Some of these devices are already commercially produced, some still under experimental verification. Amid physical fields of human that have been well studied in the present time are: a thermal field, electric and magnetic fields, torsion field, optical radiation in the near-ultraviolet region. A man is surrounded by a cloud of neutral and charged particles produced in the process of breathing, sweating and perspiration of the skin. All of these fields and particles are connected with the process of life at the basis of all forms of physiological and psychical activity.

Valerie Hunt, research scientist, professor emeritus of physiological Science at UCLA, has developed a method of instrumental confirmation of the existence of the human energy field and has dedicated her life to the study of human bioenergy fields and their links to health, disease, spirituality and human behavior.

The deficiency of standard medical devices to report on the human bioenergy field led Dr. Hunt to seek help from the scientists who developed telemetry devices for NASA. Through this avenue, a brand new research device was created for her laboratory, one which would measure frequencies from zero up to 250 KHz — much more greater than used in medical science before that time. She has called this device the AuraMeter.

Her research has shown that violations of electromagnetic fields have a stronger influence on the disease and functional impairment than biochemical misbalance. In mapping the bioenergy fields, Dr. Hunt has found that each individual has a unique human energy patterns, which can be called as: the Signature Field (Hunt, 2000).

One of the most stunning discoveries of scientists in the last century (Brennon 1988;the Hunt, 2000, etc.) is that certain talents seems to be associated with the presence of certain frequencies in the energy field. Dr. Hunt found that the frequency of the energy field are do not exceed 250 KHz. For healers or some gifted people, frequencies are in the range from 400 to 800 Hz. People, who can go into a trance, translate these "psychic" frequencies without distortion and work in a narrow range of 800 to 900 Hz.

Russian scientist - Deputy Director of Saint-Petersburg Federal Research Institute of Physical Culture (http://www.korotkov.eu) Dr. Konstantin Korotkov developed a similar to Kirlian photography technology called Gas Discharge Visualization (GDV). The Korotkov's GDV camera system consists of hardware and software to directly record, process and interpret GDV images to a computer (Korotkov, 2001,2004). GDV scanning equipment makes it possible to analyze the condition of the energy centers (chakras) as well as the energetic status of the organs and systems of the body (http://gdvcamera.com). Intrapersonal and interpersonal biofield interactions and results of the measurements of bioelectromagnetic signals that vary from low frequencies (electrostatic body-motion effects, electrocardiograms and electroencephalograms) to high frequencies (12 gigahertz microwave signals, high frequency X-rays, and gamma rays), were documented in the review of results of researches (Schwarz, 2001).

A biofield, which also has other names like as bioenergy field, human energy field, energy information field, bio-electromagnetic field, aura, etc. has two groups of components- endogenic and exogenic. The group of an endogenic component includes all types of field originated in atoms, molecules, cells, organs and tissue of the human body. An exogenous human field is a field of external origin and can include biofields of other people, techno genic electromagnetic (EM) fields (for example, radio, cell phones and TV fields), acoustic fields, torsion fields, gravitation fields and so on.

Dr. Rein Glen and others develop a hypothesis that exogenous EM fields can either induce or perturb endogenous field. Direct evidence indicates that classical EM fields can alter endogenous field associated for example with the heart (Glen, 2004, Kafatos, 2016).

The endogenous rhythms are the physiological rhythms of the body. They have a very wide range of fluctuations in time (ranging from hundreds and tens of fluctuations per second to several fluctuations per minute, hour, day, week, and month.

When the human body is in a state of homeostasis, all functional systems, and consequently their rhythms should be in harmony. The impact of exogenous rhythms or Infringements of endogenous rhythms can affect this harmony for a short or for a long time, and then the task of controlling the body systems described in the section 1.5 is to compensate for these effects, saving harmony on the another level.

According to ancient Chinese (oriental) theories, the structure of the biofield is made up of a network of meridians, which are conduits for the flow of subtle energy, and a number of chakras, the subtle energy centers whose job it is to regulate the physical functions of the body (Rich,2004).

The chakras are spinning vortices of radiating internal energy (pranic or chi) that nourish the organs and glands in the body. Chakras accumulate and distribute physical, mental, emotional and spiritual energy, on which depends the well-being of the body.

There are seven main chakras and they are located at specific points in the body, along the spine extending out the front and back of the body. The chakras are connected to each other and with certain parts of the body by means energy channels known as nadi, forming with them a system of chakras / nadis. Energy centers correspond to the seven colors of the spectrum of the rainbow, and the seven musical notes. Each of the seven main chakras affects various aspects of the body.

Usually, chakras are represented as the funnels of energy, which looks like the small spiral cones of energy. In the healthy system these spiral cones manifest the rhythmical rotation, synchronous with other, attracting to its centers energy for the use by its body. Each cone is disposed for the specific frequency, necessary for the healthy functioning of the body. In the damaged system these vortices do not revolve synchronously. The spiral cones of energy, of which they consist, can revolve quickly or slowly, jerky or unevenly. Sometimes it is possible to note breaks in the energy forms. These disturbances are correlated with certain dysfunction or pathology in the appropriate region of the physical body.

There are many reasons that can affect the functioning of human energy systems, leading to a variety of health problems. The following are some key potential issues: chemical (toxins), physical (radiation), emotional, spiritual, etc. which can create imbalances in various parts of the auras, chakras, and meridian system. Here can be also holes or tears in the auras, blockages that can be caused by any of the following: emotions that have not been felt and released; negative thought forms from other people, and so on

Underactive or overactive chakras can create a variety of outcomes. Underactive chakras can aftereffect like as fatigue, lethargy, weight problems, just a slow attitude towards life or a lack of willpower. Overactive chakras may create other types of problems of other types like as hyperactivity, panic attacks, emotional imbalances, and so on.

On the other side, the human biofield has static and dynamic components. Its static component do not change with time. Dynamic components radiate into surrounding space. We can assume that according to the above-mentioned hypothesis static component of the biofield is responsible for saving information, including subconscious.

From the point of view of classical physics, our physical body is surrounded by biofields as a mix of various fields, having different nature. Part of them like a thermal field has a boundary, some fields may not have a boundaries. It is very important, that wave of some of the fields can be modulated by amplitude or by frequency, for example, by way like low-frequency radio signal modulated for high-frequency carriers signal.

Miller I. et al. in their article "Quantum Bioholography" (2011) wrote the following: "Recent discoveries by Russian scientists Peter Gariaev & al. and later speculations by Vladimir Poponin shed tremendous light on our proposal that the human being is a transducer of universal energy and consciousness – essentially a biocomputer. The new feature of this research is the ability to physically demonstrate subtle fields emerging from the quantum foam or vacuum potential".

In the end of the 20th century in the Institute of Radiotechniques and Electronics of Academia of USSR was extensive studies of the human biofield for sensitives and normal people have been carried out (Godik,2010). Unfortunately, these studies were interrupted in connection with the collapse of the Soviet Union.

Many scientists agree, that biofields create around every living organism energetic cocoon, and all people packed in such cocoons with egg shape. From ancient times people believed that the height of this egg is equal to the length of a man's body with his arms fully extended above his head on the vertical axis, and its width is that of a man with his arms extended outwards from the center of his body along the horizontal axis. In reality for some people, the boundaries of the cocoon more in several times then the average statistical data. For example, the maximum radius of the biofield for blind people can exceed 5-6 meters. Apparently, this field somehow compensates for the lack of vision and helps to orient themselves in space.

It is generally accepted to represent the multidimensional structure of the human bio-field in the form of the physical body and its surrounding subtle bodies, inserted by one in other shells.

All human energy radiation can be divided into two classes:

- Directional - providing telepathy, ESP, telekinesis.
- Omnidirectional - the actual aura, which also provides a link to the world of the body - ambient radiation and other living beings.

Peter Fraser and Harry Massey developed NES (Nutri-Energetics System) model (theory) of human body-field and the body-field influence on health. Based on concepts of traditional Chinese medicine, Quantum physics theory, Space resonance theory of Milo Wolff, and so on they proved the comprehensive link between the human body's biochemistry and bioenergetics. They explained a comprehensive communication network in the body-field that directs information and energy to the right place in the body at the right time so that the body functions correctly (Fraser, 2008; Wolf, 2008).

The human body-field is a complex, structured network of fields that interpenetrates the physical body. The theory of the human body-field can most simply be explained as an exploration of the energetic and informational structure of the body that underlies its biochemistry.

This allows considering a man like as complex bioenergetics and information system, including his physical body and different kind of biofields. The shape and overall quality of

the human energy field is related to the energy and activity of the nervous system, organs, and glands in the physical body.

Many scientists count that human biofield in form a cocoon or in another forms plays a role of a power protection systems, which in some extend defends human body from the influence of external radiation, electromagnetic, gravitational and other fields (Brennon. 1988).

From viewpoint of hyper-dimensional theories of modern physics (Beichler, 2004) our body is a part of multidimensional university, therefore it is multidimensional too. The theory of the multidimensional man developed by Dr. Puchko L. G. allowed her to find methods for diagnosing and treating people suffering from ordinary and incurable diseases. These methods was published in many books, unfortunately mostly in Russian (see: http://www.ansmed.ru/about/catalog).

Scientific studies in recent years have shown that creativity of a human intellect is influenced by external electromagnetic fields, gravity and other fields, the ionization, the purity of the inhaled air, used for drinking water, and of course the quality of the food.

5.4. Aura and subtle bodies

One of existed definitions of aura is following: aura is a part of a person's biofield that can be seen for people with supersensory perception or after special training for other people

What is an aura, humanity has long been known prior on the East and later on the West. The concept of the aura can actually be seen in many different religions across the world: Zoroastrianism, Buddhism, Hinduism, Kabbalah, Christianism and Islamism (Longley, 2014).

Aura word with Sanskrit can be translated as the word having two components: "ay" - energy activity, "ra" - a shining light. In common, the aura is an emanation surrounding a person or object like the halo or aureole in religious art.

In 1911 Dr. Walter Kilner published in his book "The Human Atmosphere", one of the first western medical studies of aura, proposing its existence, nature and possible use in medical diagnosis and prognosis (Kilner, 1911, reprinted 2010). Innovated by Semyon Kirlian and his wife method of photography demonstrated evidences a life force or energy field that reflected the physical and emotional states of man and other living subjects (Brennan, 1988).

The following experiment of 2009 year prove the existence of human body glimmering. Japanese scientists Masaki Kobayashi, Daisuke Kikuchi, and Hitoshi Okamura successfully imaged the diurnal change of this ultraweak photon emission with an improved highly sensitive imaging system using cryogenic charge-coupled device (CCD) camera. They found that the human body directly and rhythmically emits light. The measured intensity of this light was 1000 times lower than the sensitivity of human eyes (Kobayashi, 2009).

One of first attempt to describe from position of modern knowledge an invisible part of humans was made by Cyndi Dale in her book "The Subtle Body: An Encyclopedia of Your Energetic Anatomy" (Dale, 2009).

It is known that many children can somewhere distinguish aura, but when they grow older, they lose this ability. Sensitive people, often they named as "aura readers". can see aura with or without color. Some of them can see in aura emotions. At present are created some methods, allowing to teach people seeing aura without any devices (Tazkuvel Embrosewyn, 2013). The boundary of the aura can be found easily with the help of dowsing, and any person can learn how to see aura (Webster, 2010).

Edgar Cayce perceived the aura as a series of vibrating colors that surrounded an individual and used it evaluations of person's health, state of mind, strength, weakness, desires, thoughts, and more. In fact, Cayce believed that every thought and action possessed an energetic vibration, that was essentially reflected in a person's aura (Todeschi, 2011). Today, experienced aura readers may be able to understand what the colors of aura mean on their own, and which colors comes from what and after which one (Longley, 2014).

Modern equipment, for example, Aura Video Station 7® Pro Package (www.aura.net) is capable to see aura in colors. On the market there are two types of aura cameras: the first, called a Kirlian camera, works by sending a painless electric charge into a part of the human body—usually a hand—and then captures the image of aura on a photographic plate. The second, called biofeedback aura imaging, is much more sophisticated. It works by measuring certain physical attributes—such as body temperature and skin conductivity—and then interprets the data and displays the 3D aura on a multimedia display.

Now we have the following systems like PIP (Poly contrast Interference Photography), DAS (Digital Aura Scanning system), RFI (Resonant Field imaging) etc., which claim to capture the human aura and help in detecting the subtle imbalances in our bodies.

Some people with clairvoyant ability can distinguish in aura several layers or by another words several subtle bodies, which look like Russian nesting dolls or the layers of an onion. For sensitive people the layers can be perceived in different ways that do not involve inner visualization. For example, energies of these layers can be sensed via touch, scent, or sound, and their boundaries can be measured, for example, with a pendulum.

There are many systems for determining the auric field, created by people on the basis of their observations and analysis. All these systems characterize auric layers (or auric fields) by number of these layers, and its color, brightness, shape, density, fluidity and functioning. Each system corresponds to the work that a person "conducts" with aura (Brennon, 1988; Bruyere, 1994; Stein, 1998).

The term "subtle body" is often used in a very vague sense, according to various teachings. In the different sources, the subtle bodies have different names and different functions. There are several different concepts for the number of subtle bodies - from 3 to 12 (Brennon, 1988; Stein, 1998; Gerber, 2001; Dale, 2009).

According to existed classifications here are the next components of human energy fields:

Etheric, Emotional, Mental, Astral, Celestial (Spiritual), etc. Each subtle body connects each other via chakras, which directs the energy using meridian system. On the other hand the subtle bodies hold and participate in transmission of various forms of information.

Subtle bodies distinguish from each other by a number of dimensions. Physical body - a three-dimensional, the astral body - a four (length, width, height and time), the mental body is the six-dimensional, spiritual body is seven-dimensional. Therefore, to say that they are nested within each other, it is wrong (Puchko, 2010), especially because some bodies, for example, a mental, has electromagnetic nature without any boundaries in space. But why sensitives see subtle bodies as layers with boundaries of each layer? We can provide the following explanation: our consciousness works in three-dimensional space, and the subsequent dimensions of space according to the laws of geometry give projections in the space of our consciousness.

Each subtle body has its biorhythms synchronized with the biorhythms of the other subtle bodies.

The phantom limb effect is one of a proof of the etheric body' existence. Even when the limb has been removed, sensation can be felt as if the toes were still there, like itching.

5.5. Human soul from a good sense perspective

Starting from great philosopher Aristotle or maybe earlier problem of soul existence is discussed amid religious adepts and scientists, because nobody knows where it is, but everyone knows what means "My soul is aching today". The Latin word "soul" means breath, so the soul would be associated with human or animal vital breath. Some believe that the soul is one of the subtle bodies, the other gives it a special place. In the literature can be find a hypothesis that our soul is the connection between our physical body and our spirit. Some believe that it has the shape follows the shape of the physical body, others say that it is impossible to determine the souls shape, and soul exists everywhere.

The concept of mortal or immortal soul is one of the general theological doctrines for many religions in our world: Judaism, Christianity, Islam and Hinduism.

As per one of doctrines described in the Kabbala - discipline, and school of thought of Judaism- the human soul has five levels or dimensions, namely, the "Nafesh" which connotes the physical body, the "Rauch" which relates to emotions and personality, "Neshamah" which relates to the intellect, the "Chayah" which relates to desire, will, commitment and faith and the "Yechidah" which is the true essence of the soul. The "Yechidah" is that part of god, which is in the human being. According to the Torah, the purpose of a physical life for the soul is to perform "Mitzvoth" or godly deeds. When a person performs "Mitzvoth", his soul is elevated to higher levels. When a person dies his/her soul goes through judgement after which a good soul proceeds to heaven or the Garden of Eden and a bad soul is sent to hell where it suffers.

Generally in many religious cultures the soul and its functions are inexplicable. On the other hand, if the 5-level structure of the soul, described by such an source of knowledge as Kabbalah, to compare with a description of the structure of subtle bodies (see section 5.3), then we can see a lot in common. This is certainly not enough to make a certain conclusion. Here is a reason why the hard road to evidence of a human soul existence did not finish for many disciplines of contemporary science. This science doesn't believe in soul existence, despite the fact that two scientific disciplines like psychology and psychiatry are connected by definition with soul existence because psychology translated from ancient Greek like knowledge of the soul, the term "psychiatry" means the 'medical treatment of the soul', because "psyche" in Ancient Greek means "soul". We can discuss problems of soul existence one more millennium, but today have a place a huge collection of different kind of facts, which cannot be fully explained without such argument as soul existence.

The first group of facts is evidences of reincarnation, a concept according to which, the soul leaves a body after the biological death and enters into a new body to start a new life. Among the world's scientific research of 20th century a great place is occupied by the work of Ian Stevenson.

In 1961 the chairman of the Department of Psychiatry at the University of Virginia Dr. Ian Stevenson began investigating cases of young children who claimed to remember previous lives. He traveled extensively over a period of forty years and found about three thousand cases of children around the world who claimed to remember past lives. His books became a classic in the annals of reincarnation research (Stevenson, 1980; Stevenson, 2005).

Analysis of the collected facts allowed him to draw the following conclusions:

- Most children under five years old remember past lives and experimentally confirm the details of past lives, names, place of residence, cause of death.
- Inexplicable phobias during immersion in hypnosis found their cause in the past life (the fear of water drowned, the fear of heights - in the past he/she fell off a cliff and broke).
- Phobias, affections, habits sometimes corresponded to personalities from the past incarnation.
- The man quickly succeeded in field in which a person was strong in his past life.
- Fixed cases where, after a brain injury, a person forgot his native language and began to speak in previously unknown language or acquired previously unknown abilities, for example, he/she began to play, without learning, the violin.

The next group of evidences connected with vision experience of the soul leaving a body after its death. In 1907 a French physician and parapsychologist Hippolyte Baraduc took a photo wife Nadine soul after their death.

Dr. Baraduc with special photo camera got the soul exit from body of his wife on her deathbed. He captured a photo twenty minutes after her passing which he claimed revealed her departing soul. The photo image include three misty luminous clouds over her body. In taken picture after about thirty minutes was exhibited one larger cloud. Soon after cloud left the body and floated into Baraduc's bedroom, creating an icy breeze before leaving entirely (http://www.weirdhistorian. com/thought-photography-hippolyte-baraduc/).

In the a hundred years after that Russian scientist Dr. Konstantin Korotkov used a technique called "the gas discharge visualization method" (GDV) captured the intensity of the natural electromagnetic field emitted by human body at the moment of its death. GDV method revealed that after death a physical body changed characteristics of emission of energy depending on the cause of death.

In case of the natural death of the senile organism was measured fluctuations in the parameters of luminescence from 16 to 55 hours; in case of "sharp" death, for example from accident was measured a visible jump of emission either after 8 hours or at the end of the first day, and two days after death the oscillations descend to the background level; in case of sudden, tragic "unexpected" death by suicide or mistake was measured the strongest and longest oscillations, their amplitude decreases from the beginning to the end of the experiment, the glow dims at the end of the first day and especially sharply at the end of the second; In addition, every evening after nine and until about two or three o'clock in the morning there are bursts of the intensity of the glow.

According to Dr. Korotkov results of measurement, the navel, and the head, are the parts of the body to first lose their life force, or rather, their soul. The groin and the heart are the last two areas of the body where the spirit resides before finally heading on to the great unknown.

The next group of evidences connected with a practice of out-of-body experience (OBE) and near-death experience (NED).

In 1971 a prominent researcher in the field of human consciousness Robert Monroe in his book "Journeys Out of the Body" discovered for general public the phenomenon of OBE in which a person seems to perceive the world from a location outside his physical body. At the end of the 20th and the beginning of the 21st century, studies of this phenomenon have been carried out in various laboratories and institutes around the world. Today, there is still no consensus in understanding the nature of this phenomenon, although many organizations have already emerged that teach the techniques of OBE. Here can be a scientific hypothesis which can explain OBE. Because for some people their brain can sent and receive information practically immediately for any distance this phenomenon may occur when the brain remotely controls a position of some subtle bodies using some carrier of information, for example as torsion field.

The famous an American philosopher, psychologist, physician Raymond Moody described in his best-selling book Life after Life (1975) evidences of many people who experienced "clinical death" and were revived. This book spawned the publication of a huge number of similar descriptions of life after physical death by people who faced this phenomenon in real life.

Dr. Eben Alexander in his book "Proof of Heaven: A Neurosurgeon's Journey into the Afterlife"(2012) described that while his body laid during a 7 days in coma, his entity journeyed beyond this world and encountered an angel who guided him into the deepest realms of super-physical existence". Written by Dr. Mary Neal's book "To Heaven and Back"(2012) includes recognition of being in a different dimension after death, meetings with unusual beings, feelings of being accepted and welcomed, and a realization that we are all part of the universe, and carry divine universality in us at all times.

The scientific analysis of possible reasons of NED and OBE' cases did not allow to disprove the hypothesis proposed by Moody (Blanke, 2015).

Director of the Centre of Consciousness Studies at the University of Arizona, Dr. Stuart Hameroff and British physicist Sir Roger Penrose developed a quantum theory of consciousness which they dubbed orchestrated objective reduction (Orch-OR). The theory asserts that our souls are contained in structures called microtubules which live within our brain cells. The quantum information within the microtubules is not destroyed, but just distributes and dissipates to the universe. It means that in the event of the patient's death, it is "possible that this quantum information can exist outside the body indefinitely - as a soul "(Hameroff, S.,1998a).

The famous Russian neuroscientist, former scientific director of the Institute of Human Brain of the Russian Academy of Sciences Natalya Petrovna Behtereva in the last years of her life openly said: "All my life I studied the living human brain. And are faced with a "strange phenomena"...A certain percentage of people continues to exist in a different form, as something separated from the body. For this form I would not want to give a different definition than the "soul." In the body, there is something that can be separated from that and even survive the man himself."

According to Dr. Korotkov, the navel, and the head, are the parts of the body to first lose their life force, or rather, their soul. The groin and the heart are the last two areas of the body where the spirit resides before finally heading on to the great unknown.

Described in the section 4.2 DNA phantom effect approved existence some internal maybe non material structure, which can be named as soul.

Many questions today remain open: If does soul have a memory, and can be the carrier of good or evil, then does it have kind of brain and an intellect? Where the soul is located? What energy is necessary for it and from where it does obtain?(Pandya, 2011).

5.6. The assemblage point as energy center in the human body

The concept of the Assemblage Point (AP) comes from the series of best-selling books of the anthropologist Dr. Carlos Castaneda. The author of these books had described interesting American shamanism' phenomena, one of the main legacies of the Toltec culture that developed in Central America at the end of the first millennium AD. According to one of hypothesis, the Toltecs were the descendants of the Atlanteans. The ancient Toltec believed that the universe is an infinite cluster of energy fields, similar to the threads of light, which have no end in sight, and are called emanations. A man is a combination of energy fields that are connected in some luminous cocoon. Cocoon surrounds us on all sides, and has the form of an ovaloid or ellipsoid. The area on the cocoon, in which these energetic threads unite together in a glowing wisp to get in or get out of the cocoon, called the AP. Dr. Castaneda described the AP like the luminous ball inside "the luminous egg, which is our energy self" (Castaneda, 1993). Castaneda counted that the position of the AP dictates how we feel and how we behave.

Similar to the hypothesis of the Assemblage point as energetic center exists in Chinese Qigong Philosophy. There are 3 important energy centers, called dantians that store and disperse Qi energy from taiji pole (center core) of the body (Johnson, 2000). The lower dantian, located in the lower abdomen, is connected with the Qi energy field of the physical body. The middle dantian, located in the center of the chest, is connected with the Qi field of the emotional body, surrounding the physical body. The third dantian is located inside the middle of the head and is associated with the spiritual field of the Qi that surround physical and emotional subtle bodies. Qi travels along acupuncture meridians to all organs and tissues of the body.

Robert Monroe while on one of his frequent out of the body excursions discovered that his second body was attached to his physical body via a cord consisting of hundreds of tendon-like strands packed neatly together. This cord was attached to the center of his back and the hundreds of strands or filaments spread and fanned out to form his second body. This suggests that his projected "second body" had an AP entering at the center of the shoulder blades.

Today, the reality of the AP has been confirmed by the fact that many advanced sensitives can see it as a circular spot with intense luminosity at different positions on the cocoon (aura) of any man. Anyway mainstream science in many countries no hurry to recognize its existence, as well as the existence of subtle bodies surrounding the physical body of the human being like a cocoon (Whale, 2008, 2009).

Mark Rich after 12 years of training could see energetic tubes, that are arranged symmetrically all the way around body and the cord bundle (similar to AP), where the cords are like long silvery ropes-extensions of energy coming out from the abdomen and going in different directions to fix a person's attention on different things (Rich, 2004).

Dr. Jon Whale - scientist, researcher, psychologist, analyst, and an engineer has extensive experience in the various studies of esoteric knowledge. He had written numerous articles and several books about the AP (Whale, 1997, 2008, 2009) and founded Whale Medical Inc.

His main contribution to modern science and medicine - works in the field of physical studies of the human biofield and alternative treatments with light emitters, crystals, and semi-precious gemstones.

In his studies Jon Whale discovered that every living person has an oscillating energy field with its epicenter, which can be seen by sensitives as a bright spot of high energy. This spot is the AP (Whale, 2009).

The greatest merit of John Whale is that, he based on modern science, had proven the reality of various Eastern teachings on energy, aura, chakras and, of course, the AP. Using special equipment, John Weil scientifically proved the existence of an AP and could confirm that a dip in the energy field from 2 to 5% occurs at the back and front of the AP case. He demonstrated the importance of AP in balancing the entire energy field and physical body. John Keith called the AP "a catalyst of power" (Whale, 2009).His colleague Bulgarian physicist Dr. E. Eftimova raised the hypotheses that AP is a vortex of toroidal form, which is crucial for its stability and perpetual motion (Evtimova,2010).

On one hand, the biological activity of the brain, nervous system, organs and glands determines the position of the APs and vice versa; on the other hand, any stress or emotional traumas, violence, intimidation, giving birth, accidents, as well as drug use, can shift it off center and it can even move into organs like the liver and heart; and at last in third, the location and entry angle of the AP dictates how an individual feels, behaves and how others perceive them.

In the common, AP determines our physical health and state of mind. Its location also influences the state of other energy vortices within the body, the chakras, and the state of the glands and organs they are associated with, the immune system, the posture and even the complexion.

Jon Whale counted that, the AP is a missing energy link connecting the human soul with the physical body (Whale, 2009).

Perhaps, the AP is also the energy center (epicenter) of the body, perhaps it serves as a tool for energy exchange with the world. In other words, this is the point where the energy goes to the outside world, and the energy of the outside world moves inside. The location of the AP on the body determines man's personal vibration level: the higher AP, the higher level of body vibration.

Perhaps the function of this organ is similar to the work of the television receiver: AP attunes with external information channels and builds for us a picture of the world.

Maybe the AP is a wave organ working in multi-dimensional space, available for any person and located outside the perception of people living in the normal space-time continuum, called Minkowsky space. This space named in honor of the Russian mathematician Herman Murkowski for the definition of four –dimensional time-space continuum, in which our life happens. In 2000, Dr. Angela Blaen in England founded The Assemblage Point Centre Ltd. and The Assemblage Point Associations for practitioners (www.assemblagepointassociation.com).

At the time of birth, the AP for ordinary people locate in the navel area, the place from which the child in the womb of his mother gets needed nutrients and energy.

The researches of Jon Whale and his colleagues shown, that AP should be for human in normal condition close the thymus gland in (or near) the center of the chest. The location of a woman's AP is generally, but not always, several centimeters higher than that of a man. (Whale, 2008-2009).

The AP location for the average person is to the right of the central. The central location is beneficial for mental health. It directly influences our interactions with others and the world at large and can improve our fortune. Sometimes experienced practitioners can easily find this location without any instruments.

Those of us who have centrally aligned AP prosper in health, performance and success. For instance, someone with a centered, or just slightly to the high right of center, AP at the front of the chest and centered on the back of the chest is likely to be happy, healthy, cheerful, and well balanced mentally and physically. AP is probably assisting well interactions with other people.

Each position around the center, be it left, right, up or down produces a certain feeling and thinking associated with that position. Unstable or misaligned APs affects physical and mental performance. Dropped APs, in particular, can cause unnecessary suffering and waste of human potential and resources. Long-term gross misalignment can cause serious physiological and psychological symptoms. It is often reflected in a hunched or slumped body posture. When a man approaches to time of death, his AP moves down towards the navel.

Any is below described reasons can be a cause of involuntary shift of the AP to a danger for health condition locations:

- Serious accident, disease, tragedy in family, chronic stress or depression;
- Distressed or oppressed childhood, sexual assault, violent intimidation, kidnapping;
- mutilation or poisoning, attempted suicide, drug overdose;
- Mugging, robbery, fraud, identity theft;
- War, terrorism, tortures, imprisonment;
- Betrayal, malicious divorce, bankruptcy, home repossession, and so on.

As a result of this AP shift people can undergo a serious and sometimes permanent change of their psychical or physical health. This is the reason why traditional medicine cannot effectively heal some diseases having above mentioning background (Whale, 2008).

Due to his searches and efforts through the years, Dr. Jon Whale has created a map of the AP locations of the man corresponding with our physical, emotional and psychological health, a way of thinking and perception of the world. He defined also how a position of the AP on the human body changes from various diseases according to their symptoms. Thanks his discoveries John Wale created many effective treatments for many diseases, including mental and skin. It allows a professional armed knowledge about the AP location of a patient to have more chances to heal some hard healing diseases.

With the help of the Differential Infrared Diagnostic Scanner innovated by Jon Whale, today exists an excellent opportunity to discover the entry and the exit locations of the AP. He showed how the human mind by changing the position of the AP could change its properties for the treatment of his own body.

A survey of the literature showed that the right position of the AP could give some person also the ability:

- To learn;
- Of perception of the outside world, including the Universe;
- To have an impact on the outside world.

Dr. Whale discovered that the AP remains fixed on its spot for most people, but can be moved through the use of crystals, chants, meditation, and special methods. During the dream, assemblage point moves smoothly and quietly on its own. For some people the AP can shift to new positions in a process or after book reading.

Assemblage point can be moved to any position on the surface of the cocoon, inside and even outside it. Now practitioners of AP distinguish several types of motion along 3 coordinate axises of the cocoon- outside or inside by the following ways:

- Changing in angle
- Changing in angle or location
- Shifting in depth or changing in angle, location and size.

It cannot be moved by consciousness, and sometimes it demands for the shift a power of the other person. Maybe it is a reason why Dr. Whale advised for a clinical correction of patients AP actions of two dedicated and medically ethical therapists.

The practice of J. Whale and other medical therapists armed invented him technology had revealed that such unhealthy conditions as chronic fatigue syndrome, multiple sclerosis, hypertension, trauma, migraines, nervous disorders (even schizophrenia, skin diseases, viral and bacterial infections, asthma and allergies) often can be succesfully treated due to the AP shifts at specific locations. In many cases, medication will not be required after correction (Whale, 2009).

Until now, I could not find any studies of interesting us item in field of connection between the position of the AP and intellect. However, this connection must exist, because the movement of the AP changes the perception, sense of self, and the behavior - important components of the intellect.

5.7. Emotions as important attribute of intellect

The role of emotions in the structure of functions and subsystems of intellect was enumerated and shortly described in the division 1.6.

There's actually no consensus on what an emotion is (de Sousa, 2017). Our task here is not to give a description of the entire variety of the emotions, different in the different cultures (about this written many books), but to show the principal connection of emotions with the intellect in their different manifestations. The class of emotions includes the moods, feelings, affects, passions, stresses.

Emotionality is the observable behavioral and physiological component of emotion, and is a measure of a person's emotional reactivity to a stimulus.

Recently, some researchers include to the class of emotions stress. "Stress" is a state of emotional and physical tension, which occurs in certain situations, which are characterized as difficult and timeless.

Around the world, many researchers have studied how and why people experience emotion, and a number of hypotheses have been proposed. Many esoteric teachings suggest that emotions form one of the subtle bodies that have different names in different sources. Last researches showed that emotions more look like as implants than a whole subtle body. At present is not created an unified theory of emotions, but here is one, in what the majority of scientists converges: emotion - a special class of the mental conditions, which reflect the attitude of man toward surrounding peace, to other people, to themselves and to the results of his activity.

As for the kinds of human emotions, they are divided into simple and complex. Simple occur in response to light, smell. They are polar, i.e, takes one of two opposite values (pleasant-unpleasant).

Complex emotions arise in the process of getting to know a man of the world, assessing the role of various objects in people's lives. These are joy, interest, anger, fear, disgust, anxiety, resentment, and some others.

The famous American psychologist known for his contributions to Differential Emotions Theory (DET), author of many books about emotions, Professor Caroll Izard isolated 10 basic emotions (joy, interest-excitement, surprise, sadness, anger, disgust, contempt, fear, shame, and guilt). Other authors include in this list also grief, love, etc. "(Izard, 2009).

In opinion of many scientists, the role of emotions in our thinking is so great that since publication the article of Peter Salovey and John Mayer "Emotional intelligence" many began to unite emotions in a single entity having named "Emotional intelligence" (Mayer, 1990).

P. Salovey and J. Mayer proposed a model that identified four different factors of emotional functions: the perception of emotion, the ability reasoning with help of emotions, the ability to understand emotion and the ability to manage emotions.

First step for perceiving emotions is understanding situation to exclude a false alarm and made fastest action or reaction. Especially it is important for nonverbal signals such as body language and facial expressions. Understanding allows to trigger a need emotion in the right time.

Reasoning with emotions means to promote thinking and cognitive activity. Emotions help prioritize what we pay attention and react to; we respond emotionally to things that garner our attention. Emotions direct our attention to important events, they prepare us for certain actions and influence our thought process.

Understanding emotions mean that the emotions can carry a wide variety of meanings. This capacity reflects the ability to identify the source of emotions, to recognize the connection between words and emotions to understand ambivalent feelings. If someone is expressing angry emotions, the observer must interpret the cause of their anger and what it might mean.

Emotion means changes in the body's chemical profile, changes in the degree of muscle contractions, changes in our neural circuitry. In sum, emotions connected with changes of energy.

According to Synthhia Andrews emotions have direct link to subtle energy, even to be as the language of subtle energy" (Andrews, 2013).

Everybody knows that fear or anger can be connected with huge energy allowing to weak people to be able run or jump as leopard, climb on the trees as a monkey, or lift a weight as a bear, or even kill an aggressive enemy.

Our positive emotions of love, joy, excitement, compassion, create energy within us. These are the driving forces that can help us excel in our accomplishments. They enable us to create and implement more ideas and inspire us to do more. They attract other successful people to us and radiate potential.

On the other side, negative emotions such as anger, hatred, fear, worry anxiety etc. drain us of our energy, sometimes zap our abilities to fight with whatever we do not like. They can stop us in our tracks from moving forward. Fight-or-flight emotions can make it impossible to think clearly, logically, and reflectively.

Emotions have unique vibrations. These emotional vibrations also go from higher/faster to lower/slower. When person is laughing and having fun, his body's vibrations is higher by amplitude and faster by frequency. When he/she is tired and sick his/her vibrations is slower and lower. When he or she is in love, they feel "energized". That's because their emotions are literally adding voltage and power, lightening their bodies. And everything is opposite when they negative and depressed. This is scientifically measurable.

Emotions interact with each other - one emotion can be activated, strengthen or weaken by another.

The famous psychologist Daniel Goleman suggested that a human has two brains, two minds, two different thinking ability: the rational, which is sent from the mind, and emotional which sent from another place (Goleman, 2005). As we discuss earlier this place can be abdominal according to a research of Michael Gerson (Gerson,1999). Emotions play a much greater role in the thinking, decision-making, and personal success than people think. For example, positive emotions assist to brain by cutting some brain function of checking dangers in the environment and in a condition of body organs.

Usually, we have a balance between the emotional 'brain senses' and rational 'thinking brain.' Emotion feeding into and informing the operation of the rational mind, the 'thinking brain' is able to correct the functioning of 'emotional brain' not letting emotions get total control. Only in the dangerous situations, such emotions as fear and anger can work as an emergency system. For more fast reaction, these emotions can turn off a flow of thoughts in the consciousness.

Emotions as a form of mental activity, can independently distribute in space, and from person to person. They can be contagious, does not matter if the emotion of fear or anger, or if it is laughter or joy. Most likely emotion and thought-form are programs that can be run outside or inside a man, and can be blocked. They can be compared with a very complex program such as modern soft robots or viruses of such class as Trojan horse.

In these cases the emotions lose their functions as the regulation of the quality of perception, and emotions may be accompanied by an external energy transfer in one direction or in the other way. We have many evidences when a singer like Elvis Presley, The Beatles and many others could infect the huge crowded theatre halls or stadiums and creates an specific aura of audience crowd. In these cases, the group emotions can be viewed as Egregore.

In an esoteric context, an Egregore is the general imprint that encircles a group entity. It is the summary of the physical, emotional, mental and spiritual energies generated by two or more people vibrating together towards the same goal; being a sub-product of our personal and collective creative process as co-creators of our reality.

Gaetan Delaforge defines an Egregore as a kind of group mind which is created when people consciously come together for a common purpose (Delaforge, 1987).

When a group of people pray and meditate collectively towards an objective, an Egregore of protection and blessing is sent forth, as a circle of Light that shields and safeguards the objective of the prayers. An Egregore can be either negative or positive, depending on the level of vibration and the frequency.

The Egregore is maintained through the mental and psychic energies of its creators and, as an autonomous entity, it is formed through the persistence and intensities of the current emotional and mental waves. Weak emotions and feelings tend to create undefined Egregore, with short life spans. The opposite is also true: strong circles are created by strong will, emotions and determination of purpose. There are good Egregore that are positive and that bring blessings, good energies and protection against negative vibrations.

In conclusion, we need to pay attention a fact that any management of intellect including its optimization must consider levels of the emotions in their variety.

5.8. Love and intellect

About love in various forms of its manifestation is written a sea of books, probably 99 percent of fiction is devoted to love. Unfortunately, the science of love, especially the science of the connection between love and intellect remains a little explored area.

"Love is the strongest of all passions, because it simultaneously possesses the head, heart and body" wrote famous French writer and philosopher Voltaire. Hereof, in the list of feelings and emotions a love takes one of the first places on the importance, because it connected with three main functions of human organisms defining the life instincts, which deal with basic survival, pleasure, and reproduction.

In the section 1.5 we wrote that any living organism should always solve three important tasks: how to survive, how to produce offspring and how to enjoy. The second and third tasks are most effectively solved with the help of love and sex. The studies of last decades demonstrated that the love allows successfully to solve and first task. For example, in a 2009 study, psychologists found that people who were immersed in thoughts of love or sex coped better with creative and logical puzzles (Förster, 2009). A 2014 study in Oxford University showed that romantic love can improve the ability to understand the psychological state of other people. These results (for 91 people) proved that a love stimulus can enhance subsequent performance on conceptually related mentalizing tasks (Wlodarski, 2014).

At present in philosophy, psychology and other sciences are identified many forms of love: familial love, friendly love, romantic love, sexual love, divine love and so on. Power and duration of love relationship can be also different: from light attraction to strong rage, from one short meeting to all life.

A huge variety of forms of manifestation of love, their dependence on culture and on life circumstances do not allow to describe fully this phenomenon. Let consider only one kind of love: mutual love between spouses in the family. Love, it is an intimate and deep sense of a person aspiring to another person, a sense of selflessness, spiritual and cordial affection, respect for spouse interests and needs, and care not only for him/her, but also for all members of the family including children and grandparents. A genuine love includes also the distribution of responsibilities for the life support of the family, and, of course, responsibilities for the parenting of children.

If follow above mentioned citation of Voltaire, love must have three strongly interconnected parts, defined by:

1. The heart, or rather the soul
2. The head, that is the mind and the intellect
3. The body or sexual love.

The first romantic part of love is first of all spiritual intimacy, an emotional attraction that turns into a union of souls, a condition where one cannot live without each other, wants to be with a beloved person every second, live his/her life, his/her problems, and care about him/her, feel and ability to be with him/her always at the right time and place.

On the other hand, it is the ability to accept the mind of another person as he is, with all his shortcomings and virtues, to respect him, this is an ability to behave correctly with this person and his close parents or illegitimate children, be able to forgive him, understanding the causes of his misdemeanors.

And finally, sexual love has its own rules, which allow partners to know when it's possible, and when it cannot, and, if it possible, sexual love needs realization of the ability to bring sexual satisfaction and the highest pleasure to each other.

A strong interconnection of these components allows two loving people to represent a single whole, and call themselves as halves of each other.

Unfortunately, in modern society, for various reasons, true mutual love has become less and less common. If to evaluate the strength of partners' love on a scale of one to ten points, one can love, for example, on 10 points and another for 3. Life shows that intellectual abilities of both or one spouses can help to develop or compensate a lack of love up to such degree, that the extraneous person will not see the mutual love at first sight is or it is developed up to such level.

If to admit to consideration aura of the person and its thin bodies shortly described above it is possible to see, that from the point of view of esoteric, three subtle bodies of aura-emotional, mental and spiritual, concern to love. Therefore, you can consider the functions and even structure of love also complex and multidimensional as well as the whole man (Puchko, 2008).

Love can stimulate a person's creative abilities and help him to achieve success, especially if it can spark souls of both partners. Genuine love makes any person happy and allows him/her to easily overcome any obstacles, no matter how complex they are.

On the other site, other requirements must be added to love: love should not be blind, but sighted and intelligent, not only by feeling but also by responsible behavior. Love without respect is short-lived and impermanent; respect without love is cold and feeble.

Love as passion - a kind of love, where the power of attraction obscures the mind. Love as a passionate attraction is not just a strong liking. Most of us know people who like them very much, but we do not love them, and some of us even felt a passionate attraction to someone who did not particularly like.

The following conclusion must be drawn from all of this. Love, as well as another strong emotion, requires management and therefore the inclusion of the entire human intellect arsenal.

In Western cultures, more than 90 percent of people marry by age 50. Healthy marriages are good for couples' mental and physical health. They are also good for children; growing up in a happy home. It protects children from mental, physical, educational and social problems. Unfortunately, about 40 to 50 percent of married couples in the United States and other countries divorce. The divorce rate for subsequent marriages is even higher. As a rule, any divorce connected with health loss and often with broken fate of spouses.

Love must be visible not only at the level of feelings, but also at the level of intellect. The phenomenon of love is always associated with problems, when love is, when love is not enough, or when love does not exist. One of the main reasons for the occurrence of these problems is insufficient attention to the role of intellect for both partners. To control love feeling, everyone needs not only to think deeply using all intellect reserves, but to have strong wish to save the love, willpower, and skills to find and realize compromise solutions.

CHAPTER 6

Factors that affect a power of human intellect

"A mind is like a parachute. It doesn't work if it is not open."

Guitarist and composer Frank Zappa

"None but ourselves can free our minds".

Jamaican singer and songwriter Bob Marley

6.1. Levels of influence on intellect

For today, scientific disputes have not yet ceased about what more influences intellect: heredity, education or training, and how intellect can change during life span (Kaufman, 2011).

Brain function is affected daily by many factors, including the environment, nutrition, physical activity, drugs and methods of treatment, growth and aging factors, processes and results of learning, and so on. Because the human intellect as a functional part of human body, its functions and characteristics should be also under the influence of these factors.

As was shown above, the intellect is a multi-level operating system and therefore it can be sensitive to impacts at all levels, from the atomic level to the levels of subtle bodies (see previous Chapter). These impacts can be harmful and useful; they can be short-term and long-term, compensated in the bad cases by the operating system itself or requiring external influences.

On the other hand, this means that intellectual abilities are not permanent and therefore they must always be assessed, maintained at the proper level or enhanced with help of the described below methods.

Scientific studies in recent years have shown that creative human intellect is influenced by external electromagnetic fields, gravity and other fields, the ionization, the purity of the inhaled air, quality of drinking water and food, and, of course, regular physical exercises to support a needed tonus.

In this chapter we consider some factors which can affect on intellectual abilities, in the next chapter we pay attention to some features of human intellect, which can be changed or tuned by impact of these factors.

6.2. Genetic level

As mentioned above, all information determining the capabilities of intellect is recorded at the genetic level. Therefore, theoretically, the most direct way to improve the human intellectual abilities is to make changes into the human genome. Unfortunately, the existing level of knowledge does not allow it, although such a tool as the CRISPR/Cas9 has already appeared for the implementation of genetic engineering methods. However, during the last decade, two revolutionary events occurred that inspire hope that a new needed genetic tool will be implemented soon for above mentioned goal.

At September 16, 2015, in one of Bogota (Colombia) clinics 44-year-old American woman Elizabeth Parrish- one of the leaders of the scientific and medical company BioViva- got dozens of experimental gene-therapy injections with main goal to stop or decrease aging. She chose a clinic in Colombia, because such type of therapy forbidden by federal regulations in the USA.

In 2018, the world's first genetically modified woman underwent examination. Results of tests showed that her muscle mass increased, the amount of intramuscular fat decreased. And for three years after therapy, the changes remained at a good level and she looked younger.

The first known attempt at creating genetically modified human embryos in the United States has been carried out by a team of researchers led by Shoukhrat Mitalipov of Oregon Health and Science University, in Portland, Oregon. This team changed the DNA of a large number of one-cell embryos with the gene-editing technique CRISPR (Liang, 2015). Experiment was stopped and none of the embryos were allowed to develop for more than a few days.

In November 2018, an associate professor in the Department of Biology of the Southern University of Science and Technology (SUSTech) in Shenzhen (China) Dr. Jiankui He claimed that he had successfully helped in the birth of the world's first genetically edited twin girls, nicknamed Lulu and Nana (Larson, 2018). He says his team performed "gene surgery" on embryos created from their parents' sperm and eggs to protect the children from the human immunodeficiency virus, HIV, which causes AIDS, because the children's father was HIV-positive. For the first time, a scientist claims to have used a powerful new gene-editing technique CRISPR/Cas9 to create genetically modified human babies. It is very interesting, that in China this work has been criticized as unethical and medically dangerous.

The first results of NASA's Twin Studies project with astronaut Scott Kelly's yearlong mission to the International Space Station discovered an increase in Scott's methylation rate (a process that turns gene activity on and off) and that Scott's telomeres — tiny caps at the ends of DNA strands — grew longer from his time in space (Feltman, 2016). This research has confirmed the facts that strong influences on the man, such factors as raised radiation, lack of oxygen or a condition of weightlessness in the space can essentially influence the genetic apparatus of the human cells.

Cell biologist Dr. Bruce H. Lipton proposed hypothesis about influence of our consciousness and subconsciousness on the genes. He showed that both positive thinking and negative thoughts can lead to changes in genes (Lipton, 2005).

Successes in the development of technology for the creation of genetically modified objects should sooner or later lead to such technologies that can allow the human genome to be edited, primarily to eliminate hereditary diseases incurable by other means, and then, of course, to improve intellectual abilities without repetition of crimes made on the basis of eugenics. By the way, new experiments in MIT with CRISP on the mice showed that the same alteration introduced into the Lulu and Nana ' DNA, made mice smarter (Regalado, 2019).

6.3. Age and intellectual abilities

As shown above (for example, in section 2.9), any human is not born with a fully formed intellect. In normal development intellect is formed by the end of middle or high school, for gifted children earlier, for others later depending from many circumstances of nurture.

In accordance with the life course at the stages of infancy, childhood, adolescence and maturity, psychologists have suggested that there are also discrete, qualitatively different stages in the development of intellect. The Swiss psychologist Jean Piaget developed his cognitive-developmental theory that includes the four stages of cognitive development correspond with the age of the child (Piaget, 1983):

1. From birth to age 2 (the sensorimotor stage) which characterized by the idea that infants "think" by manipulating the world around them
2. From age 2 to age 7 (the preoperational stage), which characterized by the idea that children use symbols to represent their discoveries
3. From age 7 to age 11 (the concrete operational stage), which characterized by the idea that children's reasoning becomes focused and logical
4. From age 11 to adulthood (the formal operational stage), which characterized by the idea that children develop the ability to think in abstract ways.

A child's brain undergoes a very important period of development from birth to three—producing more than a million neural connections each second. By the time when became 3 years old, his brain has formed nearly 1,000 trillion neural connections, which is about twice as many as the adult brain has. Around the age of 11, the brain will begin to gradually rid itself of unnecessary and inefficient connections.

Although the brain reaches its full adult weight by the age of 21, it continues to develop for several years. In fact, a study done by the National Institutes of Health found that some regions of brain do not fully form until age 25.

Based on recent findings relating to emotional maturity, hormonal development and neurological activity, a child psychologist at London's Tavistock Clinic, Dr. Laverne Antrobus claim that adolescence can be split into three stages: an early period between 12 and 14 years of age, a middle period between 15 and 17 years of age, and a new period called "late adolescence" from 18 to 25 years of age (Mientka, 2013).

Researches have revealed an essential difference in timeframes of achievement of the maximal intellectual capacities at various people: some abilities achieve maximum and begin to decline around high school graduation; some abilities achieve a plateau in early adulthood, beginning to decline in subjects' 30s; and still others do not peak until subjects reach their 40s or later. (Hartshorne, 2014). Newcastle University scientists have found that girls' brains can begin maturing from the age of 10 while some boys have to wait until 20 until the same organizational structures take place.

Technical progress is changing this situation. According to recent nationally representative surveys of the Pew Internet Project, teenagers and young people using computers, the Internet with social networks and smartphones immersed in the technological environment. All this affects the speed and quality of brain development during childhood and adolescence (Chassiakos, 2016).

Although generally the brain reaches its final stage of development at age 25, learning continues throughout the life span. The neural connections of the brain continue to form, change and redirect when confronted with new experiences and ideas. It is known, that a person mentally develops gradually, with the development of intellect lags behind its physical development. This does not apply innate genius or indigo children.

Perhaps everybody agrees that wisdom really does come with age, older people have better horizon and can find the better solutions based on their rich experience. Usually verbal abilities including vocabulary of elders are preserved with age. Common changes met in cases with recognition of words. Besides that, for elders it takes longer and is more difficult to find the needed words in conversation or trying to recall names of people and objects.

Not as fast as a young man, the brain of a healthy elderly person wins in flexibility. Perhaps that is why in adulthood we do more accurate conclusions and make wise decisions. Furthermore, it was found that our brain with age calmer responds to the negative emotions.

Based on the researches of American psychiatrist Dr. Gene D. Cohen, who pioneered research into geriatric mental health, the brain of an healthy old man much more developed than is commonly believed (Cohen, 2006). Distinguished Professor Emeritus of Department of Psychology at University of California (Davis) Dr. Dean Keith Simonton had analyzed a human creativity in many aspects, and found that people in the 60-70 years, while doing everything a lower speed work, can be as productive as people of more young age (Simonton, 1988).

Over time, the brain increases the amount of myelin - a substance that makes the signal more quickly to pass between neurons. Due to this general intellectual power of the brain can be increased up to 3000% compared to the average. A peak of activity of myelin production accounts for 60 to 80 years of age. If up to 60 years between the two hemispheres of the brain there is a strict division of labor, after 60 years, a person can use both sides of the brain simultaneously, adjusting their inclusion slight inclination head in one direction or another.

On the other hand, it has been widely found that the volume of the brain and/or its weight declines with age at a rate of around 5% per decade after age 40 with the actual rate of decline possibly increasing with age particularly over age 70 (Peters, 2006).

Here are some historical facts: Marcus Cato, a Roman Senator, learned Greek at 80; Socrates at the age of 70 years learned to play many musical instruments, and had mastered the art to perfection; Michelangelo created his most significant paintings at the age of 80 years. At 80 years old Goethe finished his "Faust"; The German historian Leopold von Ranke his "World history" finished when he turned on 91; the famous composer and conductor Igor Stravinsky worked up to 88 years.

6.4. Man under the influence of external fields

Influence of geo-space factors on a human body. As shown by the studies of Alexander Chizhevsky, Simon Shnoll and others, human performance can be affected by the following factors: solar activity, the Earth's position relative to the planet, Moon phases, earthquakes, magnetic storms, fluctuations in atmospheric pressure and temperature, Schumann resonances, and so on (Chizhevsky,1995; Shnoll, 2009).

In the process of evolution, man constantly adapted to the impacts of the external environment: the constant magnetic field of the earth, the gravitational influences of the moon and planets of the solar system, variations in atmospheric pressure and humidity, magnetic storms, etc. On the one hand, some particularly stable rhythms could be "recorded" in the long time memory of the brain at the level of the frequencies of internal processes. It allowed to have for healthy person an harmony in all human organs and systems. It is obvious that the appearance or disappearance of an unusual vibration in human environment can cause desynchronization, which leads to dysfunction of systems and organs. On the other hand, the human body has acquired for long years of evolution compensatory mechanisms that provide self-regulation of the main systems of the body during changes in the external environment. Unfortunately, these compensation mechanisms have limited capabilities.

Therefore, with too strong impacts, or with such impacts under which compensation mechanisms does not work, the organs and systems of our body can significantly change their characteristics for the worse. For example, when a man is exposed to electromagnetic fields with magnetic strengths or voltage above the maximum permissible level, disorders of the nervous, cardiovascular systems, respiratory organs, digestive organs and some blood biochemical parameters can be developed. Here is for example a reason why the elderly, unlike the young, feel the weather changes.

The results of research done by scientists from different countries over the past few decades are quite indicative.

Numerous experiments on the effects of electromagnetic fields (EMF) on people and animals have shown that in any organism can be selected frequencies that cause dramatic changes in the functioning of organisms (Belyaeva, 2015). Such frequencies are called bioeffective. The response to them can be different: positive or negative. There are certain "frequency-amplitude windows" within which there is a detectable reaction of a biological object, and outside of them there is not. It was noted that the frequency is a carrier of information, and the types of oscillations in the body can transform into each other. Therefore, response of the organism on an externally applied oscillatory force is possible at the same frequencies with completely different types of field (electromagnetic, acoustic, gravitational, etc. (Khabarova, 2002).

A great contribution to the study of the mechanisms of artificial influences on living organisms and humans was made by Royal Raymond Rife, Georges Lakhovsky, George De La Warr, Douglas Baker, Hulda Clark, to name just a few. Very important discovery of the American innovator and scientist Dr. Royal Rife was the fact that any organism has its own resonant frequency, which Rife called the Mortal Oscillatory Rate (MOR) or deadly frequency of vibrations. By placing a live culture of bacteria under a microscope, Rife turned on a his frequency generator, which was tuned to the MOR frequency for this bacterial culture. After that, all bacteria immediately died. Rife discovered that he could use a frequency generator on people infected with certain types of bacteria, and thus heal infectious diseases (Lynes, 2009).

Georges Lakhovsky invented the Multiple Wave Oscillator, and was the first serious, scientific investigator on the use of high-frequency electromagnetic waves in biology and medicine (Lakhovsky, 1935).

George De La Warr and Douglas Baker applying magnetic fields of varying frequency to various anatomical sites demonstrated scientific evidence validating the efficacy of magnetic fields in regulating bodily functions and in alleviating a number of medical conditions of the human body (2011).

Based on the same idea of bio resonance Dr. Hulda Clark invented the Zapper device. She used this device to kills parasites, bacteria, and viruses with help of high frequency electro-magnetic energy without any harm for human tissue (Clark, 2010).

Experiments of the Russian scientists with a group of 12 healthy volunteers exposed during 15-20 min to experimentally created slight atmospheric pressure oscillations (APO) with amplitudes 30-50 Pa in the frequency band 0.011-0.17 Hz demonstrated that even slight atmospheric pressure oscillations, which frequently occur naturally, can influence human mental activity and cause significant changes in attention and short-term memory functions, performance rate, and mental processing flexibility (Delyukov,1999).

The scientists from University of Crete discovered dependence of psychotic disorders of people from seismic activity (Anagnostopoulos, 2015).

Resonance is a change in the characteristics of the oscillatory system, which occurs when the natural frequencies of this system, absolutely coincide or nearly equal with the frequency of an externally applied oscillatory force. Natural frequency, also known as eigenfrequency, is the frequency at which a system tends to oscillate in the absence of external influence. If the oscillating system is driven by an external force at the frequency at which the amplitude of its motion is greatest (close to a natural frequency of the system), this frequency is called resonant frequency. The natural vibration frequency is an inherent property of any part of the system; it depends only on the characteristics of the object itself (for example, on the dimensions, mass and elasticity of parts in mechanical systems, on capacitive and inductive characteristics in electrical systems).

In addition, it is important, that during the resonance of complex non-linear systems (which are biological objects), the frequency of exposure and response frequencies do not necessarily coincide.

Schumann Resonance is the phenomenon of the formation of standing electromagnetic waves of low and ultra-low frequencies between the surface of the Earth and the ionosphere.

They seem to be related to electrical activity in the atmosphere, particularly during times of intense lightning activity. Because in an oscillating medium, along with oscillations of the fundamental frequency, other harmonics can be excited, to one degree or another, in atmosphere they occur at several frequencies, specifically 7.8, 14, 20, 26, 33, 39 and 45 Hertz, with a daily variation of about +/- 0.5 Hertz. Scientific interest in these oscillations is due to the fact that their frequencies fall within the range of natural oscillations of the brain waves - 8 - 12 Hz (the alpha) and 16-24 (the beta) rhythms.

Research at the molecular level showed that exposure of living cells 7. 83Hz frequency increases their resistance to external influences, reduces the absorption of toxic substances. In the United States (NASA), and Germany (Max Planck Institute) conducted lengthy experiments, as a result of which it was established that the Schumann waves are necessary for the synchronization of biological rhythms and well-being of all life on Earth. Dr. Robert Beck, conducted dozens of experiments by measuring the frequency of the brain rhythms of people involved in healing. With the help of a portable electroencephalograph he measured the frequency of the brain rhythms of people dealing with various healing practices. The results of his experiments were extremely interesting: during the session, without exception healers tuned to the frequency range of 7-8 Hz.

Experiments during the period of 1 January 2016 to 31 December 2016 in the Department of Cardiology at the Hospital of Lithuanian University of Health Sciences (LUHS),. support the hypothesis that the Earth's magnetic field has a relationship between the number of acute myocardial infarction with ST segment elevation (STEMI) cases per week and the average weekly geomagnetic field strength in different frequency ranges. Correlations varied in different age groups as well as in males and females (may indicate diverse organism sensitivity to the Earth's magnetic field (Jaruševičius, 2019).

The space missions have revealed the influence of gravity on the well-being of astronauts. This is a space adaptation syndrome experienced by around half of space travelers during adaptation to weightlessness. Symptoms can include everything from nausea and easy discomfort to persistent vomiting and hallucinations.

Influence of technosphere factors on a human body.

If we consider the impact of the technosphere on humans, we can distinguish the following factors: acoustic oscillations, mechanical vibrations, electrical and magnetic influences, ionizing effects (radiation), temperature influences.

The degree of exposure to electromagnetic and other radiation on the human body generally depends on the frequency range, the nature and duration of irradiation, strength of field, the individual characteristics of the organism, and so on.

In real life a man encounters practically every day with influences caused by industrial systems, radio, TV, music and other means of show business. Influence effects for human body and brain can be both positive and negative. With a positive effect well-being, performance, mood and intellectual abilities can be provided; with a negative effect, a person can be sick a short or long time.

The degree of exposure to electromagnetic and other radiation on the human body generally depends on the frequency range, the nature and duration of irradiation, strength of field, the individual characteristics of the organism, and so on.

During our life, we are exposed to various sources of industrial vibration, for example, in cars, buses, trains, aircrafts. Many people are also exposed to other vibrations during their working day from electrical power lines, electro-mechanical machines or electronic devices, from radio or TV, etc.

Exogenous fields penetrate through human body tissues and bones, and this field can create special wave structures inside of the body as result of modulation or resonance.

According to oriental teaching, our body can receive two streams of energy. Through the legs we can receive energy, emerging in the center of our planet. Another stream - flow of cosmic energy moves downward at the top of the head and the hands-down. Exogenous field is part of universe field, maybe part of the Cosmic Web (Gott, 2016).

Influence of mechanical, acoustic or other types of vibration on intellectual abilities. Scientific studies in recent years have shown that creativity of a human intellect is influenced by mechanical vibration, external electromagnetic fields, etc..

Each person collides with mechanical vibration of a body and acoustic vibration in an actual life, for example, during walking, horse riding, driving on various types of transport, listening to music or sounds of the nature. And everyone noticed and knew as it influences its mental faculties.

Scientists and practitioners of the XIX-XX centuries have started to study seriously influence of vibration on a human organism. In the 1880s and 1890s, a French neurologist and professor of anatomical pathology Jean-Martin Charcot, M.D., was one of the first to discover and report positive effect of vibration application. Experimenting with the oscillator, Nicola Tesla discovered some physiological effects of vibration and designed a special platform, fluctuations which unusually invigorate persons standing on that. Nikola Tesla treated with vibration platforms help himself, Mark Twain and other people. Unfortunately, comprehensive research about whole-body vibration is lacking. Today it is possible to find in literature and internet a small statistic in determining the usefulness of such devices. Here is what a practitioner Becky Chambers, for example, writes in her book "Whole Body Vibration: "Hundreds of my clients report rapid and dramatic improvement in mood, energy, and sleep" (Chambers, 2013).

Entrainment technology by using sound and light vibration. For centuries people have been using outside rhythmic stimulus in the form of drumbeats, chants and singing to induce various feelings, ranging from euphoria to sadness. Exposure to vibrations of singing Tibetan bowls, Himalayan bowls or suzu gongs is the oldest oriental methods of restoring the lost energy of the person, the harmonization of the work of chakras, cleansing the energy channels and restore the aura.

Our body and especially the brain are sensitive to signals received from the external environment through the senses organs, especially sight and hearing organs. These signals can carry information in the form of colored pictures, music, speech, or unrecognized by consciousness impacts. Some of these signals may serve as a stimulant for the brain. When the brain receives a stimulus, through the ears, eyes or other senses, it emits an electrical signals and flows of energy in response. The direction and depth of the influence of vibration effects depend on the amplitude, frequency, phase, forms of signals, and duration of impact, and even program of the impact, that can include sequences of various type (by duration and quality) influences.

When the brain is exposed to a rhythmic stimulus, such as a drum beat, the rhythm is reproduced in the brain in the form of rhythmic electrical impulses. If the rhythm becomes consistent enough, it can start to resemble the natural internal rhythms of the brain, called brainwaves. In this case, the brain responds by synchronizing its own electric cycles to the rhythmic stimulus. In some new science disciplines, this is commonly called the Frequency Following Response (FFR).

The phenomenon of entrainment for physical systems was discovered in approximately in 1665 by a Dutch scientist Christian Huygens, the inventor of the pendulum clock, who revealed that two pendulum clocks which normally showed slightly different time nonetheless became perfectly synchronized when attached to a common wood beam. He called this synchronization tendency "entrainment". Later studies demonstrated that entrainments phenomenon concerns by the same way the electromagnetic brain waves.

Numerous studies performed at the end of the last century have shown the promise of using this phenomenon to improve people's intellectual abilities. Below we mention only a few of them.

In the 1970s, director of the Institute for Advanced studies in behavioral medicine, Dr. Charles Stroebel discovered that during the stages of deep meditation the brain wave patterns of meditators altered and both hemispheres of the brain were working in harmony together.

The effect of Binaural beats was discovered in 1839 by Heinrich Wilhelm Dove and further researches demonstrated that binaural beats could help induce relaxation, meditation, creativity, and other desirable mental states. This effect was reopened in 1973 by a biophisist Dr. Gerald Oster who described how pulsations called binaural beats occurred in the brain when oscillations of different frequencies were presented separately to each ear (Oster, 1973). As a result, the entire brain became entrained to a frequency equal to the difference between the two tones and began to resonate to that frequency.

From 1988, director of the Center for Neuroacoustic Research, San Diego, Dr. Jeffrey Thompson has been exploring neuroacoustics and the therapeutic application of sound. His researches have led to the development of precise protocols for using sound to modulate brainwave patterns, affect sympathetic-parasympathetic balance, and synchronize the activity of the right and left brain hemispheres. He has applied these methods in stress reduction, cardiovascular disease prevention, management of depression, and a host of other conditions.

Now entrainment method allows using for brain activation the different types of signals, described below in the sections 9.3 and 10.2

6.5. Influence of biological organisms and other people

As was mentioned in section 5.1., human cannot exist without energy and information. Biological organisms like as plants, animals and human can exchange energy and information through various biophysical pathways.

It is well known that trees can help us think better — Plato and Aristotle liked thinking in the olive groves around Athens, Isaac Newton discovered his theory of gravity when an apple fell from the tree under which he was sitting. Trees treat the psyche, contribute to the normalization of the heart, stimulate metabolic processes in the human body, relieve headaches, reduce the effects of stress, etc.

There are vampire plants and donor plants. The first take the energy of the external bio field, the second - give energy to the external bio field.

Bio fields of various plants depend not only on their type, but also on the place, time of year, day. Donor trees usually have no neighbors closer than 6 meters. Therefore, the aura in the dense forest is weaker. All donor trees also differ in the strength of the bio field, which changes during the day, and when the tree "sleeps", it is completely absent. The bio field is also different in different seasons of the year. The most powerful bio fields are usually in spring. The trees' aura cannot be seen by the untrained eye but can be felt with practice when sensitizing the hands to feel the subtle energy.

Donor trees (oak, sequoia, pine, etc) are fed with positive energy, give strength and vigor, which is simply necessary for various diseases (colds, arthritis, rheumatism, gastrointestinal disorders). Vampire trees, on the other hand, take away negative energy. Such a pumping of energy through trees can be useful for neurosis, headaches, hyperthyroidism, osteochondritis, inflammation and injuries. Vampire trees include: aspen, willow, poplar, etc.

Amid donor herbs there are ragweeg, St. Johns wort, nettle and others.

Man is not only part of the nature, he is also a social being. He lives in the company of other people, mostly with the help of other people and for other people. It means that his intellect must have a possibility of interpersonal interaction with intellects of other people.

At last centuries was created many theories of interpersonal interaction: psychoanalytic theory of Sigmund Freud, social exchange theory of George Homans (1958) and Peter Blau (1964), the theory of impression management of Erving Goffman (1959), the theory of symbolic interactionism of the J. Mead and G. Blumer (2009), just name a few.

These theories was created to understand, explain and predict beliefs, attitudes, intentions, and behaviors of individuals, couples, groups, and mass audiences.

Currently, there is a lot of evidence for the interpersonal exchange of internal energy, automatically or by special human influence including hypnosis and suggestion. Unfortunately, science has not established what is the carrier of this energy. It can be assumed that this energy can be transmitted through various media.

Interpersonal bio field interference (IBI) is influence of people on each other, or more precisely, interaction of the bio fields of people around us. However, it should be noted that the scientific evidence of that phenomena has received only recently, and most people are still skeptical about this effect.

IBI can be usually as a root of such type of human interrelations like domestic quarrels, the personal incompatibility, and even health conditions. In the last case, IBI may deform the egg-shaped shell of the bio field of man, as a result of which he/she may perceive fatigue, declines of forces, disease and even death. The negative influence of people on each other is called "jinx", "vampirism" and so on.

Some facts of our life experience approve the phenomena of induction from exogenous biofield from one person to one or several close to him persons, for example:

- yawn
- pain
- emotions
- suppression of will and desires
- schizophfrenia, etc.

One day with a Texas psychic Ray Stanford there was a funny case of perception of exogenous fields when he met Dr. Robert Van de Castle, director of the sleep and dream laboratory at the University of Virginia. Ray noticed several pink spots in the aura around the belly of the Dr.Van de Castle. Such spots he is usually associated with pregnant women; and he remarked to Dr. Van de Castle: "If I did not know you, I would say that you were pregnant." However, Van de Castle then replied that he all morning has been analyzing the dreams of pregnant women and after that even noticed that he himself began to feel like a pregnant woman (Mishlov, 1997).

According to Sofia Blanc (Blanc, 2010) on Kirlian images are registered so-called connecting cords (they often seen by clairvoyants). These cords are evidence of energy interchange between people that are usually can be negative for one of them.

In the context of this book is very important understanding some intellect functions using in practice of communication between people. The right control and management of these functions allow creating and supporting, for example, an intellect of the team, measuring or assessing by results of this team activity.

Recent studies in the San-Francisco State University revealed that the stream of consciousness is more susceptible to external stimuli than had previously been proven. It means that our thoughts are susceptible to external influence – even against our will (Merrick, 2015).

Consider a nature of psychological impact. The effect of exposure on a person depends on what mechanisms of influence were used: persuasion, suggestion or informational infection.

Persuasion is a set of methods or skills to influence of one person on others in fields of a person's beliefs, attitudes, intentions, motivations, or behaviors based either by appeals to logic and reason, or using heuristic methods by appeals to habits or emotions.

Suggestion is the process of inducing a thought, sensation, or action in a receptive person. This is a purposeful, not reasoned impact on the person that lead to the appearance against his will and consciousness, a certain state of mind, feelings, and relationships; or to change his behavior. Suggestion can affects not only one person, but to a lot of people.

Direct suggestion - the impact when one person says to another in-form certain ideas that must be unconditionally accepted and implemented. As an example of such a suggestion can be called effects of influence parents and teachers on children.

A simple example of indirect suggestion is advertising, which are most commonly used techniques that predispose to approval without hesitation. Instead of clear information on product advertising links offered goods and notions of beauty and pleasure (this is, basically, advertising cigarettes, beer, cosmetics), when advertising any beverage, where the product is associated with the concepts of youth, health and cheerfulness. The suggestion is the process of transmission of information, based on its uncritical acceptance. The main condition for effective suggestion is credible authority and respect for him. In turn, the degree of suggestibility depends on the characteristics of the personality, which is the object of exposure. The modern techniques of hypnotism and autosuggestion allow to make suggestions more effectively.

The success of suggestion depends:

- From the abilities of the suggestor, his charisma, authority, confidence;
- From the nature of the information to be informed and the place of information to be promoted in the general information flow;
- From suggestibility (conformism) of percipients.

Charisma is a certain quality of an individual personality by virtue of which he stand out among ordinary men and treated as endowed with supernatural, superhuman, or at least specifically exceptional powers or qualities.

Conformism - the degree of compliance to suggestion, the ability to uncritically perceive incoming information, is different for different people. Conformism is higher in persons with a weak nervous system, as well as in persons with sharp fluctuations of attention.

There are three main forms of suggestion:

- hypnotic suggestion (in the state of hypnosis);
- suggestion in a state of relaxation - muscle and mental relaxation;
- suggestion during active wakefulness of a person.

Studies have shown that the suggestibility and conformism are unique to every person from childhood to the end of life, but age, sex, profession, group composition, and so forth affect the severity of these qualities.

The phenomenon of informational infection was known at the earliest stages of human history and had diverse manifestations: the massive outbreak of the different mental states that arise during the ritual dance, a passion for sport, the situation of panic.

According to memetics theory developed by Richard Brodie, we can observe cases of the human mind, infected by a mind informational virus (Brodie, 2009).

The "meme" word was first introduced by evolutionary biologist, Richard Dawkins, in 1976 (Dawkins, 2006). "Meme" means s from the Greek "something imitated". Dawkins described memes as a way for people to transmit easily and fastly social and cultural ideas to each other. A meme behaves like a flu or a cold virus, traveling from person to person quickly, transmitting a gist of idea.

A meme is a unit of information like a subprogram in a mind whose existence influences on events such that more copies of this unit get created in other minds. Something goes on in the world that infects people with certain memes, and those memes eventually influence on their hosts' behavior in such a way that the something gets repeated and/or spread. This is a reason why it is looking as a virus of the mind.

According to Richard Brodie, we can observe some variants of the method of infection with this type of information virus. The process may evolve in one of the following scenarios:

- Repetition. Repeating a meme until it becomes familiar and is among the similar programs in mind
- The cognitive dissonance. Being in a mentally uncomfortable situation can lead to reprogramming of new memes that, in order to improve the situation, relieve mental stress.
- Association. Less attractive memes can be associated and combined with more attractive ones.

In common, external psychological impact on the mind can change characteristics of mind and intellect both in the best and in the worst side. It is important to know for increasing collective intellect and for its management.

6.6. Influence of feeding

The brain, which accounts for 2 percent of our body weight, consumes about 20–25 percent of needed for whole body energy, which can be renewed through regular consumption of water, air, and food. Food can make us smarter, improves memory, helps us think more clearly, and even prevent dementia; although no one "miraculous" product cannot instantly boost our brain (Wenk, 2010).

Recently, in the science of nutrition appeared a new discipline - nutritional neuroscience that studies the influence of nutritive properties of food and its component as proteins, carbohydrates and fats, vitamins, dietary supplements, and etc. on the intellect and psychic characteristics (sensation, perception, memory, feelings) of the human.

During some last decades scientists around the world has provided exciting evidence for the influence of dietary factors on specific mechanisms that maintain mental functions (Gómez-Pinilla, 2008; Lugavere, 2018). For example, a balanced diet rich in omega-3 fatty acids — found in salmon, sardines, walnuts and kiwi fruit — can give the neurons of brain a boost and even help fight against mental disorders from depression to dementia.

A normal diet should include fruit, such as orange, apple, grapes; or dry fruits, such as chestnuts, almonds, dry raisins, and so on; along with vegetables such as carrots, beef, tomatoes, and other vegetables — especially green leafy vegetables; sea foods, such as oysters rich in iron and zinc, or seaweeds, each type of which has a unique set of nutrients.

Below here are a far of complete list of some fruits, vegetables and herbs that can change condition of human mind and intellectual abilities:

- Watermelon and banana - carry many positive qualities, enhancing the overall energy background, uplifting, helping with fatigue, relieves stress
- Bergamot - helps to achieve success in the work, it helps to overcome the difficulties, neutralizes negative energy around them
- Broccoli has phenolics, antioxidant vitamins and dietary minerals needed for brain
- Citronella is an excellent remedy for fatigue, and drowsiness. Just because of its strong flavor. Citronella has beneficial effect on the internal and external space, it takes away from all kinds of negative energy
- Carnation (Dianthus) - restoring strength after a nervous and physical overstrain, fragrance harmonizes human energy space, raises vitality, makes the mind calm and clear.
- Coffee - refreshing and invigorating, it helps concentration
- Cherry helps to feel the joy of life brighter. Cherries are nature's own little anti-inflammatory pills
- Cedar - eliminates stagnant processes in the body and in the environment, purifies the atmosphere, it contributes to clarity of mind and good health. Improves performance and immunity
- Date - It is believed that dried dates have a positive effect on the brain, increasing its performance by 20% or more
- Eucalyptus stimulates mental ability, enhances memory, improves mood, relieves drowsiness. Cleans air and kills bacteria
- Green tea - it should be noted tonic properties, by which a person begins to feel the lightness and clarity of mind. It cleanses the body from all sorts of power units.
- Gingo Beloba is helpful in different conditions that include memory loss, headache, ringing in the ears, vertigo, dizziness, difficulty concentrating, mood disturbances, and hearing disorders

- Jasmine - stimulates creativity, strengthens self-esteem, a sense of well-being, helping to adapt to an unfamiliar environment, has anti-stress effect, removes fatigue, muscle tension, and mobilizes the reserve forces of the organism
- Lily - relieves depression, signs of fatigue from your body and mind
- Lotus relieves fatigue, conducive to spiritual development, helping to smooth the acuteness of depression and melancholy, sharpens intuition
- Lemon contributes to the recovery of hard work, strengthens the immune system. Provides a balance of emotions, it helps to get rid of anxiety, improves mood and makes you feel better. It stimulates mental abilities. It has antiviral and anti-infective action.
- Rose converts the negative energy of disappointment and sadness in the constructive power of self-improvement. It helps to soberly assess the situation and find the roots of the problem. Refreshes and clarifies the mind, helps to make the right decision.
- Spinach is rich in chlorophyll, which is good for blood and therefore for brain work.
- Kale contains such a phytochemical as sulforaphane which linked to possible cancer and heart benefits, as well as brain health

Chocolate makes us smarter. Conducted study found that eating foods with a high cocoa content, within five days can increase blood flow to the brain, thereby improving its performance.

6.7. The magic role of aromas, music, colors and art

Effects of the environment. Studies show that vision gives us 70% of the information, the hearing - 15% and tactile - 10%. The better working senses, the more information we get, the more opportunities for our creativity. Of course, blind Leonhard Euler or deaf Ludwig van Beethoven - these are exceptions to the rules.

Today it is known that our surroundings may influence our emotions and state of mind. Probably everyone had to be in situations where certain places, especially irritate him, or vice versa in others places where a person was relaxed and soothed. Scientific studies show that the effects of the environment in the form of pleasant and unpleasant landscape, odors or flavors, noise or music, cold or warm colors affect on creativity or workability negatively or positively.

Aromas for life and workplace. Scents can stimulate physical and mental activity and performance. Avicenna thought that the rose oil increases the possibilities of the mind and thinking speed. In the Middle Ages students used wreaths of parsley and mint in preparing for exams in order to activate the memory and remember the material. The famous English poet Lord Byron wrote that always feels a rush of inspiration with the truffle scent.

It is interesting to know what influence of perfume or essential oils scents depends on not only the nature, character and mood of the person, his own smell, but also on his biofield. Since ancient times, people, especially woman, knew that with aromas help, it is possible to receive the emotional recovery, to maintain peace of mind, harmony of relationships of couples, families and so on.

At nowadays, in many countries started flavoring offices, cinemas, apartments, industrial premises and so on. The flavors give the impression that you are not in the room but in the park. In addition to the emotional, there is another side - and the flavors clean the air from foreign odors, germs, viruses, prevent infection of visitors during the flu epidemic. Aromas help a person nourished the energy of nature, conquer fatigue. They increase overall body tone, relieve low mood, stress.

Music for life and workspace. Henry Wadsworth Longfellow wrote: "Music is the universal language of mankind". Music has become part of our lives, and for some, a profession. Numerous studies have confirmed that music affects our brains, our minds, our thoughts, and our spirit (Levitin, 2007). As a result of evolution, we have the cognitive capacity to find or chose a music to enjoy, to remember hundreds of melodies, to detect wrong notes, and to dance under some music. Individuals report a spectrum of bodily and mental sensations while listening to music, such as the feeling of a lump in the throat, the tingling sensation on the scalp, back of the neck and spine that is often accompanied by goose bumps (Panksepp, 1995). Albert Einstein considered music as the driving force of intuition.

Researchers today believe that musical exercises optimize the development of nerve cells and improve brain functions in mathematics, analysis, research, also affect memory, creative thinking, stress management, concentration of attention, motivation.

Music is perceived not only through ears but also through all analyzers and systems of the body, affects the acupressure points and meridians. She comes to every cell of every organ, it affects all levels of life: molecular, systems and organism as a whole. Music causes chemical reactions, restructuring the status and characteristics of both physiological and mental conditions.

Medicine of the ancient states of Egypt, Greece, Rome recognized the therapeutic effect of music used. In Egypt, during the first six dynasties (2900-2270 years BC) with the help of music were treated the mentally ill people.

In France the music therapy has been developed in psychiatric hospitals since about 1830 through receptive and active (band, choral) activities.

After World War II music therapy intensively developed in USA, Sweden, Norway, Finland, Denmark, Russia and other countries. In 1969 Music therapy Centre was founded in Paris, whose employees were engaged in group therapy.

According to recommendations of the American Music Therapy Association (http://www.musictherapy.org) musical therapists now prescribing music therapy for heart ailments, brain dysfunction, learning disabilities, depression, PTSD, Alzheimers, childhood development and more.

Some people need to hear music to improve concentration and focus in the workspace. Some companies organize music playing in their offices to relax employees from stress. For example, employees can hear the sounds of birds chirping when they arrive for work in the morning, and at another time hear the radio broadcasting music that accompanies an exercise program.

An expert on music psychology professor Dr. Adrian North of Heriot-Watt University, Edinburgh, UK, has carried out extensive research on the social and applied psychology of music, and the role of musical preference in everyday life. He found some correlation of personality (including intellect functions) and favorite music.

His results showed, for example, that fans of Blues, Jazz, Classical music, Opera, and Dance have creative abilities; Country and Western fans are hardworking, Chart pop fans have "high self-esteem, are hardworking, outgoing and gentle, but are not creative, and not at easy." (North, 2008).

Everybody has heard about the Mozart effect- possibility increasing intellectual abilities by listening to Mozart's music (Campbell, 1997). Numerous independent studies by scientists, doctors and psychologists around the world prove that the music of Wolfgang Amadeus Mozart can mobilize all the natural abilities of our brain. Studies by American scientists have shown that only a 10-minute listening to Mozart's piano music can increase the IQ of people by an average of 8-10 units.

As European scientists have proved, Mozart's music enhances the mental abilities of all people, without exception, who listen to it (both those who like it and those who do not like it). Even after a 5-minute musical session, listeners noticeably increase their concentration. According to the authoritative experts of the world level, Mozart's music helps to get rid of any spiritual problems, improves speech and hearing. Even hearing of Mozart's calm music daily while eating can help in disappearing of many digestive problems..

Scientists have experimentally determined that the sounds of Mozart sonatas (especially the Sonata of K 448) are able to arrest epileptic seizures (reduce the number of epileptic attacks). People with Alzheimer's disease improve their skills by regularly listening to the Mozart sonata for two pianos in C major.

In fact, it is known that usage music (not only Mozart) in the clinical therapy allowed to receive highly beneficial results in treating by Tomatis Method, for example, famous French film actor Gérard Depardieu and the most renowned and influential opera singers of the 20th century Maria Callas.

Bulgarian psychologist Georgi Lozanov, along with his colleague Dr. Aleko Novakov discovered that under the influence of baroque music (of Bach, Handel and Vivaldi composers) the information is perceived and processed by a brain more effectively. This discovery led to the development of a new method. Under its terms, the information - for example, the phrase in a foreign language - appears at intervals of four seconds on the background of baroque music to the rhythm of 60 beats per minute. Initial results showed that used this method students learn from 60 to 500 foreign words per day.

Color environment influence. Studying the effect of color have led to the creation of a scientific discipline of Color psychology. Skillfully using the color combinations, we can sometimes get truly amazing results. Scientists have conducted interesting experiments, the results of which showed that for a certain choice of color, which use for painting the rooms for sheep, cows, chickens, the latters dramatically increasing wool growth, milk and egg production. Even plants under the influence of a certain color increase the content of useful substances.

All colors perform psychophysical and energetic influence on the person, have a different semantic characteristics: warm (red, yellow) and cold (blue, indigo, violet), active (warm) and passive ((cold), heavy (dark) and soft (light), the retreating (cold) and upcoming (warm), warning (red, yellow, green). For people colors affect workability (including learning), regulate blood pressure and appetite, attention, emotion, hearing acuity and many other processes of the human psychophysiology. However, the same color for each person can act differently depending on the personality of an individual.

If to arrange a classroom in white, beige, brown, it may raise discipline and student achievement. Even the factory floor, painted in blue or beige color cans enhance productivity of workers. Here are some examples of psychological effects of colors:

- Purple color provides a nice balance between stimulation and serenity that is supposed to encourage creativity of workers. Light purple creates a peaceful surrounding, and relieve tension

- Green and blue colors create a peaceful and calming environment. These colors are typically considered restful
- Yellow or orange colors are toning colors. Orange color acts in the same direction as the red, but to a lesser extent, improves digestion, helps rejuvenate, emancipation, strengthens the will, free from feelings of depression. Yellow color stimulates vision and nervous activity, activates the motor centers, causing a joyful mood, generates energy muscles, can stimulate intellectual ability.

Forms and art influence. Forms influence the subconscious mind sometimes no less effectively than color. For example, angular figures (squares, triangles, polygons with sharp angles) are associated with threat and hostility, and forms with smooth edges (circles, ovals, flowing lines) evoke a sense of trust, peace, serenity. Combining forms can achieve almost any emotional response.

At present, art therapy is widely used in a wide variety of hospitals, psychiatric and rehabilitation facilities, wellness centers, and so on. For people, who experience illness, trauma, and mental health problems and those seeking personal growth, during individual and/or group sessions art therapists elicit their inherent capacity for art making to enhance their mental, emotional and even physical health.

Experts in art therapy can cite many examples of how some pictures of the famous artists have selective therapeutic properties: paintings by Botticelli - relieve pain, paintings by Matisse - treat kidney disease, Picasso - affect positively the activity of the cerebral cortex, etc.

6.8. Influence of education and training

This is widely known, that it is possible to improve all main function of intellect and mind: our memory, planning, spatial orientation, processing speed, reasoning, creativity- by education and special training. For a number of years, a senior professor of education at the Harvard Graduate School David Perkins and colleagues have conducted research on thinking dispositions, devising ways to transform class rooms into "cultures of thinking" that foster thinking skills and attitudes through the teaching of the disciplines in a thinking-centered way. Prof. Perkins went to the conclusion: "We can become more intelligent through study and practice, through access to appropriate tools, and through learning to make effective use of these tools" (Perkins, 1995).

The most striking example of the role of education or training on intellectual abilities is the so-called syndrome of Mowgli, concerning most of the known cases of feral children who lived in different countries and have grown up in an environment of different animals (dogs, bears, wolfs, gazels, etc) and birds.

Feral children returned to human society can not reach the level of mental development of their coevals. For example, they may be unable to learn to use a toilet, have trouble learning to walk upright after walking on fours all their life, and display a complete lack of interest in the human activity around them. They often seem mentally impaired and have trouble learning a human language. The inability to learn a natural language after having been isolated for so many years is often attributed to the existence of a critical period for language learning (Keith F. 2008).

The peculiarity of the human mind in general and intellect in particular is that these tools require a long development to form a proper level (10-20 years) and require special efforts to maintain this level during human life. Especially important, that the process of acquiring new knowledge should be continuous, otherwise the brain will start to lose its flexibility and speed of mind operations.

Many people know the following proverb: idle brain is the devil's workshop. Our brain is idle when we do not have anything to think about. This may lead to dangerous consequences. When we do not think for yourself, we allow others to influence us. Other thoughts can occupy our mind and any mischievous person can easily exploit us for its own purpose. On the other side, a new psychology studies have shown that the students can get smarter and did better in school, if they train their brains to be stronger, like a muscle.

The more knowledge we have, the easier it is to learn. Developed intellect make easier to learn also. On the other hand, higher education is not proof of higher intellectual abilities and a lack of education says absolutely nothing about intellectual abilities.

Cognitive training—better known as "brain training"—is one of the hottest new trends in self-improvement. According to researches of the cognitive scientist Cathy Price of University College London, intellect level, measured by IQ test, can be increase after training from 110 to 130, from average level to level of gifted people (Begley, 2012).

The team of neurologists and cognitive psychologists at CogniFit, TransparentCorp, and others have designed a complex of mental games, which are aimed to stimulate cognitive skills and brain plasticity in children and adults. The program automatically presents challenging exercises specifically adapted to each profile.

It is known that such games like as the oldest board game Go, chess and checkers games, card games "Preferance" and "Poker", puzzles including Japanise game Sudoku, etc. can help to increase intellect level. At present, there are very popular computer games, that, for example, in 2016 year were products of the about 100 billion dollars industry. With help of specialized computer games users can train ability to think using both hemispheres of the brain simultaneously, increasing the areas of brain that engender a sense of wellbeing, and so on (Begley, 2007).

Scientists led by Susanne M. Jaeggi in University of Michigan trained adult volunteers to perform a difficult task for the development of working memory. They revealed that the longer the participants practiced, the better were their ability to think logically and solve problems (Jaeggi, 2008). Another way to raise an intellect level - improve the brain's ability to focus. Here helps also passionate interest, if person is indifferent to the information that read, see or hear it fade from his memory. Results of post-mortem studies demonstrated that learning increases the number of neural connections, that causes an increase in brain volume and density. The idea that the brain is like a muscle can grow and be strengthened through training, it ceases to be a metaphor.

Based on theory of neuroplasticity, researches suggests that pro-social skills such as empathy, compassion and gratitude can also be trained.

According to Dr. Michael Merzenich studies (Merzenich, 2013), a radical improvement in cognitive functioning: how we learn, think, perceive and remember information, is possible even in old age. His most recent patents granted to promising techniques to develop language skills without tedious memorization. Merzenich argues that under the right conditions of training a new skill can change hundreds of millions, perhaps billions of connections between nerve cells in our brain maps. Merzenich is sure, when training takes place in accordance with the laws governing the plasticity of the brain, mental functions are improved, allowing people to learn and perceive information with greater accuracy and speed of memorization.

Numerous studies have confirmed that the memory may have an impact on intellectual abilities. For example studies of Jaeggi S. M. and Buschkuehl M., showed that intellectual abilities can be enhanced after training worked memory (Jaeggi, 2008). In any training is very important the dosage of practice (Mårtensson, 2011).

6.9. Creative environment and inspiration

If the environment helps one person or group of people (team) to feel comfortable and be more productive in developing and analyzing new ideas, then this environment can be called a "creative environment".

At present office facilities of many hi-tech companies are made in view of special requirements to working conditions. Any office space must satisfy the basic (critical) needs of the performer. It should include items such as a desk, computer equipped by smart software and peripherals, regular or smart telephone, and so forth. For example, a graphic artist may need both a small desk for his computer and a larger table or work space for his artwork. A consultant, however, may need additional space for several locking, fireproof file cabinets, and possibly a space for clients to meet with them.

Different people have a maximum of their creativity or productivity under special conditions, which may include the following:

- Special internal condition for inspiration, for example, sufficient sleeping time, a good feeding, aroma, music and so on;
- Boundary conditions like silence, temperature in the room, maybe special scent and colors ;
- The best time for work for individuals depending on their chronotypes;
- Influence of other people, like supervisors, favorite women or man.

Creative environment can be different for different people and depends from the goals of their activity, for example, ones for authors writing books, others for composers or artists in their creativity, more different for designers of complex systems or machines working in the groups.

"I decided that it was not wisdom that enabled [poets] to write their poetry, but a kind of instinct or inspiration, such as you find in seers and prophets who deliver all their sublime messages without knowing in the least what they mean." –wrote Socrates a long time ago.

Creative inspiration - the highest state of ascent, when the cognitive and emotional sphere of man connected and focused on solving creative problems. A man in a state of creative inspiration can manage and clarify his thoughts, images and emotions, understands everything in his actions and cannot always tell how much time has passed (hour, day, night). Often, with a state of creative inspiration associated insights.

Status of inspiration occurs in humans, passionately and persistently striving to creative problem solving. People take their inspiration from different sources. This can be poetry, painting, music, personal hobby, special trainings, conversations, love, and so on.

To have an inspiration is not enough for the creative process. "I write only when inspiration strikes. Fortunately it strikes every morning at nine o'clock sharp" – wrote W. Somerset Maugham. This means the need for regularity and activity planning. Sometimes it more important to have strong motivation. We will consider that below.

Here should not be confused inspiration and insight. Usually if somebody wants to explain what means insight, he remembered as an illustration Archimedes case. As the story goes, Archimedes was inspired by the displacement of water in his bathtub to formulate Archimedes Principle: "If the weight of the water displaced is less than the weight of the object, the object will sink". Historians usually quote the culmination of this story - naked Archimedes, running wildly through the streets, unable to contain the excitement screaming "Eureka" - I found this".

Insights are those moments when a person suddenly understand something, or discover something, what give a clear answer on problem, about which he thought a lot of time (Kiefer. 2013).

Development of the creative environment for any person must be an important part of intellect management described below in the Chapter 9.

6.10. Boundaries of artificial influences on intellect

In the section 1.2 we wrote about human being wonders. Achievements of many geniuses are wonderful. Here are questions. Is it possible to make a person smarter in a cheaper and shorter way without overloading him with knowledge that may never come in handy?

The famous Russian neuroscientist and psychologist academic Natalia Bekhtereva wrote in one article about case when stimulating one of the subcortical nuclei, her colleague Vladimir Smirnov saw how a patient became two times "smarter": more than two-fold increased his ability to memorize in the operating memory (Bekhtereva, 2001). Therefore answer on these questions can be - theoretically yes.

The factors of influence on intellect described above can have both a positive and negative impact on a person's intellectual abilities, because every brain in its physiology has limited resources to improve. Mind, memory, and thoughts are "flesh and blood" which are vulnerable to many abuses, including alcohol, drugs and toxins, stress, poor nutrition, lack of physical and mental exercise (Khalsa, 1997). Here it is possible to add that human brain with its content can be spoiled by information overload. What does it means.

Now we live in the age of information explosion, the rapid increase in the amount and sources of information published with help of not only tradition methods (articles, books, TV, films), but and new methods via computers, internet and other types of networks. Today many of us are immersed in a sea of information. The term "information overload" was coined by a sociologist and futurologist Alvin Toffler in his books "Future Shock" (1970), "The Third Wave"(1984), etc. "Information Overload" occurs when people are trying to deal with more quantity of information than they are able to process to make rational decisions. By this reason we can often observe the results of human activity with either delay making decisions, or the wrong decisions, on the other side- increased number of people lived under stress or practically out of mind.

Scientific and technical progress has radically changed for many people the nature of the workplace; it imposes new quality requirements for the use of physiological and mental energy of the workers. People currently have to deal with complex equipment, where sometimes the rhythm of the machine or rhythm of technological process automatically begins to control the rhythm of the workers body or mind. Under these conditions, there is often a mismatch between technology and physiological or neuro-psychological possibilities of man. The reason is that the physiological adaptation of man in some cases can be unable to keep up with the new pace of the rhythm of life, the development of technology, the progress of production. Hence neuropsychiatric fatigue, that can be further accompanied stress, neurosis, hypertension, etc. This may also be associated with a lack of motor activity, can lead, for example, to cardiovascular diseases.

As a result, at present in comparison with last decades doctors meet more cases not only fatigue, but and over fatigue, that can change a structure of brain. Over fatigue is excessive fatigue especially when carried beyond the recuperative capacity of the individual. Over fatigue represents a pathological condition of an organism, which develops under action of long exhaustion with a prevalence of a mental or physical component.

One of simplest methods to avoid conditions of information overload and over fatigue, is a brains rest. If it did not receive needed rest, a brain defends himself, and can use regime of "autopilot". In extreme cases, the brain can automatically shut down (Smart, 2013, Compernolle, 2014)). The latter can have bad consequences, if at that moment an owner of brain is behind the wheel of a motorcycle, car or plane.

Now new theories of cognition and superlearning have appeared, which developed methods how not to drown in the flows of information and use them wisely. In this connection, today very important to have methodology alloying to increase abilities of the human intellect to cope with the increasing flow of information and emotional stress load.

The Cognitive load theory has its origins in experiments conducting by Dr. John Sweller at the University of New South Wales, Australia in the early 1980s. Today this theory has grown into one of the most widely recognized sets of proven principles governing learning and instruction in the training profession (Clark et al, 2006).

Now it is very difficult to define any limits for brain and intellect functions, which allow knowing a priori symptoms of brain homeostasis condition infringement. We must take into account that for every person these limits can be different, and therefore affecting on the intellect we have to remember the main principle of bioethics - "do no harm".

According to the World Health Organization, every fourth or fifth person in the world has a mental or behavioral disorder. The causes of many mental disorders are not fully understood.

Using an optimal intellect-tuning methods, it is possible for number people to treat some deceases, for example connected to stress, but for other people this treatment can aggravate the condition of patients.

Because human brain is the super complex system, theoretically this system can ensure the existence of an effect called the butterfly effect. According to hypothesis of chaos theory father Edward Lorenz (1963), a butterfly flapping its wings in one part of the world can cause devastating consequences in another part, for example a butterfly waving its wings in Brasil can trigger an avalanche of effects that can culminate in the rainy season in Indonesia. Andy Andrews in this book cited several examples of one "little" event that changed world history (Andrews, 2010). The experiments in the University College London (2010) proved that 'butterfly effect' in the brain can make the brain intrinsically unreliable.

CHAPTER 7

Some features of human thinking

"You have power over your mind—not outside events. Realize this, and you will find strength."—Marcus Aurelius

"As a single footstep will not make a path on the earth, so a single thought will not make a pathway in the mind. To make a deep physical path, we walk again and again. To make a deep mental path, we must think over and over the kind of thoughts we wish to dominate our lives." —Henry David Thoreau

7.1. New concepts of thinking

The question of what is thinking and what role it plays in cognition and human activities, interested in humanity from time immemorial. History of thinking theories include: associationism, gestalt psychology, behaviorism, psychoanalysis, cognitive theory of motivation, operational concept of intelligence, the theory of the ontogenetic development of thinking, and at last, the theory of thinking as an activity of information processing system.

In recent years, thanks to research of Jerry Fodor (Fodor, 1983), Richard Restak (Restak, 2012), Steven Pinker (Pinker, 2009), Henry Plotkin (Plotkin, 2004), just name of few, formed new concepts of thinking. These concepts are not generally accepted, and therefore, have a status of the hypothesis.

Here are four main assertions of this concept:

- The human brain is a kind of computer-similar device, which performs many functions of information processing for various purposes
- Structure of mind had arisen during evolution of brain to overcome the difficulties, leading to the survival and development of the human genome
- The mind is a complex functional structure consisting of specialized "modules", each of which is designed to solve its local problems based on use of specific information and means of its processing
- Mostly intellectual structure of brain has an innate origin.

Famous American philosopher and cognitive scientist Jerry Fodor suggested "a language of thought hypothesis" (LOTH), which states that thought, like language, has syntax. That allows building up complex thoughts by combining simpler thoughts in various ways (Fodor, 1975). Based on LOTH, we can view the thinking brain as a syntactically driven engine preserving semantic properties of its processes.

In his book The Modularity of Mind (1983) Jerry Fodor proposed a modular theory of mind, in which initially defined module as "functionally specialized cognitive system". He defined certain properties of mind: domain specificity, informational encapsulation, obligatory firing, and so forth.

The functional modular structure of the brain was confirmed by the work of many scientists. For example, Gnezditsky V.V. could solve the problem of three-dimensional localization of energy sources in the brain detected using electroencephalography' data (Gnezditskiy, 2004). In particular, program "BranLock (Gnezditskiy, 2004), and a software for topographic mapping «Brainsys» (Koekina, 2000) has been used in studies in the Russian Research Center of the Development of Abilities and Consciousness in Moscow (http://www.rcdmc.ru).

These results are close to results of the work of Harris Georgiou at the National Kapodistrian University of Athens in Greece, who discovered that a brain works as a parallel computing machine. Harris Georgiou found that parallelism in the brain does not occur at the level of individual neurons but on a much higher structural and functional level (Georgiou, 2014).

7.2. About technologies of the thought processes

Jerry Fodor developed a computational theory of mind in its modern form, proposed by his teacher Hilary Putnam in 1961. According to this theory, the human mind is an information processing system, which is built according same principles using in computing machines (Rescorla, 2015). At present, architecture of human mind is known only in some details, but we can suppose that it can include elements of a machine of Turing, neural network, quantum computer and so forth.

A computer analogy makes it easier to understand algorithms of the brain's activity, to find out that generates analytical and creative capacities of the person, to discover how the ability to make decisions can be realized in an uncertainties, etc.

In the light of new evidences, the human mind can never be absolutely thoughtless or quiet. The brain can work in the four main regimes (modes) of thinking:

1. On duty - to maintain the functions of the body and react to solving appearing problems under control of Body Control Unit and Psychological Control Unit, described in section 1.5
2. Creative mode for solving problems defined by consciousness or sub consciousness
3. Complete or partial relaxation mode including a regime of autopilot, unconscious (sleeping) mode, etc.
4. Alarm mode for the fast life-or-death decisions in the very dangerous situations.

For the first two regimes, we can assume that our thoughts in mind are organized in some way into streams, one or more. Moreover, such streams may include not only thoughts, but also semi-finished products of thinking, including fragments of thoughts, music, images, etc.

For many years, it was believed that the primary source of great achievements and discoveries in all spheres of human activity is a sudden insight (illumination) in a dream or in reality. Mentioned in the section 1.7 works of Alex Osborn, Genrikh Altshuller, and many others proved that creativity and other functions of mind in the most cases has a technology, which can be using approaches of a computational theory of mind described with help of special algorithms and programs.

Let us look at some functions of human thinking that might be helpful in further consideration the basic concepts and features of the thought processes.

Thinking process in the creative mode involves several operations: comparison, analysis, synthesis, abstraction. Comparison is an examination of two or more items to establish similarities and dissimilarities. Analysis is mental differentiation of the object into its component elements and their connections with subsequent comparison by many criteria. Otherwise, it can be the process of evaluating data using analytical and logical reasoning to examine each component of the data provided. For example, a medical doctor needs to consider all complaints of his patient and test results to make diagnoses of disease.

Comparison is a one of main operation of brain. Any information received from the sense organs is remembered in short-term and long-term memory, and then compare with the existing in the mind the model of the world. If the comparison shows that this corresponds to the model of the world, this information is remembered as an addition to the model or corrects what is available. If the comparison shows the danger of the situation, then other intellect functions are started. This procedure is used many times a day when choosing what to wear, what to eat, where to go, and so on. It means that comparison function must be programmed in the mind using different algorithms for different people, and executed automatically.

Synthesis is the combination of the individual components in the unit. Usually in process of development of something, our brain must have a structure and mechanism for ideas generation. Examples of synthesis process here are when a psychologist, after analyzing several tests, builds a generalized psychological portrait of a man, or when a designer created some project.

Abstraction is the act of considering in something general characteristics, apart from concrete realities, specific objects, or actual instances. As a result of abstraction can be formed some concepts. As an example, consider the term "reliability" as a low probability of failure of any variety of cars or household appliances.

Generally, products of thinking include the following: concepts, judgments and conclusions. Concept - the idea, which reflects the general, essential attributes of objects or phenomena. Judgment - a reflection of the relations between objects and phenomena, or between their properties and characteristics. From two or more judgments a man can construct the following form on the complexity of thinking - reasoning. Conception is a complex product of abstract or reflective thinking or the sum of a person's ideas and beliefs concerning something. Conclusion is such a relationship between concepts (conception) and judgements, which resulted in a new judgment or a new concept.

Described above functions of analysis, synthesis, and so forth as well the comparison functions can be programmed in the brain and executed automatically. According to Chris Walton, 95-99% of all our thoughts, feeling and actions each day is governed by programs in our subconscious mind. Almost everything that we think and do in life is driven from level of mind we are not aware of (Walton, 2010). It is obvious that when our conscious part of mind is busy thinking about past experiences or future intentions and desires, our subconscious part is also at work, driving the behaviors that we show to the world.

The term "train of thoughts" for description of streams of thoughts was introduced and elaborated as early as in 1651 by English philosopher Thomas Hobbes.

The American philosopher William James coined the term "stream of consciousness". He wrote that consciousness - is the stream in which the thoughts, feelings, memories, sudden association constantly interrupt each other, and bizarrely, "illogical" intertwined "Stream of Consciousness" is the ultimate degree of "internal dialogue", there is an objective connection with the real environment is often difficult recoverable.

Usually a thought generator is observed as an out-of-control system. Mediators and people, who tried to clean their mind, are familiar with the 'monkey mind' phenomenon. Many Buddhist teachers refer to this phenomenon because it is similar to the way a monkey will swing from tree to tree tasting a banana from each one before dropping it and moving to the next tree. Like these monkeys, we often jump from thought to thought without ever really being in the present moment.

To understand how a stream of thought is generated, we can make some computer analogy, if to suggest an existence a functional group of microprocessors in our brain, on one hand connected with our memory structure, and on another hand with a commutation device, which periodically switch their connection to the field of view of consciousness, view that you see and comprehend, from one microprocessor to another one.

The existence of commutation device explains why in the stream of thoughts one thought appears after another one sometimes without any visible connection. This explains also why we can meet in the stream of thought not only a full thought form, but and a piece of thought like a word or like alone image, sound, etc. It can occur because at the time of connection one microprocessor create a thought, but another one didn't finish the process of the thought creation and works with some component of thought.

If information from senses organs or about the conditions of human body organs exceed the some threshold of perception given by different brain structures, then it will be included in the stream of thoughts. For example, when weather sharply changes, our brain includes this information in the stream of thoughts (like "it's rain, I need an umbrella), or "it's chilly, I need a sweater"). The same is, if we feel a pain somewhere. The more painful our condition, the more room takes pain feeling in our consciousness.

We have also installed in our memory different kind of alarms. In fact, studies have found that our minds are wandering periodically, drifting off to thoughts unrelated to what we're doing — did I remember to pick up from school a son, a daughter, etc.?, did I remember to turn off the light?, what should I have for dinner?

In 1975, James Webb Young, described five steps functional model of the creative process and include in that five actions like preparation, incubation, illumination – insight, evaluation and elaboration (developing) (Webb, 2009). In his book the author presents 55 techniques of creativity with algorithms and examples that allow not only to easily find new ideas, but also to evaluate them and implement them (Scherer, 2007).

There are several theories as to how mental images are formed in the mind. For example, the Dual-Code Theory, created by Allan Paivio in 1971, is the theory that we use two separate codes to represent information in our brains: image codes and verbal codes. Image codes are things like thinking of a picture of a lion when you are thinking of a lion, whereas a verbal code would be to think of the word "lion".

Both of codes allow organize the flow of thoughts as internal monologue, a conversation with oneself, an inner voice. On the other hand, verbalization of thought can be like reading words or hearing words. People can think in any language, first of all in mother tongue. If a person knows very well another language, he can think with the help of this language. Internal monologue, may, as in real life, to be colored or created with the addition of sounds.

Albert Einstain wrote: "The true sign of intelligence is not knowledge but imagination". Ability of imagination is necessary for designing, inventing, painting, writing a book and so on. Experience demonstrates that the creative power of imagination has an important role in the achievement of success in any field. Imagination is not limited only to seeing pictures in the mind. It includes all the five senses and the feelings. For some people it is easier to see mental pictures, others find it easier to imagine a feeling, and some are more comfortable imagining the sensation of one of the five senses. At last, imagination gives the ability to look at any situation from a different point of view, and to mentally explore the past and the future.

Visualization is a methods using the imagination to create a mental picture of the desired events, objects, and thinking results. Practically everybody, especially children, has the ability to visualize everything. Unfortunately, in most cases, people misuse their creative potential with age.

The ability to visualize especially developed in creative people: for example, designers, artists, sculptors. Visualization can be in 2D and 3D formats.

Nicola Tesla had ability for 3D visualization. When he got an idea for a new machine, he was able to "set it up" in his mind, keep it in the mind and even start it running to see how it would work. His capacity for this was so developed that the results that he got in his mind were incredibly accurate. This was verified when it came to building prototypes for the new machine. He would already know exactly how it would perform because of his "Mind Lab" experiments (Tesla, 2011).

Mental scene can have a boundary or cannot. For some people this scene is like a cloud. Placed any object in the mental scene, everybody after some training can focus it, analyze it, change it, add any other object and even create any interaction between these objects in space and time.

7.3. Creativity during the sleep

The history knows many examples how people came up with an innovative solution like insight during the sleep. Here is a list of the some great discoveries made during the sleep:

- Russian scientist Dmitri Mendeleev saw his famous the Periodic Law as a table where the elements are arranged as needed.

- German organic chemist August Kekule saw in a dream an aromatic benzene cycle.
- German-born pharmacologist and psychobiologist Otto Levy saw in dream the principle of chemical impulse transmission in the frog heart.

Experimental studies by Ulrich Wagner and other scientists have proven that sleep makes it easier to extract obvious knowledge and insightful behavior (Wagner, 2004).According to a new knowledge, the three types of memories consolidate in sleep: declarative (fact-based information), emotional, and operational, which includes skills like learning to ride a bike or play a song (Magaldi, 2015).

Because dreams often alert us to our problems, and guide us toward a resolution, today can be useful the lucid dreaming—the ability to know you are dreaming while you are in a dream, and then consciously explore and change the elements of the dream (Tucillo, 2013). After a few lucid dreams, many people realized that there is much more to reality than they currently understand.

7.4. Sequential and parallel thinking mind

Thinking "correctly" means aligning our thoughts in the flow, that may include thoughts, pictures, sounding complexes and emotions. Depending on age, temperament, responsibilities, speed of thinking (a number of thoughts per unit of time) in humans can be in a wide range. The wider human horizon the greater flow of his/her thoughts. Number of random thoughts can vary from a couple thousand to 50 000 per day. People often dwell in the past or in the future, obsessing about mistakes they might have made, thinking about grievances or revenge offenders, planning or analyzing what was done.

Any stream of thoughts can be productive when it is organized. If a person has one stream of thoughts, it means he has sequential mind, if he has several streams of thought, then –parallel thinking mind.

History gives us many examples of people who could thinking in parallel. According to legend, Julius Caesar could do at the same time seven unrelated cases. We also know that Napoleon could simultaneously dictate to his secretaries seven critical diplomatic documents.

In section 5.10 we wrote about Laurence Kim Peek who can read by scanning the left page with his left eye, and simultaneously the right page with his right eye.

Francisco Maria Grapaldo was born in Parama (Italy) in 1464. He used to write his poems with both the hands simultaneously – in two adjacent pages of a notebook. Mr. Joseph Burnheart Duncan had nine secretaries for different languages when he was the director of the national library of Munich. He used to dictate official letters and other deliberations in nine different languages to nine secretaries simultaneously. He had also memorized the Bible by heart and was able to reproduce the corresponding sections completely from any page of this Holy Scripture. Recalling the location of the million books in the library was so easy for him as if the catalogue was kept open before his eyes (Shriram Sharma Acharya, 1996, 2014).

By other words "thinking in parallel" means a known from computer science term of working in the regime of multi-tasking. For example, a good driver while driving can simultaneity evaluate the road situation around the car, talk on the phone, listening to the radio, speaking with passengers in the car, and thinking about something else.

Nowadays any office worker, high school student or even stay-at-home mom could successfully make plenty deals practically at the same time. In our life, multitasking is not considered a rare gift anymore, rather a necessity.

In real life we can meet 3 types of multitasking: concurrent multitasking (doing multiple tasks at once), sequential multitasking (interrupting and resuming tasks), and multitask skill acquisition (learning and practicing multiple tasks).

By terminology of computer science multi-tasking can be based on either number of processing centers, or principle of switching between programs. Human body has many processing centers, for example 2 hemispheres in the brain.

Multi-tasking in technological environments was formerly a skill developed primarily by professionals (supervisors) working with plenty electronic displays of electronically acquired data in the center of control of complex system, like a big ship, or airplane, or electric power station. For example, airplane pilots have to develop the ability to multitask in order to be able to make the fast life-or-death decisions in his super complicated environments. But how such multiple streams can effectively be dealt with by our minds and brains has not been fully understood.

Anyway we can assume, that parallel processing can be realized with the help of the commutation function of higher level, which allow periodically to switch from one stream of thoughts or information packet to another.

7.5. Thought-forms

Most of us have an intuitive idea about what thought is. Today many people use a term "Thought-form". This term is not new. Theosophists Annie Besant and Charles Leadbeater introduced the concept of thought forms in their book "Thought forms", first published in 1901 (Leadbiter, 1901) and brought the three basic principles underlying the creation of thought forms:

- Quality of thought determines color
- The nature of thought determines form
- Certain thoughts determine the clarity of outline.

In their book Besant and Leadbeater explain how thought forms are generated from thoughts and emotions, or to use theosophical terms, the subtle mental body and the subtle desire-body. They wrote:" This thought-form may not inaptly be compared to a Leyden jar, the coating of living essence being symbolized by the jar, and the thought energy by the charge of electricity. If the man's thought or feeling is directly connected with someone else, the resultant thought-form moves towards that person and discharges itself upon his astral and mental bodies. If the man's thought is about himself, or is based upon a personal feeling, as the vast majority of thoughts are, it hovers round its creator and is always ready to react upon him whenever he is for a moment in a passive condition".

Edgar Cayce was also one of the few people insisting on the reality of thoughts long before it was fashionable to accept it. He wrote: "thoughts are things and take form as they are dwelt upon".

According to postulates of Thought-forms theory of Peter Baum, thoughts imply the existence of mental models (Baum, 2012).

It seems to me, this term can be used as a definition of a thought, which has a strong meaning. For example a word like "flower" is not a thought, but "flower is created for beauty" is a thought, because has content. In other words, the thought form - is formulated thought with content.

Every thought should consist of some components. If it appears like words and phrases, then these components are letters or words; if it appears like images or pictures, then its elements are parts of images and colors; if an idea or mood is implemented in the form of music, components are the different sounds in the different progression, and so forth. As experience shows, many people (not only the sensitive) can read minds. Cause of mind reading can be explained simply because the thought-form is emitted by the brain into the environment as a set of modulated signals.

We will understand that the thought-form is a set of thoughts components that, at first, makes this thought understandable to other people, if it uttered aloud, or understandable for sensitives (or any devices of artificial intelligence in the future), if it is not pronounced. It is obvious that in order for the idea become thought-form, it must have a certain semantic completeness, as well as having a wave function form that can be emitted by the brain and recorded, for example, by EEG or fMRI. In principle, the thought-form may convey also emotions. The nature and strength of the emotions associated with the thoughts, can give the thoughts shape, color, intensity and strength, to make them more clear and understandable.

Thought-form can include several thoughts united, for example, in a conception, and save as one package of information. English poet John. G. Byron had memorized all his works. In my opinion, it means that he saved all his poems as thought-forms. At the end of his life famous artist Van Gog could write his pictures for couple hours. Obviously, he had his pictures as one file in the brain, i.e. as a thought-form.

Physically thought-form is a wave packet generated by a man's brain. That wave impacts on the person and spread to the external space. The more often generated the same thought, the thought-form will gains greater its potential. Power of the thought-form depends on conviction and belief. Many clairvoyances can see thought- forms and use it for healing goals (Murphy Brendan, 2012).

Each definite thought produces a double effect - a radiating vibration and a floating form. The thought itself appears first to clairvoyant sight as a vibration in the mental body, and this may be either simple or complex.

In addition to the theoretical understanding, there were attempts to detect the thought-forms not only by the physical apparatus like fMRI (Crew, 2016), but and with the help of photography.

The term "thoughtography", also called projected thermography, psychic photography, etc. was first introduced at the beginning of the twentieth century by a professor at Tokyo University Tomokichi Fukurai. He published results of experiments with sensitive Nagao who had the capability of telepathically imprinting images on photo plates, which he called "nensha". Later these experiments were repeated by American psychologist Jule Eisenbud (1967), and by Russian psychiatrist Dr. Gennady Krokhalev.

Researchers at Japan's National Institute of Genetics believe they've captured a world first video — images of a thought making it's way through the brain of a zebrafish (Muto, 2013).

All these studies proved that thought has material nature, can be radiate from a brain and receive by the brain.

The phenomenon of transmission of mental images is called telepathy, and the person who reads the thoughts is called a telepath. History knows many examples of telepathy or physics reading. Some of them was mentioned in the section 3.7. Here are Edgar Cayce, Sonia Choquette, Jeane Dixon, Linda Howe, Caroline Myss, to name just a few.

7.6. Dominant motivation, mental blocks and mental barriers

Any creativity is impossible without a dominant motivation. The theory of dominant has been developed by Russian physiologist Aleksey Ukhtomsky (Ukhtomsky, 1996).

The main postulate of the Ukhtomsky' theory is that at every moment the living creature is engaged in a particular task. These activities are provisionally dominant and, at some point in time, any external impact will be either 'work' on the target or ignored. After some time, this type of activity may be exhausted or displaced by others, and this new task will otherwise organize the activities of the body. The previous activity remains a trail that may 'be remembered' in part or in full under suitable conditions. The experience of the dominants, from the recollection of our own reactions to an external situation, largely determines our continued behaviour, our ability to respond to the new external situation, the new task.

Formed dominant motivation can become the main system governing human behavior, ensuring the satisfaction of need, which in the given period of time turns out to be the main one, and all other needs therefore appear to be secondary and subordinate, and the corresponding reflex activity will be suppressed. It means that there is an urgent mobilization of all body systems, past life experience and perception of the environment for the sake of achieving a dominant need or goal.

A concept of "flow" developed by Hungarian psychologist Mihaly Csizkszentmihalyi as a a a highly focused mental state is close to concept of the dominant motivation. His studies revealed, that people who have learned how to consciously control their concentration, know how to turn off all the mental processes, except for one. 'Flow' is a state of concentration so focused that it may be accompanied by a state of satisfaction and happiness (Csizkszentmihalyi, 2002).

A mental block is a some infringement of cognitive functions of a brain (intellect), which can have as a result serious glitches. A mental block can be described as a psychological obstacle that prevents a brain from performing a particular skill. The reasons for the appearance of mental blocks can be very different: traditions and rules from childhood, habits, emotions, etc. For example, sportsmen can have mental block when a fear becomes overwhelming and create a resistance to perform the needed skill (Martin, 2016). These blocks can cause a rise in uncontrollable neuropsychic tension, i.e. stress.

Stupor is bright and representative case of the mental block for the healthy person after the gone through shock, for example after accident. Other example is a writer's block, which is a condition, primarily associated with writing, in which an author either has an inability to continue or complete a train of thought, or loses the ability to produce new work, or feels a common creative slowdown. The condition ranges in difficulty from coming up with original ideas to being unable to produce a work for long time (months or years).

Brainblock is a synonym of mental block. According to the theory of American neuropsychologist Dr. Theo Tsaousides, the brainblocks can transform the motivation into inertia, the desire to act in apathy, and minor obstacles to insoluble problems. They can cause an array of health problems, ranging from fatigue to serious clinical problems, like depression and anxiety (Tsaousides, 2015).

Identifying and overcoming a mental blockage is very difficult task. Although very often mental blocks can be caused by physical problems, they can be usually be cleared up by taking a break from the work or by some methods of psychotherapy.

A mental block can be represented in our subconscious as an irresistible wall, or as a mental barrier to be able to cope. The mental barriers, for example, can arise from following reasons:

- Painful or difficult unresolved, buried deep in consciousness experience
- Lack of knowledge or bad experience creates a lack of confidence
- Fear of the unknown, which formed paralysis and incapability to cope.

Bulgarian psychotherapist Georgi Lozanov substantiated some of these barriers. The first one he called "critical-logical", and the second he called "intuitive-affective". These barriers are due to the existing system of education when human development protects the brain from unnecessary overload him with information.

Breaking down the mental barriers of our subconscious is no easy task also. For each of these and other barriers, is needed to identify the factors causing the mental blockage. Lozanov in his practice use some of innovated him methods (see section 8.6). To eliminate barriers usually used method of personal intellect management allowing to work with subconscious (see Chapter 9).

7.7. Mirror mind

In 1992 year Italian neuroscientist at the University of Parma Dr. Giacomo Rizzolatti had discovered mirror neurons, unique cells of the cerebral cortex, that are fired when we observe the actions of another person, and automatically reproduce someone else's behavior in our head, allowing to feel what is happening as a personal action. In other words, this phenomenon reveals the veil of secrecy over one of the main abilities of the human brain-capacity for empathy (Blakeslee, 2006).

If viewers see on the screen match of the boxers, they tense up muscles and clench their fists. If mother smiles baby, he instantly overreacts respondent smile. Mirror neurons without mental analysis, capturing only facial expressions and gestures, contributing to sympathy or, conversely, avoid.

Any child captures from its parents different emotions - joy, anger, compassion, etc.. Throughout his life he would smile, frown, ashamed of the way his relatives did.

Italian researches also found that our will and consciousness are able to partially reduce the impression caused by the fact that we see, or what is intended. Interestingly, the study discovered another fact - women have a much more of mirror neurons than men do. This explains their tendency to emotional relationships, more sympathy and understanding.

Mirror neurons allow us to adopt (to repeat) behavior of surrounding people. In particular, they allow an understanding of very important feature of emotional part of intellect - capacity intuitively to perceive that others feel.

Mirror Mind ability is very important for intellect management, for example, inside of team.

7.8. Mindset and Mind tuning

Mindset. Power of mind depends on a mindset - a fixed attitude of mind or disposition that not only predetermines a person's responses to any manifestations of environmental processes, but and a control or interpretations of situations.

Mindset is a type of mind tuning as a result of education, training and life experience. Mindset is the sum of a person knowledge, including beliefs and thoughts about the world and himself in it. Human mindset based not only on innate ability, but and on the result of everything, which happens in the human life. Generally, mindset is based on a collection of all beliefs, tendencies, feelings, emotions, habits, attitudes, mental models of rounding us world, etc. Mental models are based on our assumptions, opinions of the people whom we trust, as well as on the specifics of education and experience. We are not always aware of their impact on our behavior. Nevertheless, it determines our interpretation of the outside world, and the reaction to it.

Mindset can be connected with mental image of what an individual wants in his life - his desires, aspirations and the resulting action plan that will propel him to succeed.

At presents many experts is distinguishing two main opposing categories of the mindset for healthy people, all other categories can be intermediaries. First category is the positive mindset with the underlying positive beliefs, values and attitudes necessary to affect success in chosen area of endeavor. The second type of mindset- negative mindset, which belongs to a person with negative mood, hopeless, being afraid everything, not taking actions at right time, blaming others, liking to cheat and so on.

Fortunately, our mindset regarding certain aspects can be changed. The decision to develop positive thinking, supported by the joyful expectation of only the best, can allow to put yourself in the right state of mind.

Any mindset depends on type of mind. American psychologists Robert Kegan and Lisa Lahey defined three types of mind: the self-authoring mind, the socialized mind, and the self-transforming mind (Kegan, 2009). The self-authoring mind is an independent mind and perceives the world through its belief filters. The socialized mind is a dependent mind, because it strongly influenced by opinions of other people, media, etc. The self-transforming mind is an interdependent mind. The way a self-transforming mind responds to a situation or request is by seeking out more information to further its need to find meaning, make a difference, and be of service.

On the other hand, depending on what function of thinking is more developed for individual, scientists distinguish several styles of thinking, for example: analytical, creative, critical, inquisitive, insightful, open minded, systematic, according to beliefs or common sense, and so forth. This does not mean that an individual cannot have several thinking functions more developed. The style of thinking as well as a humor sense are supplements that allow describing an individual mindset better.

According to theory of Professor of Psychology at Stanford University Carol Dweck, here are else two categories of mindset: fixed mindset and the growth mindset (Dweck, 2006).

People with a fixed mindset believe that their intellect and personal traits are constant. They spend their time only by exploiting their intellect, and not developing it. Fixed mindset is usually associated with such things as protection, avoiding difficult problems, competition, and difficulties, minimization of investment in own development and the development of other people.

People with a growth mindset believe that their most basic abilities can be developed through dedication and hard work— innate intellect and talent are just the starting point. This view creates a love of learning and a flexibility that is essential for great accomplishment. Virtually all great people have had these qualities (Dweck, 2006). In general, a growth mindset stimulates, taking on challenges, welcoming feedback, putting in effort, persisting, seeking cooperation, and focusing on improvement.

The multiple researches approved that the growth mindset allows having more success in any field of knowledge. Studies have shown repeatedly that with training, supplements, and time – intellectual abilities can always be increased. We need to remember, that many geniuses did not were wunderkind in their childhood.

Psychical self-control subsystem of our brain, mentioned in division 1.5, works based on the mindset characteristics and willpower.

Mindset must include goals of life. For some people this goal – to live as all around us, for others to became a movie star, or famous sportsmen, for most people - to create a happy family.

Depending on the goals of life and other purposes, mindset must include the following operations, described in the next Chapter:

- Compensation of existed homeostatic imbalance
- Mind liberation from mental blocks and barriers
- Building protection against extraneous interference.

We have the ability to adapt our mental activity to environment without even being aware of it. But this adaptation very often is not enough in our life overwhelmed problems and stresses. When we are babies, mothers may rock us and sing lullabies to calm and soothe us. This can help many adults too, but we need more strong methods to tune our personal activity to a calmer, more harmonious condition. These methods must be in the arsenal of Mind tuning.

Mind tuning and control. Numerous studies show that by analogy with an orchestra or any musical instrument human brain or human mind can be effectively attuned for solving different tasks.

If a piano tuner uses tuning forks to strike precise tones and calibrate the keys of the piano to those tones, tuning the human brain needs other methods and tools.

The processes of changing growth mindset can be natural or artificial. Praying, love, sport, different kind of entertainment and so on include many means for modifying our mental, emotional, and physiological states, therefore and for natural change of mindset. Artificial tools for impact on the brain with help of for example different kind of oscillators, and techniques of mind activation are an essence of mind tuning.

Goal of the mind tuning must be balance and harmony between mind, body and soul, which are essential to overall well-being of our body and needed cognitive, emotional and physical performance (see 7.2). Nobody has ideal mind traits, but the special methods and tools of mind tuning allow to have above mentioned traits the better or the best for any person.

In the ideal case, a mind tuning means the development of the following:

- Compensation of homeostatic imbalance (see Chapter 7)
- Optimization of intellectual abilities

- Enhancing a capability to harness the power of knowledge
- Strong ability to keep an attention a long time
- Improvement of flexibility and fast learning
- Boosting an ability to interpersonal connection with the people around.

At present existed arsenal of mind tuning include mainly different types of meditation, some methods of suggestion and entrainment.

Methods of optimal tuning intellectual abilities of mind will be described in the Chapter 8.

CHAPTER 8

Problems and some methods of enhancing intellectual abilities

Two qualities are indispensable: first, an intellect that, even in the darkest hour, retains some glimmerings of the inner light which leads to truth; and second, the courage to follow this faint light wherever it may lead.

Carl von Clausewitz

8.1. Classical individual methods of mind activation

Above, we described some factors for mind activation, including the creative environment, an influence of motivation, will, memory and so on, which affect an intellect. Analysis of these methods makes it possible to divide them into two groups: one is artificial stimulation of a brain using special techniques and devices, second is a management of intellectual functions with help of some algorithms and more complex methods.

The history has saved up a lot of become classical simplest methods of the creative activity' stimulation used by some famous people in the past.

Aristotle method

Aristotle liked to think and teach while walking, and he communicated with his disciples on the covered path in the Lyceum. This was called the "peripatus", from which the word "peripatetic" derives, meaning something that wander or related to walking.

Body cooling and warming methods

A German composer and pianist Ludwig van Beethoven has a ritual before composing. He liked to dip his head in cold water before getting creative. A French novelist and playwright Honore de Balzac wrote his novels, standing barefoot on the chilly stone floor. A philosopher, writer, and composer Jean-Jacques Rousseau made his brain work harder, standing in the sun with uncovered head. A German poet, philosopher, physician, historian, and playwright Friedrich Schiller working on his compositions, always kept his feet in cold water.

Method of Tomas Edison and Salvador Dali

Edison worked on his inventions in a very busy schedule. When he came to a dead end in the process of searching a problems solution, usually he sat down in a chair, took a metal ball in hand. Falling asleep, he released the ball from his hand, and loud sound of the falling to the floor ball would wake him. He confessed that he had been visited by new and bright ideas about the project he was working on.

The favorite technique for Salvador Dali was that he would put a tin plate on the floor and then sit in a chair beside it, holding a spoon over the plate. The moment that he began to doze the spoon would slip from his fingers and clang on the plate, immediately waking him to capture the surreal images.

Hypnosis

Two great classical music composers Mozart and Sergei Rachmaninoff were hypnotism devotees. Mozart opera "Cosi Fan Tutte" was entirely composed in a hypnotic trance.

Albert Einstein used the hypnotic trance state to develop many of his theories and formulas. Thomas Edison practiced self-hypnosis regularly. Sylvester Stallone used self-hypnosis tapes to boost willpower and creativity. During each day of filming Rocky, Stallone worked with well-regarded hypnotherapist Gil Boyne. The basketball legend Michael Jordan was hypnotized before every game to enhance his mental focus (Morcan, 2014).

Mozart effect

Don G. Campbell in his book, published in 1997 described scientific studies which proved that listening to specific Baroque and classical music – especially Mozart – allows humans to study and remember information better (Campbell, 1997). Besides Mozart, researchers found other composers whose works are most conducive to learning include Vivaldi, Bach, Pachelbel and Handel (Morcan, 2014).

When Einstein met difficulties while working while working on a serious problem, he would play Mozart until he could dive to what he called the harmony of the spheres.

Rotten apples.

Some geniuses use very strange methods of brain activation, for example, famous German writer Friedrich von Schiller kept rotten apples in his desk, claiming he needed the scent of their decay to help him write.

At last decades, for brain stimulation were created many new advanced and more effective methods:

- Changing the position of the assembly point
- Transcranial brain stimulation
- Synchronization both hemispheres
- Activation of subconsciousness
- Self-mind management
- Meditation and trance techniques
- Activation of channels for connection to Universal field of knowledge
- Mind machines
- Brain boosting supplements and medical treatment
- Etc.

Below we will describe some of these methods.

8.2. Compensation of homeostatic imbalance

The motto "A healthy mind in a healthy body" ascribed to Thales of Miletus, Greek philosopher, mathematician and astronomer from Miletus in Asia Minor. Today this phrase sometimes has mutation like "healthy body – healthy soul", "in healthy body is healthy spirit and mind" with explanation "A healthy spirit is the basis for a healthy mind. A healthy mind is a requirement for a healthy body!" and so forth.

Scientific studies of last decades allowed more understanding a relationship between mind, body, soul and even spirit. Brain balance and harmony are essential to the overall well-being of our body and needed cognitive, emotional and physical performance (Blackett, 2014).

As previously remarked (see 1.5), intellect must support equilibrium of homeostasis state of physical body and all subtle bodies, shortly described in 5.4. When asking yourself what is homeostasis, it refers when temperature of body, blood pressure and many medical tests are normal, everything is running smoothly, body and mind are in harmony and we do not feel our bodies. In reality, homeostasis is a complicated state that is very difficult to adequately describe and especially measure quantitatively. Unfortunately, the equilibrium that our body is trying to achieve through the many mechanisms usually cannot be absolute.

First of all, physiological and psychological states of organism are dynamic, and all organisms systems and organs continuously vary their characteristics in accordance with the circadian and other rhythms described above. So, even being in homeostasis, body temperature, blood pressure, heart rate and most metabolic indicators are not always at a constant level, but change over time within the permissible range. If some parameters of body or mind functions exceed permissible range, then both control units BCU and PSCU, described in section 1.5, will execute actions to compensate imbalance. When the resources of the body do not allow doing this, or something is preventing the human system from adapting or correcting, we get an uncompensated balance that can lead to illness. Here are many reasons for this kind of imbalance beginning from mind-body dissonance and ending with aging or psychosomatic disorders. Mind-body dissonance – is the ability to display bodily expressions that contradict mental states. The word psychosomatic refers to physical symptoms that occur for psychological reasons. Psychosomatic disorders are physical diseases, for example, as psoriasis, eczema, stomach ulcers, and some heart diseases that have a mental component derived, for example, from the stress or strain of everyday living (Franz, 1950).

It is very interesting that the psychological and physiological factors characterized psychosomatic disease is almost interdependent, i.e. psychological factors can influence the appearance of medical diseases (called psychosomatic), and the severe or continuing chronic diseases inevitably to one degree or another can affect the mental state of the human being. Very often a person who has been ill for a long time spoils a behavior, which characterized by appearance hysterical reactions, sleep deteriorates, appetite changes, and other negative manifestations.

Dr. Lyudmila Puchko (see 3.7) developed a method for measurement of level of homeostatic imbalance based on radiesthesia technique using a pendulum and scale from 0 to 100 percent similar the Fig.3-2. (Puchko, 2008, 2010). Used for measurement a simple pendulum consists of a bob with mass m hanging at the end of a string of length L. This technique helps a person to measure his/her levels of homeostasis for all physical and subtle bodies. This approach is a part of developed by Puchko the diagnostic method for evaluation of mental or physical health.

The solution of the problems of compensating a homeostatic imbalance is what Western and Eastern contemporary and alternative medicine have been doing during several thousand years. Our task is to activate the intellect to provide self-treatment in some cases and thereby make one more step towards improving the physical and mental health for any person.

Currently, there are over a thousand different classical psychotherapy techniques and new continuously developing methodologies of alternative medicine based on self-suggesting, hypnosis, meditation, anti-stress exercises, Neurolinguistic Programming (NLP), radiesthesia method of vibration series of Puchko, just to name a few that allow treating psychosomatic illnesses and eliminate imbalances that are not yet diseases.

At present, these methodologies and tools can reprogram the subconscious mind to change the unwanted feelings, beliefs, and behavior to the desired positive outcome of the client.

Using methods of imbalance and level of stress measurement (see 3.7) allow sometimes simplifying the task of eliminating above mentioned homeostatic imbalance, because knowledge of abnormality of body-brain functions stimulate a search the best methods and tools of imbalance compensation and mind tuning.

8.3. Mind liberation from mental blocks and barriers

Above we considered a man as the complex bioenergetics and information system (CBIS). It is obvious that brain-body balance or harmony concern the whole CBIS, which unfortunately can have emotional and energetic blockages.

Breaking down the mental barriers and blockages in mind is no easy task. Humanity in its history has accumulated a lot of simple and complex methods of getting rid of all this. For example, it is possible to list a little from used strategies:

- Coaching - a relatively new profession that really became popular from the 1980s. Before this decade we had some kind of coach often in the sporting arena but now, coaching may help us to reach our highest potential in our careers, our relationships, with our money, community, spirituality and more;
- Emotional Freedom Technique, that combines mind-body medicine and the ancient practice of acupuncture without the needles;
- Deep meditations, developed by many oriental practices;
- Hypnosis based for example on ideodynamic concept in therapeutic hypnosis (Cheek, 1994);
- Aromatherapy (essential oils) usage. Essential oils that are derived from flowers, leaves, roots of plants have been used for thousands of years;
- Neuro-linguistic programming (NLP).

Emotional Freedom Technique was created by the American engineer Gary Craig (Craig, 2008) in the 90s, based on the techniques of Dr. Roger Callahan named "Thought Field Therapy" (Therapy Mind Fields) (Callahan, 1985, 2001)) and used the principles of traditional oriental medicine (acupressure) and western psychology.

Dr. Larry Phillip Nims developed some years ago a new method, called him "BE SET FREE FAST™ ". This method can rapidly eliminate unconscious blocks, negative emotional roots and the self-limiting beliefs which affect numerous everyday problems of human adjustment, personal ineffectiveness, emotional pain, dissatisfaction, and personal unfulfillment, and can embed new functional programs if necessary (Nims, 2003).

NLP is an approach to communication, personal development, and psychotherapy created by Richard Bandler and John Grinder in the 1970s (Bandler,1979, 2014). The primary focus of NLP is how to operate your own mind and body to create what you want in life. NLP is the methodology and technology of human communication and its changes. NLP is based on studying the structure of subjective experience, not paying special attention to the content of this experience in the semantic and value aspects. Based on the traditions of the American school of psychology and psychotherapy NLP has developed a wide range of methods. These methods are designed to improve human performance.

NLP practice allows to do the many things, for example, the following:

- Build instant rapport with anyone – anytime as a prerequisite for good communication and cooperative relations
- Understand how to access emotional states that are appropriate and useful for any given situation
- Adopt empowering beliefs
- Develop intuition; know how others are thinking
- Build confidence and motivation
- Release negative emotions and limiting decisions
- Reinforce and accomplish goals
- And so on.

Described below the Silva Method can be also useful for solution of this task.

8.4. Accelerated learning

Job-hopping is a common phrase that is used for workers who jump from job to job. According to a report of the U.S. Bureau of Labor Statistics, the median number of years that wage and salary workers had been with their current employer was 4.2 years in January 2016, down from 4.6 years in January 2014. It means that time of sticking with a single profession for one's whole working life is going away. Changing jobs frequently has become the norm rather than the exception. According to CareerBuilder, 45 percent of employers now expect new college graduates to stay less than two years. Therefore, the stigma of being a job-hopper is quickly becoming outdated.

New technologies demand new skills. The labor market is changing at a fantastic rate. For example, now more and more employees are working outside of an office. Many companies are moving to a partially or even fully distributed team model, which can minimize overhead costs. Having a roster of talented contractors to draw on allows for employers access to a wider range of specialized skills without paying additional full-time staff salaries.

At present, applying in an educational institution, a person cannot know whether his professions will the claimed through 4-5 and more years. Thanks to above mentioned information, learning process become practically for most people a constant process. Therefore this process must be right organized and takes not too much time. This is a reason why accelerated learning (AL) is needed for students, businessmen, actors and representative of many profession.

AL means not only to receive more information in the shorter time, it means to receive a professional skills in the shorter time with ability to have a media to support these skills during a given time, for example for time of new job with new profession. Because this is very serious problem, on the market appears a new teaching and training professions: facilitator, instructional designer, organization development professional, etc.

The methodology of AL must be based on:

- The best in people and organizations
- Encouraging discovery, reflection, and learning
- Ability to think critically, systemically and innovatively
- Reducing the unnecessary expenditure of time for learning unnecessary things
- The right physical environment for learning process.

Learning must involve the whole brain- body system with all its emotions, senses, and receptors. It engages the mind, body and spirit.

All programs of AL have to provide guided experiences in the classroom, online and on-the-job through observation activities, simulations, case studies, and experiments that allow participants to embody the learning, recognize the key elements of profession or discipline, and develop their skills. During the learning process, students observe their own results and recognize the role they play in technology of learning. Positive results enables a shift in thinking, the creation of a new plan of action, a correction in behaviors or a strengthening of key behaviors that will help them achieve consistent success. Obviously the result of training depends on:

- What stimuluses used for learning
- How relevant and significant were the topics
- What relationship was with other participants of learning process –students (members of team) and teachers
- What equipment used for learning process.

In 1966, Dr. Lozanov founded the Suggestology Research Institute in Sofia, Bulgaria. The name Suggestopedia derives from two English words – "suggestion" and "pedagogy". Unlike most classroom learning, Dr. Lozanov developed a holistic method of teaching involving role-playing, games, the visual arts, and music. Dr. Lozanov's early program focused on the teaching of foreign languages using relaxation, pictures, and music. Using this method, students were able to learn between 100 and 1,000 new foreign words per day with a retention rate of 98% or even better. The method of Dr. Lozanov was described in the book "Superlearning" (Ostrander, 2012).

In 1978, Dr. Ivan Barzakov with a group of teachers and psychologists formed the Barzak Educational Institute in Novato, California. The Barzakov team have since trained more than 10,000 people in 17 countries.

Prof. Ivan Barzakov calls his method OptimaLearning. While Lozanov used his techniques mainly for foreign language learning, Barzakov applies his principles to any subject. He developed a careful selection of music tapes, which are used to enhance learning, creativity, problem-solving and decision making. He carefully blends different types of music together for contrast, "because variety stimulates our minds and keeps us alert". He also changes the "texture" of the music, from violin to flute through to mandolin, clavichord, and piano.

By the late 1970s, a psychologist Dr. Donald Schuster of the University of Iowa at Ames founded the Society for Accelerative Learning and Teaching (SALT). The name of this society was changed to its current name of IAL, International Alliance for Learning. The new name reflected a cultural preference and a desire to demonstrate the adaptations over the years of the original methodology developed by Lozanov. Now similar societies appeared in the different countries, for example, the Society for Effective Affective Learning (SEAL) in England.

As proven by numerous applications, a very promising way to accelerate learning and training is applying such artificial intelligence systems as virtual reality systems (VRS). Here are just a few of the possibilities of VRS that allow accelerating learning by:

- Studying any discipline in virtual environments which differs little from the real environment
- Stimulating imagination and creativity in bigger and bolder ways creating an imitation of reality-similar gaming tasks or projects (Iserlis, 2009).

Many methods and tools for accelerated learning allow to learn anything two-to-five times faster, and described in many books, for example (Meier, 2000; Dryden, 2005; Cobb, 2013).

Many of today's leading organizations and educational institutions are benefiting from the power of accelerated learning.

One of main and very important problem of AL is regulation and acceleration of information input into the brain. Let consider this in more details.

8.5. Acceleration of information input into the brain

If before the second half of the twentieth century, we lived in an atmosphere of lack of information, there was no computers, Internet, and cell (smart) phones, no multi-program TV, there were few books in comparison with nowadays, and there were few devices for the reproduction of information like copy machines, printers or players.

In the 21st century, we live in a world with such excess of information that we can talk about information overload. This is a reason why today a problem of information perception and information processing takes the first place in the information technologies.

If earlier efforts were needed to get some knowledge somewhere, now here is a gigantic surplus, any person needs only to learn how to correctly and quickly find relevant and reliable information, and then be able to input (remember and understand) it quickly and correctly.

There are several ways of information perception - through the consciousness, through the subconscious and combined. Through consciousness, we can assimilate generally the information having a lot of time and bearing meaning. The subconscious mind perceives easier emotions, but to grasp the contents it requires special technologies, including hypnotic or non-hypnotic suggestion, the use of certain types of music, and so on.

Acceleration of perception of the information through consciousness is provided, for example, by developing methods of fast reading. Only approximately 1% of readers can read at speeds of above 1000 words per minute (wpm) with near 85% comprehension. Average readers can reach only around 200 words per minute with a typical comprehension of 60%. We can calculate a reading efficiency as multiplication of reading speed on the percentage of comprehension defined by special tests. For example, if reading speed equal 200 wpm and comprehension equal 60%, reading efficiency will equal 200 x 0.6 = 120 efficient words per minute. It is obvious, that a speed of reading must be equal or less a speed of comprehension.

Very few people know that Presidents John F. Kennedy, Richard M. Nixon and Jimmy Carter sent many high-ranking members of their Administration to the Evelyn Wood Reading Dynamics Institute in Washington. Evelyn Nielsen Wood was an American educator, widely known for creating the fast reading methods, allowing increase a reader's speed by a factor of two to five times (Wood, 1958). She could herself read 2,700 words a minute, her students -1400 words a minute and more and about 95% comprehension (Van Gelder, 1995).

A method developed by Evelyn Wood included some innovations:

- Reading down the page rather than left to right
- Reading main parts of sentences rather than whole sentences
- Avoiding involuntary rereading of material.

At last decades, was created several fast reading systems like as The PhotoReading Whole Mind System, developed by Paul R. Scheele (Scheele, 2007), and Speed Reading, based on 37-speed reading and photo reading/downloading techniques and developed by Jan Cisek and Susan Norman (Cisek, 2012). Frank Waldman, Maik Maurer, and Jamie Locke founded in the Boston the company Spritz Inc. and first released to the public their product in February 2014 at the Mobile World Congress in Barcelona. With the Spritz reader program, "spritz" people must read words, as they appear one at a time, in rapid succession. With Spritz the user can read anywhere from 250 to 1,000 words per minute. This super-fast reading technology may activate natural ability to quickly and easily absorb information, three and more times faster.

One of the modern ways to increase the speed at which information is entered in the mind is the audio books that at first allow adjusting reading speed and at second to use extra reading time— for example, when driving a car. Now the second-generation of the Amazon e-reader Kindle and the Kindle DX have an "experimental" feature that allows to convert any text to speech and reads it to listener in any circumstances.

8.6. Memory training

Memory training is one more approach to accelerate learning.

The three main conditions for fixing in the long-time memory the ideas, facts, events or data are the impression, association, and repetition. This determines the order they should be put in place.

Around 500 BC, the Greek poet Simonides left a dinner party to receive a message. When he returned, he found the roof caved in and every other guest crushed among the ruins. Worse still, the corpses were so uglified that it was impossible for family members to identify their dead. But Simonides intervened and offered to name each person. How? He remembered where they were sitting at the table.

Thus was born the Simonides' method of loci (from the Latin for "place"), the worlds oldest and arguably most famous memory technique. This method is based on one simple brain ability: people have a far better memory for the tangible information (physical spaces, images) than they have for the abstract info (numbers, words, ideas). To employ that, it is needed to pick a physical space and correlate it with vivid representations of whatever must be remembered.

For the next thousand years, the method of loci remained a fundamental pillar of intellectual life. But in the middle of the second millennium, it disappeared from popular use.

Here are some practices for training visual and hearing memory:

- Photographic memory training, for example by Darkroom method (http://www.menprovement.com/how-to-develop-a-photographic-memory)
- Loci method training
- Reading aloud
- Memorizing poetry
- Computer and online games like as Trolley Dash, Pattern Memory and so on (http://www.memory-improvement-tips.com/brain-games.html)
- Attention concentration - Turn on the radio, then gradually reduce the volume; set the lowest volume limit, when you can understand what they say. The low intensity of the sound will make you focus. Do not continue this exercise for more than 3 minutes. Work with special programs.
- New language learning. It is good for improving and maintaining the memory.

8.7. Changing the position of the assemblage point

Here are some methods on how to move the Assemblage point (AP) to its best location (the higher the better) and angle. The Assemblage Point can be re-aligned with the use of a large crystal by a therapist who has been properly trained (Whale, 1997). According to Jon Whale, the quartz crystal suitable for the AP correction must have the minimum length 18 cm, minimum width and depth (diameter) 3 cm, and weight more than 200 g.

Some therapists and patients may have a strong energy, which can be enough to realign the Assemblage Point use some exercises and without using a crystal.

The Assemblage Point can also be realigned using sound frequencies by a method developed by Simon Heather. From his website (http://www.simonheather.co.uk/media/men.mp3) it is possible to choose the male or female frequencies and listen to this 8-minute MP3-format track morning and evening. Everybody can listen to the sounds while sitting or lying down. The frequencies may sound quite discordant at first. Discordant sounds tend to move energy while harmonious sounds have a balancing effect. Dr. Jon Whale developed the Gem Lamp Therapy (Whale, 2009) realized with help of the electronic apparatus Caduceus Lux (the Lux IV and Stellar Delux). For more information, see http://www.whalemedical.com or http://www.theassemblagepoint.com.

8.8. Meditation

Meditation has been used throughout history by adherents practically of all the world's religions like as Buddhism, Hindu, Judaism, Christianity, Islam and Taoism. Meditation traditions include a plenty different practices, which are very difficult to classify, because they are closely related to the cultures that gave birth to them. According to Stephan Bodian, more than 10 percent of adults in the United State meditate regularly (Bodian, 2012).

Under the umbrella of this term there are a broad variety of mental practices, ranging from techniques developed to promote physical health and positive emotions, to exercises ensuring empowerment of mind control and relieving the stress. On the other hand, meditation can include the process of unlocking the door to subconscious and superconscious parts of mind.

Meditation allows to change mindset and correct mind abilities in the different circumstances and the environment. Using various meditative practices, a person gets the opportunity to develop new, more positive ways of thinking and understanding the essence of the occurring phenomena.

Generally, scientific information gathered from different sources indicates that meditation improves:

- the will power with the ability to focus, attention, work under stress
- the speed of information processing
- ability to think clearly, mental strength and flexibility
- emotional intellect
- ability faster learning, memory and self-awareness
- mood and psychological well-being
- mindfulness and executive control
- creativity abilities
- the healing effect, for example, mental improvements for old age
- and so on.

Meditation helps to get rid of mood disturbances, anxiety, depression or increased aggressiveness, can alleviate loneliness, hopelessness, and despair.

In the process of meditation, our brains can shift into more stable, stronger brain frequencies (called alpha and theta) normally reached during sleep. Changing brain frequencies to these levels while staying awake, a mediator can be able to bring the unconscious mind to the conscious level.

Brain studies in Buddhist monks have shown that the human brain in process of meditation has the ability to significantly change its functions (Danzigo, 2011; Josipovic, 2012). Meditation reduces frequency of waves of brain activity and allows us to concentrate our thoughts better.

In the last years, the use of meditation became more popular firstly because the increasing stress and anxiety of modern life stimulated a growing need for new effective methods of treatment and secondly because the Western world became more familiar with the centuries-old proven Eastern methods of influence on human health and mood.

There are many factors that indicate that meditations have a beneficial impact on the health of people in different age groups. Meditation not only reduces stress, decreases anxiety, enhances self-awareness, lengthens attention span, may reduce age-related memory loss, but and promotes emotional health, can generate kindness, may help to fight addictions, improves sleep, help control pain, and even can decrease blood pressure.

The studies at the University of California (Willer, 2012), at Harvard University (Hölzel, 2011), at University of Wisconsin, at EOC Institute (http://eocinstitute.org), etc. found that meditation actually changed the circuitry of the brain and beneficially enhanced the functions of some the brain regions associated with such qualities of the mind as attention, memory abilities, and even brain power. Meditation has been scientifically proven to synchronize left and right hemispheres of the brain, neural chemistry, and many intellectual abilities. This has shown also that meditation naturally and beneficially increases the neural mass (gray matter) of the brain by harnessing the brain's "neuroplasticity" potential. In particular, researchers from University of Oregon found that with as little as two weeks of regular meditation, the brain begins to build axonal density, which means a greater number of signaling connections. After a month of meditation, the number of signaling connections continued to increase, while an increase in myelin (a protective tissue around the axons) also began to increase (Schwarz, 2011).

State of relaxation occurring during meditation contributes to the regulation of various physiological processes in the body and to achievement of the inner peace of mind. It is known that people who regularly meditate are much happier and healthier than who doesn't do it. And they have greatly extended life spans because meditation dramatically reduces, and even reverses some diseases of the different types.

According to classification of Lutz, Travis and Shear, the many meditation techniques, which exist today, can be united in the three categories: Focused Attention, Open Monitoring, and Automatic Self-Transcending (Lutz, 2008; Travis, 2010).

Quality of meditation depends on its depths (or level) and goals. Depth of meditation is connected with frequency of brain waves. It is true that people who have meditated for many years and mastered advanced techniques, are able to slow down their brain waves to a deeper level than beginners. Usually meditators can only reach alpha waves the first years, but eventually they can master techniques that produce theta waves, which are rarely dominant in waking states.

Numerous meditative techniques can include the following components: repetition of any expressive word or phrase known as a mantra; breath tracking; concentration of attention on the flow of sensations; concentration on some geometric figure or for example, focusing solely on the flame of a lit candle; contemplation of the picturesque image of a sacred object or a saint;

cultivation of love and goodwill, empathy, forgiveness and other positive and healing emotions; hearing special music, and so forth.

Today, are widely used plenty meditation systems and methods such as Vipassana meditation, Zen, Taoist meditation, Transcendental Meditation, Silva Life System and others, which have similar benefits.

The most popular meditation technique in the West is known as mindfulness meditation, which is a combination of above-mentioned concentration and open awareness. The practitioner focuses on an object, for example, the breath, bodily sensations, thoughts, feelings, or sounds. This practice is often extended to daily actions like as housework, eating, or walking.

Harry W. Carpenter recommends in his book (Carpenter, 2014) Transcendental Meditation (TM), that demands use some mantras, which is used to interrupt the chain of chatter that goes on in mind. The results of serious studies demonstrated that transcendental meditation has an unprecedented impact on all aspects of human life - the development of mental potential, health, psychological state.

The Silva Method is the name given to a self-help program developed by José Silva. His techniques allow using the relaxed, healthy state of mind that occurs during meditation to solve your day-to-day problems. It is called "active" meditation. According to book (Silva,1977), this method allows students to receive insights and practice extrasensory perception (ESP). He also developed techniques to achieve what many people call "tapping the super-conscious" to solve problems, reach goals, come up with new ideas and solutions to problems. According to statistics published on the official website of the Silva method organization (http://www.silvalifesystem. com), at present, the method is being taught in over 30 languages in 110 countries, over 13 million people have graduated and benefited from the Program. The method used in sports, business, creative and many other areas has helped many people gain wealth, health, success in various fields, tolerance and many other positive changes. Among the famous followers of Jose Silva can be noted singer Madonna, a political star Margaret Thatcher, and American writer Richard Bach.

The choice of meditation method is determined by the individual characteristics of each person and his concrete tasks. Meditation can be practiced at any time, some types anywhere and even in the process of walking. Best of all every day to adhere to the same time. Like any exercise, meditation is effective only on the condition that is performed regularly. Many experts recommend the best time for meditation before sleep or right after awakening. Meditation eliminates the automatism of consciousness, the perception of the world becomes more profound and clear, it improves intuition, clears consciousness.

The main advantage of meditation is that using relatively simple exercise it allows everybody to combine and customize the efficient use of the three main cognitive components such as imagination, intuition, and emotion.

Meditation helps to reprogram the internal evaluation of past experiences (for example, grievances, disappointment, self-flagellation, etc.). This is possible because the brain does not distinguish the events of the external world from those that occur in our mind (Dispenza, 2012).

Meditation can be organized as a multi-stage process (Dalai-Lama, 2003), which according to different authors can include a different number of stages: 3, 5, 7, and more.

Apparently that technical progress has affected applied methods of meditation, first of all by use as a feedback of devices and programs for an assessment of activity of a brain before and after meditation. About it there will be a speech in following section.

Because the practice of meditation requires some dose of determination and will-power, many types of meditation are often practiced with in-person guidance at first, and then later with decreasing need for guidance.

CHAPTER 9

Intellect tuning

> It doesn't matter how beautiful your theory is, it doesn't matter how smart you are. If it doesn't agree with experiment, it's wrong.
>
> Richard P. Feynman

9.1. Intellect optimization as unused opportunities for enhancing quality and power of brain functions

The greatest portraitists of the 16th century German artist Hans Holbein the Younger was the King's Painter of English king Henry VIII in the period 1536-1543 years. According to legend, at one day Holbein got into a fight with an anonymous English earl and threw him down the stairs. The injured and offended earl complained to Henry VIII about his injustice, but instead of punishing Hans Holbein, the king admonished the whining courtier by reminding him, "I tell you, earl, that if it pleased me to make seven dukes of seven peasants, I could do so, but I could never make of seven earls Hans Holbein or any as eminent as he."

Over the past 500 years a lot has changed in the world, and today thanks to technical progress there are created some methods of teaching and training which can allow making from a simple peasant a good (sometimes outstanding) artist, engineer, scientist or actor. Unfortunately, now not exist methods to create geniuses from ignorant people.

Anyway, the idea to create a super brain for any person is very popular in the scientific world a long time. Unfortunately, different people put different content into this concept. Some people believe that super brain can learn faster, resist stress easier, focus and concentrate better, stay energized longer, and so on using diet, meditation and training (Chopra, 2013; Hutchison, 2013; Wood, 2017; etc.). Some scientists believe that a super brain can be created artificially with some implants in the human brain (Lebedev, 2018). Some suggested a method to multiply mental power in 21 days, with 21 simple and enjoyable exercises, others don't limit a time for enhancing intellectual abilities.

Unfortunately, existed learning technologies, meditation and hypnosis do not include the methods of direct enhancing of intellectual abilities of brain. Especially it concerns cognitive performance, including ability to focus/attend, retain, and recall information requires the brain to engage in a complex series of tasks.

Relatively recently, the possibilities of using brain neuroplasticity and genetic engineering have been discovered, the facility of epigenetics seem promising, but the problems of reliability of the methods used to affect the brain today are largely unresolved.

The super brain creation methods used to improve a person's mental abilities do not allow using the main brain reserves, simply because they cannot be measured using conventional direct measurement methods. Many studies have shown that our brain (mind) has a number of unused capabilities that can allow to improve the efficiency of its work many times over.

The measurement method used by the author allows not only to measure individual intellectual abilities of a person, but also to carry out their optimization on local or general mathematical models.

Currently, the most important unused opportunity is the possibility of directed optimization of the characteristics of the brain or mind (thinking), which can be implemented based on existed mathematical methods of optimization theory.

In the simplest case, the optimization problem is to maximize or minimize the real function by systematically selecting input values from the allowable set and estimating (measuring) the value of the function (optimality criterion). In the process of optimization of complex systems, there are three problems:

- Construction of the mathematical model of the system, which allows predicting its behavior in any given situation
- Formulation and formation of optimality criterion for the given system
- Choosing a mathematical method to solve the optimization problem.

Modern science has accumulated many methods that allow describing systems of any complexity with incomplete knowledge. For our goals, we use one of these techniques, based on the principle of black, gray and white boxes.

Principle of the black box is a term used in system engineering and cybernetics for description and modeling a system with a very complex or unknown working mechanism. Generally, to describe such a system for different environmental conditions, is used the output information as the reaction to the specific conditions and input. The term white-box is used to describe a system with known architecture or where the inner components or logic are available for inspection. Gray-box is a commonly used term to describe an object about which we do not have full knowledge. Based on the results of feedback, using principle of black and gray–box allows in the processes of optimization to achieve needed results.

The correctness of this approach is confirmed by the fact that in the realized systems of artificial intelligence many functions of the human brain have been simulated using black and grey-box principles. We can find examples of this realization in different devices and systems, including:

- Expert systems
- Sense and recognition systems

- Intellectual control systems
- Robots, etc.

According to the theory of optimization, several ways of mathematical modelling of complex systems with unknown structure and unknown functioning law are possible;

- Building a mathematical model of the system as a black box based on regression analysis of experimental data or other methods of statistical analysis
- Construction of a local mathematical model in the vicinity of the optimum search area allowing to define a direction of enhancing for the selected criterion of optimization (see method of Nelder-Mead, described below).

9.2. The goal of optimal mind tuning. Criteria and Target (Goal) functions

In our era of technical progress optimization usually are considered as procedures of modification of the system due to the variation of the certain parameters of the system for improvement of its efficiency. Optimization of design or characteristics of different systems are actively used practically in all areas of a science and technology, including medicine (Alves al, 2008; Hassanien, 2016). The number of parameters of objects of optimization may be various, and parameters can be changed continuously or discretely (discrete optimization).

The efficiency of a process of optimization of systems is estimated (if it is done on mathematical model) or measured (if it is done by means of the step-by-step algorithm on the actual object) using some certain criteria or functions called in mathematics the functions of the target (as synonym -the goal functions). For example, if the task is optimizing the car engine, the goal or the target function can be a power of engine or specific fuel consumption (per unit of output).

The problem of optimization of intellect cannot be stated in the form of the task: to optimize intellect on all parameters. It is the formulation of the multicriterial problem because the intellect is expressed by means of an assessment of efficiency of many intellectual functions: creativity, intuition, attentiveness, etc., and each of these functions maybe evaluated with help of one or several criteria. Usually multicriterial problems have only compromise solutions. Such solutions are called Pareto efficient or non-inferiority solutions; what means that if none of the objective functions can be improved in value without degrading some of the other objective values. Because of multicriterial problems complexity (Sartini de OliveiraI, 2010), we will not consider this task in its entirety. Below we suggest two simpler task formulations for multicriterial problem solution:

1. Transfer multicriterial task to one criterial task for example if in target function different criteria combine in a weighted linear function

$$\text{Target F= opt} \sum \alpha_i \cdot F_i(x_j)$$

Where:
Target F – target function
α_i - weigh coefficient

F_i – optimality criterion for intellectual function

x_j – varied parameter in the optimization process

2. For each specific person to choose as target function the most important for the current moment intellectual function

$$\text{Target } F = \text{opt } F(xj)$$

The goal of optimization includes not only a demand by varying the variable parameters to find a maximum or a minimum value of the goal function but and a condition to have a solution inside of the area of admissible decisions. To define boundaries of this area, the optimization task must include also a collection of constraints given usually at the beginning. Among the constraints for optimization of mind can be included admissible diapason of health parameters, mental characteristics, physical, chemical or other parameters, and even economic factors. For example, it makes no sense any optimization which requires unreal funding or huge amounts of time. The goal of the optimization process is considered to be achieved as soon as all these constraints were satisfied and value of optimizations criterion become optimal.

To formulate the problem of optimizing the human intellect correctly, it is necessary to take into account the following. Each person receives intellect from birth and developed this in the process education and upbringing. It is possible to improve or worse a human intellect, but nobody today even using genetic engineering can create an optimal intellect. Optimization is a very broad term, and it has been taken in many creative tasks related to the creation of new objects and systems.

For a human we cannot create a new optimal brain. Hence, in our case, it makes sense to talk about optimal human intellect tuning, and this definition we will use in next chapters.

From a mathematical point of view to optimize some characteristic (or function) of the brain, for example, general intellect ability, we can use such criterion, for example, as g-factor (G-factor), measured with help of pendulum. Because we do not know the latent unit, and real properties of a measured factor (for us is important comparative value of this factor), it makes sense to add to letter G an index p (from word "pendulum) – it will look like as "Gp".

The method described below for the optimization of G –factor of intellect is a step-by-step process of influencing the brain in one way or another, measuring intellectual functions as a brain reaction at each step of the optimization process

This allows to optimize not only G-factor of intellect, but and many components (functions) of intellect, for example, creativity, intuition, and even such condition of mind, as a stress.

Generally, depending on situations, the goals of the optimal intellect tuning may be different, for example:

- Increasing G-factor of intellect with help of Audio-visual entrainment (AVE) systems, optimizing such parameters, as color and pulsation frequencies of light signals, frequencies of sound, amplitudes, volume, phases and forms of signal in the various brain (mind) states
- Enhancing some intellect features such as creativity and intuition by dosed classical music hearing

- Minimizing a stress level
- Maintaining intellect and some its functions at a peak condition at long period of time.

Unfortunately, maybe, fortunately, our hardware –brain has not fixed structure because the number of brains cells and links between them can be increased or decreased depends on different influences (internal or external). This means that by acting on the brain (or body) one way or another, we encourage the work of the body to change the structure of the brain, for example, the growth of its cells and change the quantity and quality of links between axons.

On the other hand, the human brain is the open system, constantly exchanging energy and information with its environment. Beside it, the system can reorganize itself at another order depending on the circumstances.

This means that it is impossible to perform the optimal tuning the mind once and for all, it must be repeated periodically. On the other side experience of the Institute for the Awakened Mind, The Centerpointe Research Institute and others with regimes like as Awakened Mind or Super Awareness proved that we can after long time brake return to optimal combination of values varied parameters for last time tuning procedure and prolong optimal tuning again and farther.

According to hundreds of studies, EMF/RF radiation is a major culprit in a wide variety of health issues ranging from cancer, miscarriages, and birth defects to depression, dementia, and fibromyalgia. Therefore, when optimizing the tuning the brain, we must think about serious possible health disorders especially of the following population groups:

- Pregnant women
- Children and teenagers
- People with some chronic diseases
- People with mental illness or with unstable psyche.

We will return to this one more time in the section 10.2.

9.3. Means and parameters for the mind optimal tuning

In Chapter 6, we considered many exogenic factors that affect a power of human intellect: geo-space factors; biofields of plants, animals and other people; artificial sources of different kind of vibration, and so forth.

In recent years, it became clear that among these factors, the most promising were used in entrainment methods, realised with help of special devices.

A review of 20 studies on brainwave entrainment found that Audio-visual entrainment (AVE) systems, descibed in sections 6.4, is effective not only in improving cognition and behavioral problems but and, for example, in alleviating stress and pain (Huang, 2008). To describe "Audio and Visual Entrainment", it is also commonly used the following terms: "brain entrainment," "audiovisual stimulation (AVS)," "auditory entrainment," or "photic stimulation (Siever, 2000, Huang, 2008)."

The use of flashes of lights and pulses of tones purposefully guide the brain into various states of brainwave activity. By stimulating the brain with pulsing auditory tones transmitted through headphones within a different frequency range and/or flashing LED lights built into special eye glasses, it is possible to enhance brain function. Some users sure that audio or visual entrainment changes brain waves within seconds or minutes.

Although glasses (goggles) are the best way to entrain the brain using light, screen flashing can be useful in the absence of glasses. Screen flashing uses the computer screen to deliver there required pulses of light.

The AVE application "Mind WorkStation" amid systems developed by Transparent Corp. was chosen by the author as the most convenient means for demonstrating the possibilities and efficiency of brain tuning with the help of entrainment systems.

It should be emphasized that this does not mean that other technologies cannot be used for the same purposes.

To better understand the results of the researches and experiments, let consider some definitions below.

Types of vibrations include:

- Binaural beats
- Monaural beats
- Isochronic tones
- Modulations and audio filtering.

Binaural Beats. In recent decades, the use binaural beats has become very popular esp thanks to the research of the Monroe Institute and Hemi-Sync. The word binaural means "having or relating to two ears". From oscillations and waves theory we know, that beats is a phenomenon with the regular increase and decrease in the amplitude of, for example, sound waves caused by two waves of slightly different frequencies being superimposed in the same medium.

As a result of hearing two tones of different frequencies which sent simultaneously to the left and right ears, the brain perceives a new tone, based on the mathematical difference between the two frequencies. The brain then produces brainwaves at the same rate as a new tone, becoming entrained to that frequency. For example: If a sound of 200 Hz frequency is sent to the left ear, and a sound of 210 Hz frequency to the right ear, the brain will process those two frequencies and perceive a new frequency at 10 Hz. Observation of the electromagnetic waves of the brain shows that brain activity changes with the same frequency of 10 Hz in both hemispheres.

It is known that different auditory vibrations affecting the ears should be below 1500 Hz, and the difference between them should be no more than 40 Hz. The lower tone is called the carrier, and the upper is called offset.

In telecommunications theory, a carrier wave, carrier signal, or just carrier, is a waveform (usually sinusoidal) that is modulated (modified) with an input signal for the purpose of conveying information. This carrier wave has usually a much higher frequency than the input signal.

Many studies have proved the serious effect of listening to binaural beats. Binaural beats in frequency 1 to 4 Hz (the delta) and 4 to 8 Hz (theta) ranges have been associated with reports of relaxed, meditative, and creative states, and used as an aid to falling asleep. Binaural beats in frequencies 8 to 12 Hz (the alpha) have increased alpha brain waves and binaural beats in frequencies 16 to 24 Hz (the beta) have been associated with reports of increased concentration or alertness and improved memory.

Monaural beats are a type of oscillation when two oscillators tuned to different frequencies are combined and sent to a loudspeaker (or two separate loudspeakers) to mix them in open air. They produce a very precise beat that can be heard with both ears or with one ear.

Isochronic Tones. At its simplest level, an isochronic tone is a tone that is being turned on and off rapidly. They create sharp, distinctive pulses of sound. They are an effective auditory entrainment method because they elicit a strong auditory evoked response via the thalamus and most people find them tolerable. Widely regarded as the most effective tone-based method, isochronic beats produce very strong cortical responses in the brain. Many people who do not respond well to binaural beats often respond very well to isochronic tones.

Modulations and Audio Filtering. Modulating sound is a way to produce brainwave entrainment using something as complex as a musical track. In effect, this is "embedding" brainwave entrainment into the audio. Any sound can be used, from nature sounds to a classical symphony.

The method allows separate left and right hemisphere stimulation.

Audio wave simulations are not the only one way the brain can be entrained. In fact, the brain is affected by any kind of rhythmic stimuli. Clicks, drum beats, light flickers, and even physical vibrations or electric pulses have all been proven to effectively entrain the brain.

Harmonic "Box": This is a technical term for combining multiple forms of auditory entrainment, usually monaural beats with binaural beats. There is generally a difference in the rates of the beats from the right and left side. Some belief this protocol results in faster brain waves in one hemisphere and slower waves in another. This is a concept invented by James Mann (Stefanelli, 2014).

Usually AVE systems have an interface, that allow to change oscillations of sound and light (separately or together) by the following parameters: value of frequency, amplitude, volume, and phase for one or several sources of vibration to create entrain brainwaves like binaural beats, monaural beats, isochronic tones, and harmonic "Box".

By combining sound with light, it is possible to utilize eight different phase settings: in phase, out phase, cross phase, sound out phase with lights in phase, sound in phase with lights out phase, clockwise motion, counterclockwise motion and front/back phase.

Another method of brainwave entrainment is combination in one soundtrack brainwave entrainment tones within the audio of music. With this type of entrainment, nearly any sound can be utilized for the embedding of the tones. It could pick out the favorite pop song, nature sounds, or something simple like white noise. Typically, any forms of music modulation entrainment incorporate adjustments in rhythm as well and in some cases volume adjustments so that the music does not overshadow the entrainment stimuli. Generally, there can be some distortion of audio, but distortion is relatively insignificant. In the MindWork Station, there are two types of tracks: Content Tracks, and Entrainment Tracks. While Content Tracks are used to contain sound files, tones, and effects, Entrainment Tracks have a purpose: to specify brainwave entrainment frequencies across a timeline. If a content track is associated with an entrainment track, the content will be used as a carrier to stimulate the brain. For example, an entrainment track is associated with tone content, and then binaural beats, monaural beats or isochronic tones will be the result. The same can be done even to sound or white noise tracks - Mind WorkStation uses a filtering process to apply entrainment to all types of audio and visual content.

Glasses equipped with light-emitting diodes (LED), usually named as LED Glasses are able to generate quick pulses of light that entrain the brain. Certain types of glasses can be used with eyes open, but most were developed to be used with the eyes closed. Usually, these glasses included with a "mind machine" but may be purchased to enhance auditory entrainment. LED glasses allow to change the following visual parameters: color, contrast, intensity, and so on.

Flashing screen/light: now are created special computer programs that may produce subtle visual entrainment effects by creating a "flashing screen." In other words, your computer screen is capable of flashing at a certain rate to entrain your brain waves. Some software even includes 3D visualizations that can be used with open eyes. Various visualizations at "pulsed" rates may also be effective for entraining the brain to the desired frequency. Further research is needed to verify visual pulsation formats of entrainment.

The following devices allow having together light and sound brain stimulations: Kasina and Procyon systens (MindPlace.com), AudioStrobe (http://www.audiostrobe.com), etc.

Biofeedback is carried out with the help of special devices that measure the parameters of the human body functioning before and after different types of impact. Used in these devices sensors attached to the different parts of the body allow to monitor various physiological signals. Some of the most applicable biofeedback today are:

- Electrical and thermal skin resistance measurements
- Human body temperature measurement system
- Respiratory feedback that monitoring the respiration process
- Heart Rate as is the speed of the heartbeat measured by the number of beats of the heart per minute (bpm). (Can be measured for example by Fingertip pulse oximeter)
- Electroencephalography, electrocardiography, electromyography, electrooculography, etc.

Restrictions during optimal tuning process

Restrictions can be two types: restrictions of varied parameters, connected with possibilities of measuring or impact devices, or connected with physical or psychological conditions or reactions of the person (patient).

For example, there are lower and upper limits of audibility for ears. The lower limit of hearing is called the auditory threshold (or the absolute threshold of hearing), the top - the threshold of pain. The pain threshold - the maximum sound pressure, which is perceived by the ear as sound. The similar lower and upper limit exists for any sense organs.

Although experience shows that entrainment methods has been found useful in the treatment of a wide range of problems, including learning disabilities, anxiety, depression, tinnitus, headaches, pain management, and sleep disturbances, some procedures of search for optimal tuning can be a source of temporally headache and other disorders for some people.

Effect of the brain entrainment can be created using two and more oscillators, that can form binaural and monaural beats, isochronic tones, and so on. Here can be used also modulations and audio filtering systems.

9.4. Math methods and results of optimal tuning an intellect by G-factor

When experimental optimization of a complex system has place with multiple parameters influence, can be used two main strategies:

1. To change parameters in turn (to reach an allowable optimum of the target function by changing a first parameter, then at the received optimum value of this parameter to pass to another and so on, until then there will be used all parameters.
2. To change at the same time all varied parameters according to algorithm of optimization. The application of both methods will be shown in this and in the next sections.

The second strategy is called Design of experiments (DOE), and many researchers used this as a cost-effective way to solve serious problems afflicting their operation, because DOE does not require the testing all possible values in the different combination directly.

Described below results of experiments represent differents variants and possibilities of optimal tuning the intellect. Optimization was carried out by two ways: first for one parameter of optimization, for example –frequency of the audio signal, second when a number of parameters were 2 and more. In the second case was used the mathematical method of Nelder-Mead in its simplest modification.

Unfortunately, the volume of this book does not allow describing the possible mathematical optimization techniques. Obviously, the choice of the best method will depend on how to set a task of optimization of the brain function and what are chosen criteria. The fact, that the brain can change its structure in the process of search of an optimal solution, greatly complicates the process of optimization. Therefore, to find a solution most likely need to spend several iterations.

During my engineering and scientific practice for purposes of multi-parameter optimization I many times used the method Nelder-Mead due to its simplicity and ability to give quick results (Iserlis, 1985,1986).

The Nelder–Mead method (named also as "downhill simplex method", or "amoeba method") that was proposed by John Nelder & Roger Mead, is a commonly applied numerical direct search method used to find the minimum or maximum of an objective function in a multidimensional space (Nelder, 1965). In mathematics, a multidimensional system is a system where are several independent variables can be varied.

Multi-parameter optimization techniques allow significantly reduce the search time (therefore and cost of search) of optimal values of the function, particularly if such function has several extreme. Here one more reason. Any impact on the brain in process of optimization needs a time for changes in the structures and cells of the brain, theoretically different for different peoples.

Remerbering the admonition "The more a book has formulas, the less readers it has" we try to explain shortly a gist of this method.

According to the Nelder–Mead method, user constructs a pattern of n + 1 points in dimension n, which moves step-by-step across the surface to be explored, sometimes changing size, but always retaining the same shape. According to Wright, "it remains remarkably popular with practitioners in a wide variety of applications. In late May 2012, Google Scholar displayed more than 2,000 papers published in 2012 that referred to the Nelder– Mead method, sometimes when combining Nelder–Mead with other algorithms (Wright. 2012). Nelder-Mead has one disadvantage- it cannot find a global extremum of optimized function. In this case is needed combination with other optimization methods.

As research has shown, this method has shown a positive effect in the process of optimizing the intellect and its functions.

The main goal of described below experiments was to show only possibility for brain tuning by changing one, two and more varied parameters.

To calculate the relative effect size (RES) of the optimal tuning, for any target function a simple relative change can be estimated by using the difference in the values, measured at the start and on the finish of measurements.

By analogy with formula (3-2) from section 3.8 we will evaluate any result of optimization as

$$RES = 100(F_{opt} - F_0) / F_0 \ \%$$

Where: Fopt – optimal (or maximum in the series) value of criterion (target function), F_0 –value of criterion (target function) on the start, RES- relative deviation of function value.

Using RES allows to decrease errors of the operator working with a pendulum, and correctly estimate effectiveness of influence of all varied parameters change.

Optimization of one parameter via a regular search of versions.

As we wrote above, the intellect can be estimated in the first approximation with the aid of a generalized G-factor (Gp-factor), which was measured with the help of a pendulum and the new corresponding scale Fig. 3.15 calculated on the basis of the R-scale (0-1000) or, if necessary, on the basis of any other R-scale that can be developed, for example, based on the R-scale (0-10) -Fig. 3-1 or R-scale (0-100).

In experiments represented on the Fig. 9.1 are demonstrated the results of optimization of Gp-factor via the impact of monophonic pure-tone audiosignals with different frequencies, taken online from sound generator (onlinetonegenerator.com).

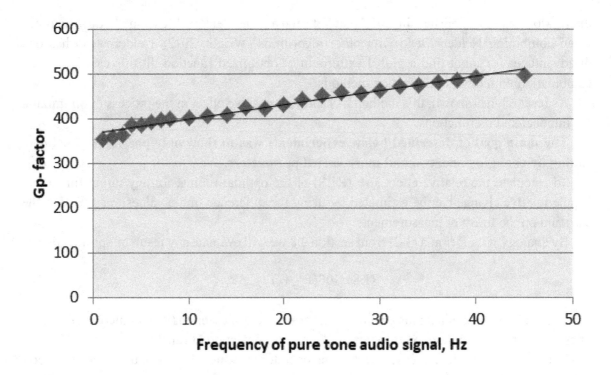

Fig. 9.1. Optimization of Gp-factor by changing frequency of pure tone audio signal with triangle waveform (volume-average). Any session equals 20 min. RES= 40%

Sometimes for optimization of intellect and its component can be used an approach of the brute-force search or exhaustive search, also known as "generate and test approach". It is a often used problem-solving technique that consists of systematically enumerating all possible variants for the solution and checking whether each variant satisfies the problem's statement. This approach used for optimization of creativity level.

In experiments, represented on Fig. 9.4, demonstrated the result of optimization of Gp-factor by changing a frequency of offset for given frequency of carrier, equal frequency of Solfeggio (Ascension)– 528 Hz, one of six Solfeggio frequencies, the use of which has become fashionable at last decades to affect the health and psyche of people (see: https://attunedvibrations.com/solfeggio).

Optimization by 2 parameters – frequencies of offset and carrier for binaural beats with help of Mind Workstation (Fig. 9.2).

Impact by binaural beats. Variable parameters: Frequency of carrier Fc and Frequence offset beats Fo. Volume =4, waveform –sin. Variable parameters at the beginning are: Fc=8 Hz, Fo=180 Hz, function Gp=37, predetermined increments for parameters: ΔFc= 2Hz, ΔFo=200Hz, parameters of optimum: Fc=1.25 Hz, Fo=468 Hz, optimal value of target function Gp=50. RES =35%. Optimum Gp=50 got on the 10th step. One session by 20 min, one time per week.

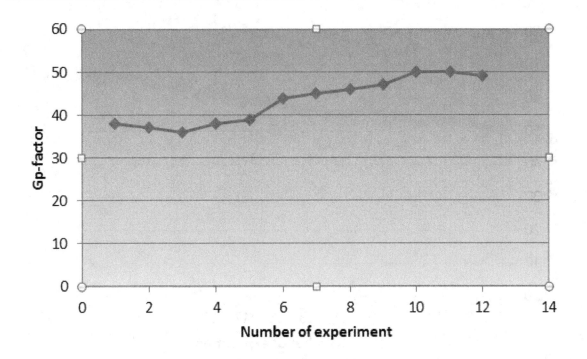

Fig.9.2. Optimization by 2 variable parameters – frequencies
of offset and carrier for binaural beats.

Optimization by 3 parameters-Volume (V), phase (Ph) and intensity of audio signal (In) with help of Mind Workstation represented on the Fig.9.3. This graph demonstrates an impact by binaural beats. Here are waveform –sin, variable parameters at the beginning are: V=5, Ph=180, In=50, Fc=12Hz, Fo=600Hz, predetermined increments for parameters were: ΔV= 1, ΔPh=40, ΔIn=30; parameters of optimal point were: V=0.1, Ph=243, In=85, Gp=69. RES =21%.

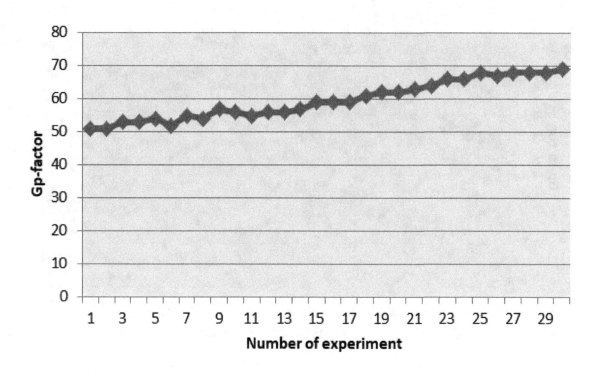

Fig. 9.3. Optimization by 3 variable parameters

Optimization by 4 variable parameters using a Harmonic box of Mind Workstation represented in the Fig. 9.4.

Here are created optimal combination of tones forming a chord. Variable parameters at the beginning are: V=2, Ph=250, In=100, frequencies F_1=10Hz, F_2=100 Hz, F_3=200Hz, F_4=300Hz, on the start value of target function Gp=78, predetermined increments for parameters: ΔF_1= 5, ΔF_2=50, ΔF_3=50, ΔF_4=50 ; parameters of optimal point: F_1=0.25Hz, F_2=400 Hz, F_3=131Hz, F_4=378Hz, Gp=114. RES = 46%.

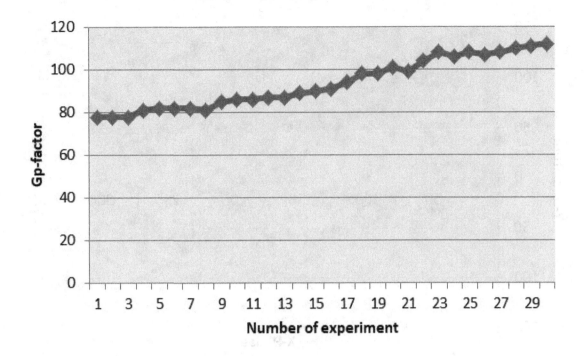

Fig. 9.4. Optimization by 4 variable parameters

Optimization of waveform for binaural beats by 3 varied parameters for any semi wave (Fig.9-5.)

Alpha wave frequency= 12Hz, tone-binaural beats, carrier frequency =200 Hz, volume =2, varied parameters: y-coordinates for 3 nodes of the positive semi wave and symmetrical negative semi wave. The initial form of a wave was the sine. Criterion – Gp-factor. The experiment included 10 sessions (30 min) with defined values of varied parameters, one per day. In the end of optimization, I received the waveform that looks like "optimized square for Autopan". The value of intellect level increased from 72 to 81, RES= 12%. Variable parameters of sine waveform at the beginning demonstrate on table 9.1 and Fig. 9.5:

Table 9.1

X	Y
36	60
90	100
141	57
216	-60
270	-100
324	-57

203

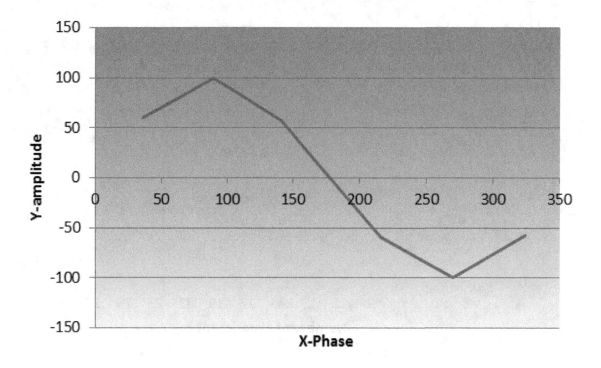

Fig. 9.5. Approximation of sinusoid by discrete form

After optimization parameters of waveform was received a form, demonstrated in the Table 9.2 and the Fig. 9.9, and results, demonstrated in the Fig. 9-10.

Table 9.2

X	Y
36	95
90	99
141	110
216	-100
270	-99
324	-95

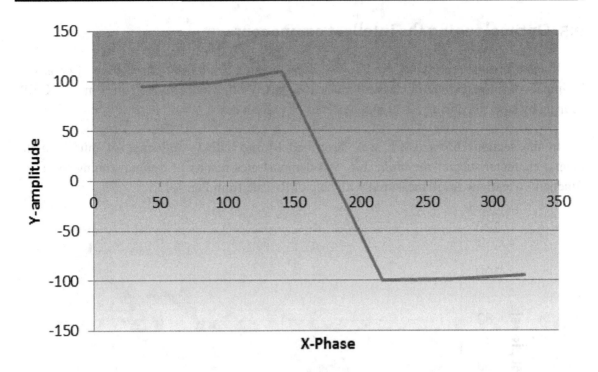

Fig. 9.6. The optimal wave form

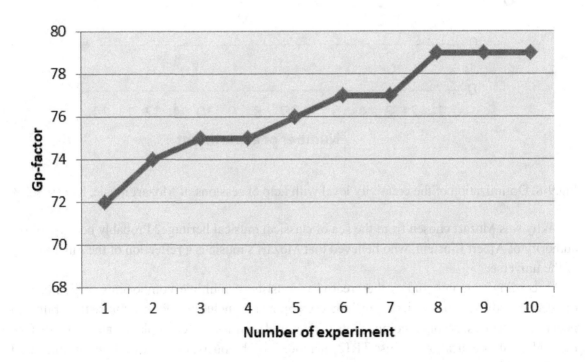

Fig. 9.7. Optimization of waveform for binaural beats by 3 varied parameters. RES =16%

9.5. Optimal tuning the intellect components

In experiments, represented on Fig. 9.8, demonstrated results of optimization of creativity by impact of some sessions of classical music (Mozart. Violin Concerto No.5 in A major, K.219) hearing by a person during 20-30 minutes ones or twice a day.

In the sections 6.7 and 8,1, was described Mozart effect –influence of music on the mental characteristics of the brain. Fig. 9-8 demonstrates results of optimal tuning by target function –Creativity level, measured with help of R-scale from Fig. 3-13.

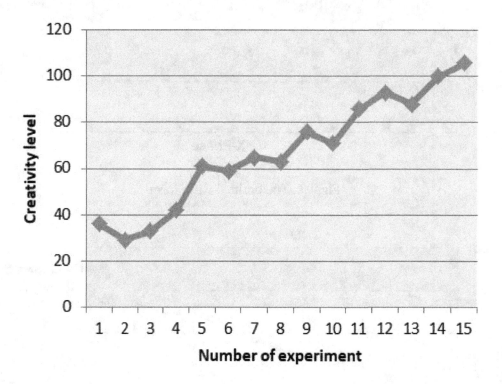

Fig.9-8. Optimization of the creativity level with help of sessions of Mozart music. RES = 863%

Why was Mozart chosen from the sea of classical musical heritage? Probably because of the authority of Albert Einstein, who believed that Mozart's music is a reflection of the inner beauty of the universe.

According to many studies the creativity consists of multiple components, which can be divided into the 2 groups: first - an innate components including, for example, the ability to generate new ideas; second - components of acquired abilities, for example the ability to analyze these ideas after education and use TRIZ method. It is obviously, in process of our optimization are changed innate components.

Fig. 9.9. demonstrates results of optimization of the intuition level with help of sessions of Mozart (Violin Concerto No.5 in A major, K.219) and Beethoven (Violin Concerto in D. major, Op.61) music, measured with help of scale from Fig. 3.12.

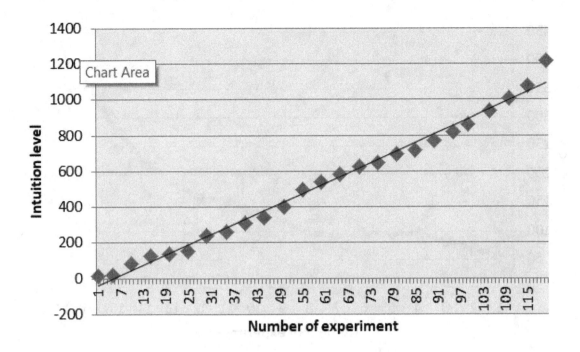

Fig. 9.9. Optimization of the intuition. RES = 8,046%

In both experiments, the maximum values of the target functions could not be found. The works were stopped because in these and some of the experiments below, the main goal was to show the possibilities of optimal brain setting.

The results of the measurement of the intelligent component like creativity and intuition, also as stress, willpower and so on, described below, proved also 3 main qualities of used measurement method: objectivity, reliability and validity.

9.6. Optimal tuning a willpower

Method for the determination of the access code to the subconscious mind allows using measuring ability of the brain to measure certain characteristics of a person or his intellect which only recently seemed impossible. For example, this method allows to determine the access codes for the measurement of willpower and vital energy index.

Experiments showed that willpower (Wp) can be increased by impact of festal or some types of pop-music, and especially revolutionary songs like Anglo-American Yankee Doodle, Italian partisan Bella Ciao, French La Marseillaise, and so forth. Fig.9.10 demonstrates relative change in willpower (Wp) in percent in process of optimal tuning by hearing Bella Ciao or Marseillaise songs in various performances once a day for 20-30 minutes.

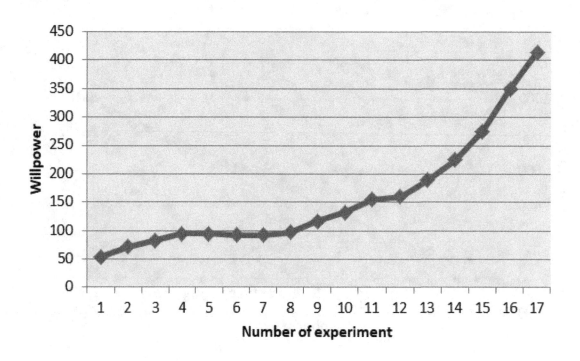

Fig. 9.10. Willpower optimal tuning. RES=1085%

The results of this experiment do not cancel and contradict the theory of Walter Michel, the essence of which is that the behavior of a person depends to a large extent on the circumstances. Moreover, it is obvious that the ability to increase the willpower under the influence of music or through special training is determined by flexibility of a person's gene system. Anywhere any person has a some reserve for increasing his willpower.

The described way of increasing willpower opens the prospect for many people to maintain their willpower at the proper level throughout life. Many people do not know and are unaware of the existence of this reserve. It is worse when they decrease this reserve accepting their life circumstances in the role of victim. About this the Nobel Prize-winning poet Joseph Brodsky said in his speech to graduates of Michigan University in Ann Arbor in 1988 year: "At all costs try to avoid granting yourself the status of the victim....No matter how abominable your condition may be, try not to blame anything or anybody: history, the state, superiors, race, parents, the phase of the moon, childhood, toilet training, etc. The menu is vast and tedious, and this vastness and tedium alone should be offensive enough to set one's intelligence against choosing from it. The moment that you place blame somewhere, you undermine your resolve to change anything;" (Brodsky, 1997).

9.7. Measurement of vital energy index

No need to explain that intellect as an operating system cannot work without energy. Without pretending to give a correct name, let's call it Vital energy (see 5.1). And if we want to somehow improve the intellect, tune it or control it, we must be able to measure this energy and evaluate the available reserve before carrying out certain procedures.

As far as I know, in Western science no one could find a formula for this energy[5], sources of this energy are known in general terms, unclear the main carrier of this energy and potential, such as voltage or pressure, which determines the energy flow direction. Described above vibration series method allows to assess a lever of this energy. As the results of the experiments showed, this method so far allows us to measure not the life energy itself but the radiation correlated with it, let name it as Vital energy index (VEI). The basis for this conclusion are the next facts: the readings do not depend on the mass, the radiation of the solution does not depend on the concentration of the radiating component. Because VEI can change in the very wide diapason was used logarithmic scale and such unit as decibel.

Researches with VEI measurements demonstrated the following:

- Vital energy has electro-magnetic, informational (see table 9-3) and sexual (for human and animal) components
- All alive and non-alive objects have different VEI
- We can find on the surface of the Earth places with high level of VEI (Place of Power) people, plant and animals can exchange vital energy between them
- For transferring energy from one to another person sometimes it is enough a hug, touch erogenous zone, handshaking, sometimes a glance.

The term Place of Power was introduced by Mexican Indian Juan Matus, described in the books of Carlos Castaneda. Places of power are geographic zones, sometimes large, sometimes small, which possess high level energy fields.

According to esoteric literature, places of power are subdivided into positive and negative ones. To the first category can be assigned the places which give their energy to people, to the second — the places which take energy.

---

[5] Einstein's energy formula $E = mc^2$ provides very little information for estimating human energy

Table 9.3 Vital Energy Index (VEI) for different signs

N	Type of symbol	Symbol	VEI in Db
1	Letter	**W**	10
2	Number	**9**	20
3	Geometric symbol	◯	35
4	Platon solid		40
5	Yupiters sign	♃	55
6	Ancient sign		65
7	Oven (Zodiak) sign	♈	75
8	Cho Ku Rei (Reiki) symbol[6]		90
9	Fingerprint of God - Rodin Vortex Mathematics basic Diagram[7]		600
10	Marko Rodins torus		900

VEI can be increased in Place of Power, charging energy for some trees like oak, birch or sequoia, bearing items like locker with signs from table 9.1, etc. Vital Energy Index (VEI) can be increased also as willpower by impact of festal or some types of pop-music, and especially revolutionary songs like Bella Ciao, Marseillaise, and so forth, by a hug and so on. Fig. 9.11 demonstrates experiments of assessment of impacts of same type of music on the deviation of the Vital Energy Index (δVEI), which was measured using a special chart, combined with access code and shown in Fig. 9-12. In the process of an experiment a person heard this music one time per day during 25-30 minutes in same time for evaluation of Willpower.

---

[6] Cho Ku Rei, which is considered one of the first symbols used in Reiki teaching, means "Placing all the powers of the universe here".

[7] Marko Rodin – American mathematician not recognized by official science, but having followers

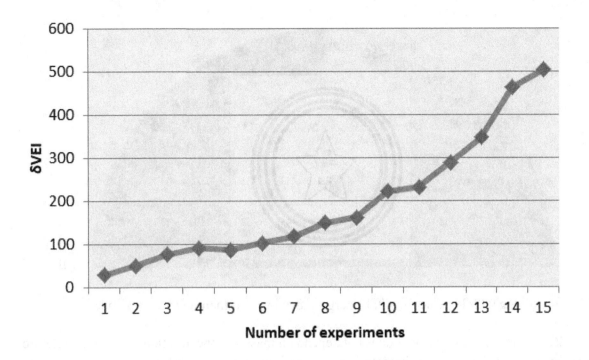

Fig. 9.11. Optimal tuning a Vital Energy index. RES=500%

$$\delta VEI = 100 \ (VEI - VEI_0)/VEI_0$$

Where: δVEI – relative deviation of Vital Energy Index, %

VEI - current measured value of Vital Energy Index,

VEI_0 - initial measured value of Vital Energy Index for the first measurement.

Experience has shown that this energy in reality can reach very high values. Therefore, it is more convenient to measure it using scales in decibels. Universal logarithmic units like decibels widely used for quantitative estimation of parameters of various audio and video devices. In radio electronics, in particular, in wire communication, in the technique of recording and reproducing information, decibels are a universal measure.

Therefore VEI index can be expressed in decibels according to the formula:

$$VEI \ dB = 10 \ log \ (VEI/VEI0),$$

where VEI – measured value of a parameter, VEIo- initial value of this parameter, dB- decibel

The use of the logarithmic scale and decibel as unit may be justified by the fact that inside the brain this scale is used according to the Weber–Fechner law which relates to human perception and states that the subjective sensation of a human is proportional to the logarithm of the stimulus intensity.

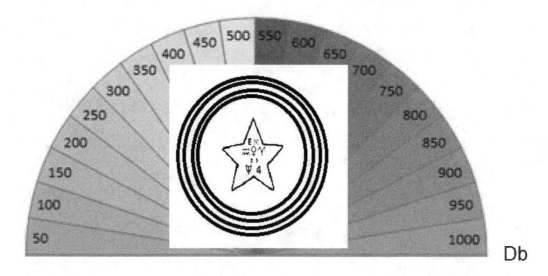

Fig.9.12. R-Scale for Vital Energy Index measurement with access code

Studies have shown that VEI does not depend on mass or concentration a solution, therefore we can assume that this index characterizes and measures energy not by quantity but by radiation level.

VEI is very sensitive to the influence of external fields, described in Chapter 6. For example, in the table 9.4 is demonstrated influence of choral karaoke singing songs on VEI during about two hours, measured with help of R-scale (Fig.9-12).

Table 9-4

N	Person initials	Before singing	After Singing
1	B.S.	120	300
2	I.T.	50	300
3	V.S.	100	300
4	L.I.	400	600
5.	N.N.	50	150
6	L.G.	250	350
7	D.N.	40	250
8	V.S.	50	150
9	L.K.	30	150
10	Z.G.		150

On the other hand, several characteristics of measured biological and not biological objects are very similar with characteristics of orgone energy discovered by Dr. Wilhelm Reich, an Austrian psychiatrist, who researched this energy in the earlier half of the 20th century. While conducting his research, Dr. Reich found that orgone is attracted and hold by organic substances, while orgone is attracted, but then immediately re-radiated, by metallic substances.

Nicolas Tesla seemed to agree with the orgon hypothesis. He in his last interview with a journalist called John Smith in 1899, said the following "… I have nothing against Mr. Einstein. He is a kind person and has done many good things, some of which will become part of the music. I will write to him and try to explain that the ether exists, and that its particles are what keep the Universe in harmony, and the life in eternity. Here meaning of ether is very close to meaning "orgone energy".

In any case, this radiation and method of its measurement requires additional studies.

9.8. Aging and optimal tuning

It is known that the intellectual abilities of any person are not permanent during the day, season and even life.

In section 6.3 was shortly considered influence of age on intellect. But today there is very little information about changing intellectual abilities during the day or week. In section 6.5 we described ultradian and infradian rhythms of the brain and human body. It was interesting to check how ultradian rhythms impact on characteristics of intellect. Results of these experiments, executed during about 10 hours daytime, demonstrated on Fig. 9.13 - 9.15.

Graph on the Fig 9.13 proved that creativity level (Cl) can vary during the daytime within 20 percent depending on many reasons.

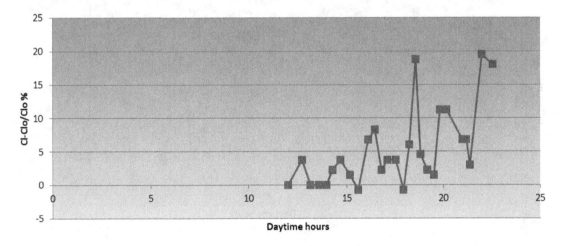

Fig. 9.13. Relative deviation of Creativity level (%) from the start point during the daytime

Graph on the Fig 9.14 demonstrates that stress level (Sl) can vary during the day time within 20 percent depending on many reasons.

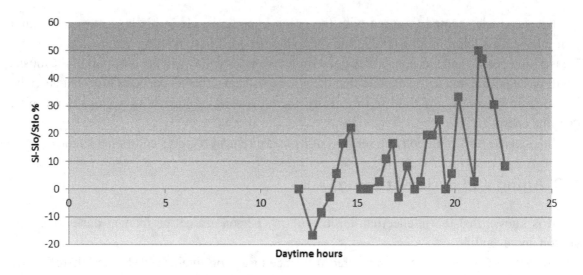

Fig. 9.14. Relative deviation of Stress level (%) from the start point during the daytime

Graph on the Fig 9.15 proved that intuition level practically does not change during the daytime (about 1%).

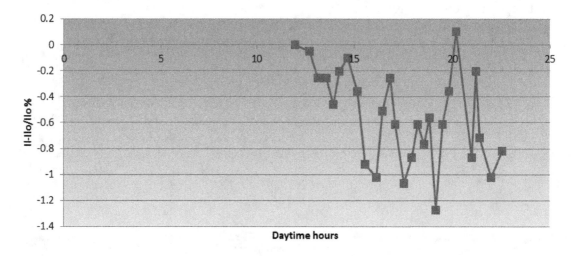

Fig. 9.15. Relative deviation of Intuition level (%) from the start point during the daytime

The same is concern the intellect, such parameter, as Gp did not change during the daytime.

It is very important to know how long the optimal adjustment is maintained in the course of a person's life.

The Table 9.5 shows how parameters of tuning are kept for a long time (from 7 to 13 months).

Table 9.5

Name of criterion	Symbol	Data of measurement	Value
General intellect factor	Gp	11.9.2016	496
		01.05.2018	445
		08.09.2019	500
Creativity level	Cl	11.16.2016	100
		08.09.2019	70
Intuition level	Il	05.17.2017	3196
		01.05.2018	3060
		08.09.2019	5500

According to table 9.5 results of optimal tuning a mind practically kept a long time. Anyway here is needed more observations.

9.9. Some words about random errors of measurements

Here makes no sense to do a rigorous mathematical analysis of random errors of measurements, including calculation of the random-variable distribution law, assessments of standard deviations, correlations, regression formulas, etc. for measured parameters or the factors, because this would make it difficult for the majority of readers to understand the results of the measurements.

On one hand dowsing method has disadvantages because don't allow to compare results with standard unit, on the other side this result don't depend on any rules of evaluations as for psychometric tests.

If to think about it, dowsing method by its accuracy locates somewhere between methods of physical measurements and less accurate methods of psychometric assessment.

Analysis of results of applying the method for determining some parameters of the intellect demonstrate that these measurements met the demands to objectivity, reliability reproducibility and validity, because of: measurements of trend g-factor of intellect in the process optimization by one parameter (for example a frequency) during about 3 moths (August-November 2016), intuition during about 2 months, Creativity level during 2 weeks, etc. laid on the one curve, different for different parameters of measurements. See: Fig. 9-9, and Fig. 9-10.

In experimental science, there is a numerical measure of confidence in the results, according to which the measurement can be trusted if the measured value exceeds the measurement error at least five or six times. Results of measurements above presented showed that these results can be trusted.

According to the Table 9.5, measurements of g-factor made on November 11, 2016 and on August 9, 2019, practically the same. For creativity and intuition levels the difference are more.

In real life subjective measurements allow optimize main functions of intellect and its components.

Described result of experiments demonstrated, that graphically functions of intellect behavior as a rule looks like a mathematically smooth functions. This indicates low random measurement errors.

CHAPTER 10

Intellect and mind control (management)

Your mind is your instrument. Learn to be its master and not its slave.

Remez Sasson

A wolf tells her cubs: "Bite like me" - and that's enough; a rabbit teaches her cubs "run away like me" - and this is also enough; but when a man say his child "Think like me", it is a crime.

The Strugatsky brothers

10.1. The necessity for the mind (intellect) control

The methods of optimal mind tuning can allow to adjust a brain as the tool for the decision of practical problems, but they are not enough to solve the diverse challenges that any person in life and at work encounters.

First of all, the complexity and interconnectedness of modern problems often does not allow one person to fully understand them due to lack of knowledge. This can happen when a person alone or as part of a group finds himself in a difficult life situation threatening with loss of property or even death. His mind must help him. On the other hand, designing many machines or systems in diapason from kitchen appliances to spacecraft requires knowledge of engineering, applied sciences, information systems, marketing, human psychology, economics, and so on to create a product competitive in the market. If add, that way from idea to end-product can be very long and dangerous. In this case, this person also needs a good mind. Unfortunately, our brain is very delicate instrument which not always can work as we want, depending on the set of both internal and external causes and circumstances.

According to a computational theory of mind, which accepts that mind/brain is completely analogous to software/hardware of computers, we can assume the following. Because in described in the previous chapter elements of technology of the intellect optimal tuning we used only physical impact on intellect (mind) like lights and sound oscillation or music, and the resulting changes in intellectual abilities persisted for a long time (a year and maybe more), so you can think that the optimal tuning refers to the hardware of the brain (increasing a number of brains cells, and their interconnection). To solve the problems described above, in some cases it can be more effective to use methods of influencing on the mind (software) as well as it is done in computer technology-by its reprogramming.

Below we will not consider one more conception of mind control which refers to a process in which a group or individual uses unethically manipulative methods to persuade others to conform to the wishes of the manipulator, often to the detriment of the person being manipulated (Biden, 2017).

The concept of Mind Control was suggested and developed by Jose Silva in his book "The Silva Mind Control Method", published in 1991, and other books published later. According to Silva Mind control method it is self-control method. According to John Biden, here are two different things, Silva Mind control method is method of self-control.

At present, if a person has a powerful computer and uses it once or twice a year, his friends or coworkers can find him very weird or even crazy. As we have already found out the human brain is more powerful than any computer and yet most people do not know how or cannot use their brains at full capacity. George Bernard Shaw wrote: "Few people think more than two or three times a year. I have made an international reputation for myself by thinking once or twice a week."

Here is one more a real story transferred into a legend. One student in the lab of the famous British physicist Ernest Rutherford was very hard working. Rutherford had noticed it and asked one evening: "Do you work in the mornings too?"

"Yes," proudly answered the student sure he would be commended.

"But when do you think?" amazed Rutherford.

Ernest Rutherford actually restricted the hours his students worked, and thanks to him, many of them went on to win Nobel prizes.

The moral of this story, cited in many articles and books, is following. A stream of thoughts is not a thinking. People do not always think. Usually a person thinks in three basic cases when:

- performing analysis to select from available options
- trying to understand (comprehend) the information received
- creating something new.

In other cases of life, a person may not think, but only looks through thoughts from the stream of thoughts as usually do users browsing on the computer the flow of postal messages or information in social networks, like as Facebook or Twitter. If he has to act to meet his needs, then he can use programs of various actions recorded in the brain, received at birth (instincts) or as a result of life experience or training. Among these programs can be a programs of eating food, hygiene, driving a car, and even program of activity for a certain kind of professions. If it is not enough he can use a brain of other people asking for example an advice.

Therefore it is understandable why Rutherford asked his question to the student. For him it was obvious that the student does not know how to think.

In the process of thinking, almost all the components of the intellect described above, and especially such cognitive components as analyzing, comparing, assuming, inferring, questioning, contrasting, evaluating, etc., should be involved.

If we again compare our brain with a computer, it's easy to make sure that when we turn on our computer we can use all of its functions in any sequence, but we cannot get it from our brain even if it's really wanted. Of course one of the main reasons is the lack of good instructions for managing the brain.

Today there is a huge demand for the skill to operate the brain or brains of members of a team. In connection with the advent of robots many professions have already disappeared or will disappear in the nearest future, because these professions didn't demand abilities to think. Therefore, there is an urgent need to ensure the possibility of solving the following problems:

- How to force the person to think?
- How to think faster and better?
- How to overcome fear, anxiety or other strong emotions, if they impede to think
- How to develop critical thinking skills
- How to read faster and comprehend better at the same time?
- How to remember all needed, etc.

Today, the science of brain control is still in the stage of formation and development, in this field thousands of scientists and specialists of different professions work, sometimes simultaneously building their own business in this fashionable direction.

It is already clear that it is necessary from childhood to teach a person to think in any situations, and our schools do not yet know how to do this. We can only talk about some attempts to solve this problem with the help of charitable foundations (for example, the Goldie Hawn Foundation). In California were founded The Center for Critical Thinking and Moral Critique and the Foundation for Critical Thinking — two sister educational non-profit organizations, which work closely together to promote educational reform.

The general task of mind (intellect) control is to force the brain to think effectively as long as it is necessary and using all reserves of memory and sub and super consciousness.

For realization of this goal should be used such methods that allow organizing the brain's work so that a person can ensure maximum productivity of his thinking without harm to his health. This means that the effect of improvement of the mind may be received due to the best organization of work of a mind – right distributions of functions between parts of the brain and right organizing effective interoperability. By analogy with a musical instrument like the piano, brain-boosting technology is pianos tuning, mind (intellect) control concerns playing practice according to piano rules and skills.

The process of mind control is a difficultly formalized process of improving the work of mind for a particular person whose results depend on many factors, including intellectual abilities, willpower, character traits of this person, situation and goals of management.

The cognitive component of the mind is one of main parts of mind. If we consider only this function of thinking, then the task of the intellect control is maintenance of concentration of the attention, extraction or delivery of the necessary information from resources of memory or from external sources, realization of an assessment of reliability of the information, an assessment of resources of time, etc.

The human mechanism of thinking is created so, that a part of thinking is some emotional functions which informs us of how we are doing (well or poorly) in any given situation or set of circumstances. It is impossible to forget about our ultimate driving force - willpower (volition) which play a key role in determining our behavior.

The more developed the intellect, the more it requires adjustment. If an individual cannot manage nor his thoughts nor his wishes and emotions, he can have many troubles in the simplest life situations. With the big degree of reliability it is possible to approve, that if the person has no intellect under the control, he has a great chance of being either in a madhouse or in prison.

New time demands new methods of thinking. Most people, when common sense cannot help, used to solve the serious problems either trial and error method, either intuition or the advice of other people. Today here are many possibilities to break this habit and establish a new way of thinking, although this is not an easy task, and there are no simple decisions to implement it.

Our brain is very complex system and beside of operating systems its software (by analogy with the computer) includes a multitude of knowledge obtained from birth and acquired during training, learning or through experience. Some of this knowledge is programmed very similar to computer programs. By methods of creation, these programs can be divided on the model-driven or data-driven. Model-driven is the way everybody learned to do in the school or other institutes, data-driven is programs created in the brain basing on the experience. Thanks to these programs, the minds of twins may have a big difference. The older are these twins, the more difference.

Most of these programs locate in the subconscious and their code does not have direct access to the change. The older a person the more he/she has obsolete programs, which don't meet the demands of new circumstances of his/her current life period and must be corrected. The best way for these correction is use meditations, allowing to access to different levels of subconsciousness.

The modern science has accumulated methods that make possible intensification of work or increasing quality of some brain programs. It can allow to any person at his/her strong desire to be more productive, more efficient, wiser, happier, and to make smarter decisions.

Here it should be emphasized that 90% of our mental processes work subconsciously and have decisive impact on our subjective reality. The other 10% are part of our consciousness, and that part makes our daily grind by dealing with things like terms, sensual perceptions and other conscious information.

Because the brain does not distinguish the events of the external world from those that occur in our thoughts, it is possible reprogramming many mind programs, including past experience (Dispensa, 2012).

To reprogram the existing information it is possible only by introduction and replacement with the new information. This is an essential difference between methods of optimal tuning intellect from methods of mind control.

If for optimal tuning tasks used physical or chemical factors to maximize the effectiveness of mind, then for the purposes of mind control, information exposures should be used to a greater extent. Here is very important problem. Our subconscious with Body Control Unit (BCU) and Psychic self-control Unit (PSCU) have maybe several levels of information protection and only special methods can overcome all levels of this protection.

This is obvious, that one of major task of mind control is to determine for each individual the best combination of methods the most effective for a given state of his/her body and the intellect.

Intellect control can be like self-control or control inside of the team from leaders or coworkers. By use described below methodologies of intellect control, we can organize and intensify a job of the mind for any person.

Mind control or self-control may include the following operations:

- Meditations[8]
- Procedures of mind (intellects) hygiene
- Cleaning the mind and Mindset
- Provision and maintaining a creative environment
- Setting brain altered state of consciousness, if necessary
- Choose regime of one task execution or multitask execution
- For one task stopping unneeded flow of thoughts
- Attention filter and cognitive control
- Usage visualization techniques
- Information impact on the subconscious by suggestion or reprogramming
- Usage problem solving, analysis and forecasting methodologies
- Emotional control
- Regular mind and body condition support, recovery and training.

Jose Silva gave the following definition of his method: "This is Mind Control – going to a deep meditative level where you can train your own mind to take charge, using its own language of images reinforced with words, bringing results that become more and more amazing, with no end in sight for the person who keeps in practice" (Silva, 1977).

Described above in the section 8.9 methods and systems (machines) of mind technology allows to carry out listed operations of control for a brain more quickly and more qualitatively.

In this chapter, an attentive reader will not find ready receipts, at first because theory of mind control is in stage of development, and at second, because all people are different, and each of them is needed a special approach.

10.2. Mind Technology for enhancing the brains performance

Trends. Decades ago, American neurologist and neuropsychiatrist, Prof. Richard Restak in his numerous books predicted how modern computer technology can be used to enhancing the brain's performance including speed of response, working memory, imagining ability, reasoning, calculation, abstraction and mental endurance (1995, 2007, 2012).

At present, in the field of science and on the market have been formed several trends in the creation of the brain stimulation techniques including the following ones based on:

---

[8] The idea of controlling thinking through meditation has its roots in the teachings of the Buddha, who wrote: "Meditation brings wisdom; lack of mediation leaves ignorance. Know well what leads you forward and what holds you back, and choose the path that leads to wisdom".

- Bio stimulants for brain
- Transcranial electrical or magnetic stimulation
- Methods of influence on the conscious and subconscious by using techniques of hypnosis or meditation with the special devices as feedback or measurement tools,
- Audio-visual entrainment (AVE) systems and Mind machines
- Feedback systems for brain control,
- Cognitive training videogames.

Most of these technologies based on the high-tech technologies, last achievements of hi-tech and the modern science (Fernandez, 2015).

Bio stimulants for the brain. Millennial human experience selected number of biological stimulants improving the condition of the body and thus improving brain function. It is known, that tea and coffee can improve concentration and focus. The main difference between the effects of these two beverages is that coffee is able to quickly shake the brain, but a relatively short time. For tea, an invigorating effect lasts much longer.

Our brain demands in day ration some quantity of the omega-3 ingredients. Omega-3 fatty acids and Medium-Chain-Triglycerides help the brain work better.

Now existed a class of remedies coined as nootropics (or smart drugs), which includes memory enhancers, neuro enhancers, cognitive enhancers, and intellect enhancers. These remedies usually issued like drugs, and supplements, that improve one or more aspects of mental function, such as working memory, motivation, and attention.

Amid them are used BrainPill, Smart Pill, OptiMind, Ginkgo Smart, and so on. Every year on the market of nootropics appears hundreds of new products. For example, Amazon.com has on the market such product as "SharpMind Solaray", "True Focus", "Brain Hacks", to names just a few.

Everyone reacts differently to different chemicals and drugs. Each person has a unique organism, and what works for one person does not work for everyone. Therefore everybody has to keep learning and experimenting and/or get the advice of his/her physician and trusted advisors.

Performance-enhancing substances are also used by military personnel to enhance combat performance.

Cranial electric or magnetic brain stimulation. Cranial electrotherapy stimulation (CES), (also known as "electrosleep", "transcranial direct current stimulation (tDCS) " and by many other names), involves a form of treatment that sends low-intensity microcurrent (under 1 milliampere) to the brain (Zaghi, 2010).

For example, tDCS utilizes weak, direct electric current to modulate the activity and excitability of neurons which enhance brain plasticity, the ability of the brain to reorganize and transform itself. tDCS also improves attention, memory and learning for healthy individuals and patients with disabilities. tDCS is performed at some clinics, it can be self-administered with training and supervision through some programs, for example, as the Home-Use Program (Looi, 2015, 2016).

The tDCS procedure is safe and easy to do. Sponge electrodes are positioned over targeted areas of the brain. A portable, FDA certified stimulator provides a tiny, imperceptible direct current which modulates nerve impulses.

tDCS allowed to relieve also a chronic pain of central origin, which includes a migraine, fibromyalgia, painful neuropathies and complex regional pain syndrome. Beside it, psychiatric conditions, which respond to tDCS, include treatment-resistant depression, bipolar disorder, addiction, borderline personality disorder, schizophrenia, and hallucinations. Neurological disorders improved by tDCS include such diseases as tinnitus, epilepsy, stroke, insomnia, age-related cognitive decline and the dementias. Many doctors approved clinical benefits that are achieved as a result of the long-term potentiation or inhibition induced by tDCS.

Early studies have shown it enhances motor skills, memory recall and concentration. As a result, the US military now employs tDCS to assist fighter pilots, snipers and other personnel (Morcan, 2014).

In an experiment at the Air Force Research Laboratory, Wright-Patterson Air Force Base in Ohio, researchers have found that tDCS of the brain can improve people's multitasking skills and help avoid the drop in performance that comes with information overload (Nelson, 2016).

Transcranial magnetic stimulation (TMS) and Repetitive transcranial magnetic stimulation (rTMS) are magnetic methods used to stimulate small regions of the brain. During a TMS procedure, a magnetic field generator, or "coil", is placed near the head of the person receiving the treatment. The coil produces small electric currents in the region via electromagnetic induction. The coil is connected to a pulse generator or stimulator.

Researches in this field are in theirs early stages. A number of studies suggest that it may improve learning, vigilance, intellect, and working memory, as well as relieve chronic pain and the symptoms of depression, fibromyalgia, Parkinson's, and schizophrenia.

The methods using techniques of meditation and hypnosis with the special devices as feedback or measurement tools. Today any person can choose for himself/herself any effective method from the following ones, for example:

- "Silva Mind Control Method" of Silva International Inc.
- "OmHarmonics" of Mindvalley
- "Awakened Mind Training" of Institute for the Awakened Mind.

Jose Silva's techniques allow using the relaxed, healthy state of mind that occurs during meditation to solve your day-to-day problems. It is called "active" meditation. According to book (Silva,1977), this method allows students to receive insights and practice extrasensory perception (ESP).

OmHarmonics based on Multivariate Resonance Technology, and allow to activate physical, emotional, mental and spiritual components of the intellect.

The Institute for the Awakened Mind (IAM) develops methods controlled by an electroencephalogram (EEG) special awakened mind meditation as the means to personal evolution.

Brain-boosting technology. Michael Hutchison's books Megabrain (1986) and "Mega Brain Power" (1991) revolutionized the field of brain-boosting technology. With the publication of these books, the use of brain-boosting technology exploded worldwide. The term "Mind Technology" was coined by Michael Hutchinson as definition of the technology based on using special computer program and special devices (mind machines) for enhancing mind abilities.

Audio wave stimulations are not the only one way the brain can be entrained. In fact, the brain is affected by any kind of rhythmic stimuli. Clicks, drum beats, light flickers, and even physical vibrations or electric pulses have all been proven to effectively entrain the brain.

In the 40-ies of the twentieth century, an effect similar to the effect of monaural and binaural beats, described in the section 6.4., was found on people's perceptions of light flashes, repeated with some frequency. The frequency of dominant electromagnetic waves of the brain tends to be synchronized with the frequency of light flashes.

These discoveries and achievements of the science and industry of the computer era laid a foundation for direct brain-boosting technologies.

Here are the group of companies developing AVE technology: The Monroe Institute, The Exploration of Consciousness Research Institute, Centerpointe Research Institute, Transparent Corp.(now it is out of business), just to name a few.

The Monroe Institute has a rich history in the development of Hemi-Sync® technology which was been scientifically and clinically proven. This technology is the process of listening to a set of complex audio rhythms (special blends of Alpha, Theta, and Delta frequencies) in combination with music, various kinds of noise and special exercises for concentration. In most cases, this procedure also includes breathing and relaxation exercises and training, aimed at increasing the suggestibility and development of visual quality enhancing intellectual performance in various mental disciplines, the enhancement of creativity, and the acquisition of "extraordinary" abilities such as nonlocal knowing, etc. (Skip Atwater, 2012).

The Monroe institute created Remote Viewing powerful program that united state-of-the-art Hemi-Sync® technology along with scientifically tested perceptual tools and techniques. This program allows expanding consciousness, focusing the mind on receiving information, and ultimately, remote sensing of the unseen.

The brainwave audio technology EquiSync developed by EOC Institute includes such program packages that allows enhancing the body's natural ability to maintain optimal brain functioning, improve blood flow to the brain and normalize its chemical and electrical activity. The technology of Holosync, developed by Centerpointe Research Institute, includes several packages, Awakening Prologue, Autofonix, and so on. Autofonix, for example, is an incredibly powerful method for delivering affirmations directly to subconscious mind. The technology was used by the US Army and it has only recently been declassified for the use by the general public.

The technology of Holosync allows:

- Meditate as deeply (actually more deeply) than an experienced Zen monks, literally at the touch of a button
- Virtually eliminate stress
- Boost mental powers
- Eliminate most so-called "dysfunctional" feelings and behaviors
- Attain a level of happiness and inner peace.

Transparent Corp. founded in 2003 is developing such tools as Neuro-Programmer, Mind WorkStation, Mind Stereo, to name a few. Mind WorkStation could use AudioStrobe Glasses and be easily integrated with biofeedback and EEG devices. AudioStrobe glasses are used to deliver rapid flashes of light to entrain the brain, and allowed the user to adjust the intensity of the flashes, the pattern of the flashes, and to select any frequency from extremely high beta to very slow delta waves, simply by turning some switch or knob. Light system can turn flashing lights of all colors at the same time, (first in right, then in the left eye) together with fast switching of audio signals.

The screen flashing uses the computer screen to deliver there required pulses of light. Although glasses are the best way to entrain the brain using light, the screen flashing can be useful in the absence of glasses. In Mind WorkStation of Transparent Corp., the flashing could also be used on top of 3D visualizations, creating a subtle flash that is comfortable even with eyes open.

Many companies use different types of glasses. The Mind Machines employ the unique Tru-Vu Omniscreen glasses that were carefully designed to provide clients with the most effective and safe light and sound experience. The Tru-Vu Omniscreen glasses use eight high-efficiency white bulbs mounted over a silver reflector behind a light blue-tinted translucent screen. This disperses the light evenly, protects the bulbs and removes any red light, which may be produced by the bulbs. White light has proven most effective in imagery, relaxation, and visualization.

Most AVE systems users notice a feeling of mental clarity and sensory acuity that lasts many hours after an AVE session. This can be explained by the continuing elevation of certain neurochemicals associated with heightened consciousness, and with the continuing presence of slow brainwave activity.

As demonstrated by the first studies performed by the founder of Mind Alive Inc. Dave Siever, AVE exerts a powerful influence on brain/mind stabilization and normalization. At the end of an AVE session, the user may realize that he/she has not felt so relaxed for years - perhaps not since childhood (Sieger, 2003)

Mind machines. Usually Mind machine (MM) consists a microprocessor-controlled device, a light/sound synthesizer, stereo headphones, and stimulation glasses with pulsating lights. Built-in light/sound subsystems direct the stimulation synthesis, creating a carefully composed sequence of rhythmic light and sound patterns. The tone, frequency, pitch and other parameters of the light and sound stimulation vary the effect, just as different styles of music do.

MM's sessions typically aim at directing the average brainwave frequency from a high level to a lower level by ramping down in several sequences. Target frequencies typically correspond to delta, theta, alpha or beta brain waves, and can be adjusted by the user based on the desired effects.

MM are often used together with biofeedback or neurofeedback equipment in order to adjust the frequency on the fly.

Modern MM can connect to the Internet to update the software and download new sessions. When sessions are used in conjunction with meditation, neurofeedback, etc. the effect can be amplified.

During the last decades was developed several Mind machines (sound and light machines), for example: the DAVID Delight Pro (www.mindalive.com), Kasina Mind Media system (www. mindplace.com), Nova Pro100, Lumina (http://www.photosonix.com/), to name a few.

The benefits of the above mentioned systems are wide in range. Some people use them to relax, to boost the brain, to prepare to sleep, to feel better and so on, while others use the practice of controlling their to aid in meditation or in healing. Mind machines can lead to the desired state of mind quickly and reliably. In addition, the brain wave activity, achieved by means of mind machines, can be observed for many hours or even days after the sessions. In some cases, mind machines can help people achieve mental states, that they simply cannot achieve on their own, by conventional methods. MM have capacities toward specific tasks and applications, such as accelerated learning, sports training, weight loss, etc.

According to Hutchinson, many people use mind machines for very specific purposes: stress relieving, lowering blood pressure, increasing sexual pleasure, eliminating insomnia, improving memory, etc. (Hutchinson, 2014). For example, mind machines like DAVID Delight Pro (Mind Alive Inc.) allow to energize a body in the morning, meditate at any time, to boost the brain, to prepare to sleep, to feel better and so on. The sessions on the David Delight Pro help in stress reduction, insomnia, improved mood, mental sharpness and balance (reduced risk of falling) for seniors, and reduced worry plus improvements in concentration and memory for college students.

Stimulation by brainwaves is forbidden for people with epilepsy, arrhythmia and other heart diseases, people with heart stimulator, with severe mental disorders, people under alcohol, pregnant woman and children.

At present, various companies use described above brain-boosting systems to improve their staff's focus, memory, concentration, intelligence, critical thinking, problem-solving abilities, reduce stress, get a better sleep, increase energy levels, and so forth.

Feedback systems for brain control. At present, many high-tech companies, including Medtronic, Neuropace, St. Jude Medical, etc. are developing systems to actively monitor brain activity and respond in real-time with appropriate treatments.

Cognitive training videogames. Many researches including work of Adam Eichenbaum, Daphne Bavelier, and C. Shawn Green (Grey, 2015) demonstrated long-lasting positive effects of video games on basic mental processes–such as perception, attention, memory, and decision-making. Companies like Posit Science, Lumos Labs, ELM, etc. develop and issue this type of intellectual games.

10.3. Hygiene and prevention of the intellect and brain diseases

"Diseases of the mind are more common and more pernicious than diseases of the body" –wrote famous Roman philosopher and orator Marcus Cicero. More than 2000 years have passed, this statement is true today. For this reason, any mind (intellect) intellect control should begin and end with hygiene to prevent brain overload and side effects.

Psychologists know that it is very easy for people to cause neurosis with the help of several words. For example, scientists in experiment offered to people simple at first glance task that did not really have a solution. However, people did not know about it. They undertook to solve this task but failed. Moreover, experimenters friendly reproached them: 'What the problem? Such simple task cannot solve. Try a new one (unsolvable too). It is simple too". After a while, the people, who underwent experiment, received stress, ran into depression and neurosis (Sheremetiev, 2013).

Our mind as any part of our body must have a service and support. First of all, it must have to feed, energy, sometimes vibration and magnetic field. It can live in the safety condition within the permissible temperature, humidity and atmospheric pressure range for a living, without excessive overloads, shocks, strokes, emotional, and sometimes unsatisfied sexual hunger, etc.

The existence of the emotional hunger was discovered in Russia in the process of scientific experiments connected with the problem of long-term stay of astronauts in space.

Modes of mind operation (work, training or optimization (see chapter 9) should alternate with the rest plus 7-8 hours of a dream. Session of optimization can be used between them for a period of the rest (about 2 hours), needed for neuroplasticity work in the issues of the brain.

As for each complex system, it is very important to detect serious maladies characterizing the brains overload. Early signs of mental overburden can include:

- Recurring headaches without apparent causes
- A palpable fatigue that does not disappear even after a night's sleep
- The skin of the face changes color (becomes pale or grayish), under the eyes there are stable bruises
- Fluctuations in blood pressure
- Redness of the eyes
- Insomnia or difficulty falling asleep, and so on.

These types of symptoms is often considered as the usual weariness and people try to restore their forces by a simple dream or rest on nature. But doctors knows, that in some cases such change of activity will be not enough, it is necessary to pass adequate treatment.

As many scientists and medical doctors recommend, for any person in any age very important to have hobbies and interests outside any business. Use those of them that are not connected with blows to the head (for example, boxing) and injuries. Hobbies provide relaxation and may inspire creative ideas that can be used in the business. Now too many people skip vacation time. It is harmful. Vacation time along with hobbies and other interests provides helps to restore depleted brain and first of all creativity.

It is obvious, that to keep brain physically healthy, everyone should keep his body physically healthy: eat a balanced diet, get regular exercise, get plenty of sleep, don't smoke, and don't use too much alcohol or recreational drugs.

Our brain requires a lot of energy and if that energy is not enough, the brain can be overwhelmed.

In the overwhelmed cases can be used some strategies, like the following:

- Step away from the task or change a type of activity,
- Change over-whelm-inducing thoughts
- Embrace stress as a challenge
- Rest and complete (deep) relaxation.

For a state of complete relaxation, some people need a long time rest. Other people need a holotropic therapist or to pass a course of meditations. Very useful can be "Floatation (isolation) tanks" therapy, which is the most effective method for stress relief and relaxation. (Hutchinson, 1984, 2003). Isolation tanks were invented in 1954 by American physician and neuroscientist John C. Lilly, who proved that brainwaves are considerably altered while in this deeply relaxed state, making the floating participants very receptive to new information (Morcan, 2014).

10.4. Mindset management

The difference between successful and non-successful people usually starts in their mind, with their predominant mindsets. Few people know that many losers can be turned into successful people by changing their mindset. Hereof, very important part of mind control system is a mindset cleanse and reprogramming, because the burden of mental worries, stresses, positive and negative emotions, biases, bad habits and wrong beliefs defined by our background may block or decrease our intellectual abilities.

The definition of mindset and basis for mindset management were described in section 7.8. Analysis of literature about mindset revealed a total absence of consensus in definition of this concept, borders of applicability and specific methods of realization.

Therefore, it makes sense to accept that the mindset (let call him as a right mindset) is a frame of mind, i.e. it is a way of doing business, a behavior, an approach, an inclination, a disposition, confidence, the drive, or motivation to solving serious complex problems.

Individuals with a right mindset must believe that they can solve any problem and no challenge is too great. They approach problems with the attitude of optimism, persistence, confidence, and resolution to improve the situation.

Usually mindset is collection of thoughts and beliefs that shape thought habits. Main task of mindset management to transfer any given mindset to condition of right mindset by changing some knowledges, beliefs and thought habits by cleansing or reprogramming subconscious mind.

Strategy and methods of this management depend on personal psychological characteristics, based on Carl Jung's theory of personality types. These characteristics may be formalized with the help of Myers-Briggs Type Indicator (MBTI), that was constructed by Katharine Cook Briggs and her daughter Isabel Briggs Myers (Myers, 1980,1995). Some companies used MBTI test to make organization of teamwork in the workplace more effective. Test results indicate personal preference in four areas:

- Extravert/Introvert — How a person interacts with others and is stimulated,
- Sensing/Intuition — How a person prefers to gather information,
- Thinking/Feeling — How a person prefers to make decisions,
- Judgment/Perception — How a person prefers to orient their life.

A hierarchy of motives usually drives human activity. Motive can derive from necessity, objects of interest or concern. Depending on the type of personality, motivation can be two types: extrinsic or intrinsic. Extrinsic motivations arise from outside of the individual and often connect with the promise of rewards such as trophies, money, social recognition or praise. Intrinsic motivations are those that arise within the individual.

Human motivation can be conscious and subconscious. Conscious motivation associated with intention. Merriam-Webster defines an intention as "the thing that you plan to do or achieve: an aim or purpose. The intention is a plan to made decision to reach a definite purpose with a distinct concept of means and ways of action. In intention are united prompting to action and its conscious planning".

The internationally renowned author and speaker in the field of self-development called the "father of motivation" by his fans, Dr. Wayne Dyer wrote: "I know that intention is a force that we all have within us. The intention is a field of energy that flows invisibly beyond the reach of our normal, everyday habitual patterns. It is there even before our actual conception. We have the means to attract this energy to us and experience life in an exciting new way" (Dyer, 2005).

An intention is a powerful force, it sends a strong message to the subconscious mind. By creating an intention, a man activates his mind and willpower.

According to the definition of Deepak Chopra, "An intention is a directed impulse of consciousness that contains the seed form of that which you aim to create. Like real seeds, intentions cannot grow if you hold on to them. Only when you release your intentions into the fertile depths of your consciousness can they grow and flourish" (Chopra D., 1994). Numerous studies show that for every person in the process of any project development, is necessary to keep in mind, especially in our subconscious mind, tirelessly of our intentions until we resonate on the deepest possible level with the idea, which has inspired us as to banish all doubt within us and concentrate inner energy for this idea embodiment. This is a gist of the power of intention.

In 1981, George T. Doran, a consultant and former director of corporate planning for Washington Water Power Company, published a paper called, "There's an S.M.A.R.T. Way to Write Management's Goals and Objectives." In the document, he introduces S.M.A.R.T. goals as a tool to create criteria to help improve the chances of succeeding in accomplishing a goal (Doran, 1981).

The acronym SMART stands for:

- S (Specific) - Goals should be simplistically written and clearly define what you are going to do,
- M (Measurable) – Goals should be measurable so that you have tangible evidence that you have accomplished the goal,
- A (Achievable) – Goals should be achievable; they should stretch you slightly so you feel challenged, but defined well enough so that you can achieve them.
- R (Relevant) – Goals should measure outcomes, not activities,
- T (Time-Bound) – Goals should be grounded within a time frame.

The right mindset provides the formation of a strategy of thinking. The word "strategy" has Greek roots and means skillful leadership. The strategy of thinking in many respects is how a person manages his resources to achieving any goal. As an example of such strategy, we can remind a favorite strategy of the founder of the Soviet space program Sergey Korolev: "You can do it quickly, and it can be bad, but you can do it slowly, and it will be well. After a while, everyone will forget that it was fast, but they will remember that it was bad. And vice versa".

Any mindset management can be carried out only with the help of the correct organization of the work of Psychological Control Unit (PCU) and sufficient amount of willpower. What is willpower we described into many previous sections. Because of according to McGonigal (2012), willpower of the person during his life may decrease; any person needs always to maintain that at a high level for the realization of his/her intention.

To use successfully mindset self-management, everybody needs to learn and have in his arsenal some special strategies, which fit only for him to struggle with some emotions, temptation, addiction, distraction, and procrastination. How to find these strategies partly described in the book of Kelly McGonigal (2012).

Meditation actually may train a brain to become a self-control machine. Even simple techniques like mindfulness, which involves taking as little as five minutes a day to focus on nothing more than breathing and some senses, improves self-awareness and brain's ability to resist destructive impulses.

The main goal of mindset management is to create for a person an easy way for a state of awareness and inspiration. People of art and creative people know that inspiration and creativity are not constant throughout all day, all week or month.

Therefore, very important goal of mind control to make so that new ideas appear by the wish of their owner in the workplace during the period of work. The easiest way to get inspiration state is to have or come up with a personal trigger. As a trigger can be used gesture, object like a favorite pen, table or computer (Oleynic, 2014).

Mindset management is a great way to success and to develop the success. To build the such type of mindset can be used, for example, the following techniques:

- Suggestion or Self- suggestion
- Creation of positive habits and routines
- Hypnosis or self-hypnosis
- Meditations
- Visualization
- Subliminal messaging.

10.5. Control the subconscious mind by means of information influence

In 1963, Dr. Joseph Murphy wrote a book called "The power of your subconscious mind", in which he explained that the power behind any thought is the subconscious mind (Murphy Joseph, 2012).

As has been shown above (Section 3.4), our subconscious part of mind works millions of times faster than consciousness. If we also take into account that on one hand, conscious part of your brain makes up 10% of your brain chemistry, when the other 90% of your brainpower is derived from the subconscious mind; and on the other hand, human subconscious part of mind can include an unlimited amount of knowledge (or connect to the Universe Knowledge Base), then the hypothesis, that the great geniuses of mankind Da Vinci, Einstein and Tesla used more the subconscious part of mind than the consciousness, does not look very stupid.

Many scientists recognized the subconscious mind as a source of creativity, intuitive thoughts and feelings, inspiration, and spiritual awakening. Many people believe that it is possible for an individual to use his conscious mind to make changes in his subconscious mind, which may translate into observable changes in the life of that individual.

The subconscious mind doesn't have cognitive filters, therefore can perceive any information as events, knowledge, beliefs, and emotions in the different forms. Dr. Bruce Lipton compared work of subconscious mind with a tape recorder, which doesn't evaluate quality information, it just tapes as is. Hence, it doesn't make any difference between right and wrong, or good and bad.

Unfortunately, the subconscious mind prefers to communicate with the consciousness part of mind not by words or gestures, but by using special graphic language (images), feeling, emotions and maybe simple with acoustic and light waves. To learn how to influence the subconscious mind, at first, we must learn to listen to the subconscious. In everyday life, we constantly ignore information from the subconscious, that try to send to us very important signals, and it often ends badly. For many people possessing the developed intuition in some circumstances can decipher this information.

To facilitate access to the subconscious part of mind can be used special states of the brain, for example, awakened, or alpha state. Often to use the subconscious mind effectively, we need to learn how to go into the alpha state at will and stay there. Many special schools allow taking practice to master this art. On the one hand, alpha and theta brain waves provide the mode of operation of our subconscious; on the other hand, they perform the function of the communication between consciousness and the subconscious in a special language. Thence subconscious mind accepts suggestions and commands readily when we are in the alpha and theta states of mind. Suggestions and commands to our subconscious mind are relatively ineffective when our mind is in the beta state where conscious mind dominates.

If we want to make changes in our life, at first, we have to convince our subconscious mind that a change is needed, at second, to use some techniques to access and communicate with normally defended subconscious mind, at third, delete unneeded or harmful information or program from subconscious mind, and at last, to insert new information or new program in this part of mind through consciousness.

By this reason, the first stage of the intellect control must be devoted a preparation and maintenance of working capacity of mind system by cleaning from biases, blocks, stress, etc. and after that - harmonization a body- brain system.

A person with a cleared mind can become "open-minded" and be aware of what causes limitations located mostly in thinking. This person can become also armed by such traits as curiosity, desire to learn new things, and understand that there can be some things he/she may not understand. This person can be learned also to understand an intuitive impulses.

Above, several ways of accessing the subconscious with the help of a pendulum and access code, meditation, hypnosis, or music were discussed. These methods are quite effective, but not all of them allow changing the information content of the subconscious.

At present, many new methods of transferring, extracting and erasing information stored in the subconscious and memory are created. These methods assist to block, or at least to weaken protection of sub consciousness, and thereby help to carry out information impact on certain structures of the brain.

Let's consider in detail some techniques which can be useful for control and management of subconscious part of mind.

Suggestion for mind control. In addition to what was said in section 6.5, mention should be made of propaganda as a powerful way of influencing a large number of people at the same time.

Propaganda is a purposeful and systematic actions to form perceptions, manipulate knowledge and guide audience behavior to achieve a reaction conducive to the realization of the goal the propagandist wants.

Description of suggestion methods can be found from special literature or the Internet. Self-suggestion is one of the easiest ways we could use to consciously convict our subconscious mind what we would like to see happen. For self-suggestion can be used different kind of affirmations and mantras. For example, here is famous affirmation of Loiuse Hay: "I am in the right place, at the right time, doing the right thing" (Hay, 1994).

In its essence, affirmation is a statement intended to provide encouragement, emotional support, or motivation, especially when used for the purpose of autosuggestion. Mantra is a phrase (or word) repeated by Buddhists and Hindus when they meditate, or to help them feel calm.

One of the most powerful technologies of influence on the subconscious is multiple repetition. The more repetitions of coding information, the better. This technique is widely used in advertising.

Creation of positive habits and routines. When a person is constantly telling himself negativities, he is creating negative beliefs about himself, his environment, and others people. This can cause changes in his health, his relationships, career and every aspect of his life. Speaking positive auto suggestions with feeling will awaken subconscious mind and reboot it so that very core belief system is changed. Many studies proved that when we focus on creating positive habits in place of negative ones we can literally change our lives (Dispenza, 2012).

Hypnosis or self-hypnosis. It is known, that hypnosis is a wakeful state of focused attention and heightened suggestibility, with diminished peripheral awareness. While in this state, messages can reach the subconscious, cleansing or reprogramming certain information. This makes hypnosis very effective for relief from stress or anxiety, to overcome bad habits, to increase confidence and to unleash creativity. Hypnosis usage enables the individual to achieve a state of extreme relaxation allowing easily to reprogram subconscious mind accepting new thoughts and related information as reality (Rossi, 1993).

Meditation. Described above in the section 8.8. meditation techniques allowing to assist also in reprogramming subconscious mind (Kain, 2016). Usually a brain normally works in the 'beta' pattern. This state is associated with alertness, but also with stress, anger and anxiety. In meditation, brain patterns slow and calm down and move first to alpha and then to theta and, in deep meditation, delta patterns. During meditation practice a person may become more adept at taking the position of 'observer' with his conscious mind.

Visualization. The role of visualization in the creativity process was above mentioned. Not all people have an innate ability to visualize information not only in 3D but even in 2D format, but practically all people can use drawing, diagrams, and graphs made manually or with help of computer as a way to capture and operate with whole or part of needed information for goals of understanding or creativity. Visual images have a huge impact on the brain, both consciously and subconsciously. Of course, to become proficient in the use of thoughts visualization, a person must continuously learn and train especially because with age this ability can decrease.

In the end of 20th century was appeared many methods and computerized tools for visualization of objects, simple and associative links between them, and even for construction of graphical or functional models of these objects.

Currently available computerized methods allow to receive 3 D images of any objects on a computer and even provide a possibility to see them from different points of view.

In the 50s of the last century was invented "semantic nets", and later in the early 70s, a British brain researcher Tony Buzan innovated a mind mapping as a whole-brain alternative to linear thinking (Buzan, 2002). This method was developed and described by Michael Michalko in his book "Cracking creativity "(Michalko, 2001).

On the American Market here are in use top 5 (or more) mind-mapping software packages (Connel, 2014):

- XMind
- Mindmeister
- Mind Manager
- Free Mind
- iMindmap.

For example, XMind is a mind mapping and brainstorming software, developed by XMind Ltd. In addition to the management elements, the software can capture ideas, clarify thinking, manage complex information, and promote team collaboration for higher productivity.

Mind Manager developed by Mindjet (www.mindjet.com) can be used for:

- Brainstorming new ideas or solutions
- Planning a project or work
- Capturing and organizing requirements and notes
- Facilitating meetings
- Conducting research
- Developing presentations.

The new MindManager 2016 for Windows helps to visualize workflows with powerful new capabilities for brainstorming, planning, process improvement, decision-making, and more.

Ralf Lengler and Prof. Martin J. Eppler from Institute for Media and Communication Management (Switzerland) designed a specific guide of visualization methods named as "A Periodic Table of Visualization Methods". They divided all visualization methods into 6 visualizations categories: Data visualization, Information visualization, Concept visualization, Strategy visualization, Metaphor visualization and Compound visualization (http://www.visual-literacy.org/periodic_table/periodic_table.html).

We will not analyze here also the theory of creative visualization developed by experts such as Shasti Gawain (Gawain, 2010) or Nick Farrell (Farrell, 2013). This theory asserts that creative visualization can improve human life and attract success and prosperity.

Subliminal and similar messaging. Subliminal stimuli (literally "below threshold"), are any sensory stimuli below an individual's threshold for conscious perception. Subliminal messaging is based on the theory that by briefly flashing a word or image in a film or video, the viewer will subconsciously pick up on the message and act accordingly. The problem subliminal messaging is discussing in literature and in Internet a long time. The United States does not have a specific federal or state law addressing the use of subliminal messages in advertising, but both the Federal Trade Commission and the Federal Communications Commission have issued policy statements declaring some advertising technique to be prohibited. Britain and Australia ban subliminal advertising for any reason. Fortunately here we are talking about the coordinated influence on subconscious mind of each specific person at his will or by practitioner after signed agreement.

Therefore some techniques and tricks similar to subliminal messaging can be used for control of the subconscious part of mind. Here are some of them.

Overload. The main principle of overloading is: to submit one of the channels of perception a large amount of information, such that the consciousness does not have time to properly process it. At that moment, encoding information is sent through another channel (usually in a hidden form). There are several options for using this technique:

- Direct model: the supply of overload information on the visual channel with the simultaneous supply of coding information through the auditory canal.
- The inverse model: the overloading information is fed through the auditory channel, and the encoding by the auditory channel.
- Reversible model: with a certain periodicity or in a random order, the channels of coding and reloading information are exchanged.

The cognitive dissonance. The essence of the application of the cognitive dissonance method in the case of audiovisual coding is that diametrically opposite information is supplied simultaneously through two channels of perception - auditory and visual. For example, we show a red square, but we say that it is a green circle. It is dissonance. The consciousness is confused and cannot adequately filter the coding information that follows immediately after the appearance of a contradiction or in parallel with it.

Marking or coding. This is one of the most effective technologies for influencing the subconscious. Information is labeled according to certain rules, as a result of which, its importance for the subconscious mind increases manifold. It is possible to mark both visual information and sound. In this case we have a very effective technology of cross-labeling, when the encoding information concerns a visual nature, and the marker is sound, or vice versa.

Management of intellect and mind is a difficult formalized process, the features and arsenal of which can be significantly different for different people and circumstances.

Techniques of the mind cleaning and reprogramming may include not only described practices, but and music therapy, prayer, hearing and exercises in the team, and so on. Mind cleaning does not mean the same as cleaning teeth or gut. Here any person needs a lot of caution; otherwise, he can clean the programs needed for life.

Financial stimuli. Financial motivations for creativity are the level of wages, benefits, bonuses, pensions, scholarships, the provision of additional benefits, including tax benefits, tax holidays, and cancellation of tax debt.

Mind technologies. According to Michael Hutchinson, mind technologies open up new opportunities for humanity. These technologies can become a "labor-saving" tool that helps us quickly switch from depressed states to states of inspiration, and change unwanted behavior to desired ones - what before required us months and years (Hutchinson, 2014). The use of Audio-visual entrainment (AVE) systems does not impose strict restrictions on the creative environment and allows a person to achieve a creative state of mind and proper efficiency in solving the most complex problems.

10.6. Cognitive control

Cognitive control system together with attention filter allow to short a volume of information in our consciousness and increase a speed of our mind.

Cognitive control is provided with skills or a set of techniques that make it possible to turn one or several thought streams, inherent in each person, into a controlled conveyor of thoughts on the stage of consciousness with the speed control of this conveyor (up to the stop), with defense from unneeded emotions, and with possibility to open or close an access to the sources of information packets for the cells of this conveyor.

Cognitive control is one of main intellect functional part, and in light of last studies (The Wiley Handbook, 2017) may incorporate the following cognitive functionalities: perception, organized communication, categorization and concepts, emotion, memory, decision making, reasoning, planning, problem solving, motor control, decision making, learning, motivation, activation of language forms, etc. Cognitive control intertwined with other functions of intellect and with working long-term memory. A primary task of cognitive control is to choose behaviors that maximize effectivity of mind activity on the basis of decision driven by emotions and desires in its scope. We can assume that our thinking or thinking machine by definition Italian physicist Eduardo r. Caianiello at its core is a functional system that manages the stream of consciousness or by other words the flow of thoughts (Caianiello E., 1961).

Described above Psychological Control Unit (PCU) executes quality-control inspection for this conveyor, allowing only the needed quality thoughts to pass through attention filters, preventing interference from other processes in the brain. Moving by belt of conveyor thoughts may include ideas, suggestions, daydreams, inspiration for action, things to do, music, etc. Among thoughts items can be different: thoughts about work or family, serious or frivolous, thoughts of love or sex, hate or grievance, and so on. PCU may focus on whatever pleases, matching our preferences while ignoring the rest. By learning or training processes we can use mistakes to get better, remembering our experience. To purify a conveyor, we get rid of the any undesirable thoughts and replace it with a new one or several, more appropriate to the task. Over time we become more efficient in these procedures, processes of cognitive control becomes automatic or semi-automatic deleting the pollutants.

The another role of PCU is regulate multitasking by opening or closing connection with some of sources of streams of own or other people's thoughts, emotions (positive or negative), and music without or with words (arias or songs).

Depending on age, temperament, responsibilities, and situations speed of thinking can be in a wide range. The wider human horizon the greater flow of his/her thoughts. One people can think very slowly, other very fast. A number of thoughts can vary from a couple thousand to 50 000 per day.

If a person has a habit of thinking he needs a little control over mind, practically he places himself on automatic pilot. To say by another words, he uses a special thinking program, that turned on by trigger either from willpower or from emotional (or intellectual) temptation. Great thinkers have always been motivated more by the pleasure of thinking than by material rewards or coercion.

Usually, without training, people cannot focus their thoughts for more than a few minutes at a time. After that, the mind goes into chaos. When we left alone, with no demands on attention, the basic disorder of the mind reveals itself. To keep reading, or writing, or hearing during a long time people must make an effort to force their attention back to the object of activity.

When we remain at rest, not including desire and will, the main disorder of the mind is manifested. In order to read, write or hear for a long time, people must make efforts to return their attention to the subject of activity.

State of inspiration intensifies the following mind processes: the generation or collection of thoughts, cognitive filters, access ability to working and log time memory, sometimes, what important for designers, artists, decorators, etc., ability to visualization, and so on. Let consider what is it and how we can manage the intellect using some of the inner functional mind algorithms.

Usually, the thoughts flow can be started or stopped by willpower, or from circumstances. The visualization also allows stopping an unorganized stream of thoughts in the needed time and transforming this stream in the conveyor system.

For many mentally working people a needed flow of thoughts starts when a person falls into the creative environment, or if it is a normal job, with a timer turned on, for example, since the beginning of the working day. Sometimes it can be accelerated with help of different stimulus. If a stream of thoughts is not productive, the best way is to change the type of activity or rest.

Altered states of consciousness. The possibility of cognitive control is largely dependent on regimes, in which there can be our consciousness. These regimes as altered states was described in the section 4.5. The task of intellect control (management) is choosing one or several states of consciousness to achieve a goal.

Attention and concentration as the instruments of cognitive control. It is obvious, that to have instead the unorganized flow of thoughts the thoughts conveyer are needed attention and concentration, which depends on mindset and willpower. Partly attention was described in section 1.9.

The process of ensuring attention is very complicated and requires manual management. Attention characteristics: its volume (the number of problems that need to be monitored), stability in time, the speed of switching the process of attention from one subject to another, or a project - are very important in the process of cognitive control.

The main tasks of attention in the process of cognitive control:

1. not to miss important information coming into consciousness or subconsciousness. The attentional filter should be configured first to notice changes in the incoming information and assess its importance
2. maintain consciousness only to address the important and defined problems, fighting with obsessive thoughts (or which are out of place), connecting with negative or inappropriate emotions. Obsessive thoughts are irrational, emotional thoughts, as a rule, always connected with our fears, doubts, resentments, guilt feelings, anger or with something important and disturbing us. At the heart of these thoughts is always an emotional charge. Our brain can include also thoughts out of place. Usually people simultaneously forced to

solve several problem simultaneously. For example, in addition to work-related problems, there are homemade problems, thoughts about hobby or love, thoughts about conflicts or unsolved problems, etc. To get rid of these thoughts, usually, are used a different kind of meditation or special techniques.

In the first case, the an attentional filter must be created to not allow an access for unnecessary information. In the second case, is needed a control (or management) for brain, which should allow focus on solving current problems in the most efficient way, not forgetting about all existing problems.

If attention is turned out, a flow of thoughts can include huge occasional and unneeded information, under attention - only useful things. Concentration allows stopping this flow to work with details. Concentration is vital for success in any field of skilled performance.

Cognitive filters in our mind allow us to consider the world also through our belief systems. These are unique to each person. Our belief systems are the result of experience, education, and culture. Based on these beliefs we actually create our reality in the world we live in, by first choosing what we want to notice, and then by assigning meaning to what we notice.

Such instruments of the cognitive control as self-suggestion, hypnosis, and meditations can affect on our beliefs, changing the beliefs' objects and rates of their importance.

Memory. In reviewing the conditions that help establish order in the mind, we cannot forget the extremely important role of memory. All forms of mental flow depend on memory, either directly or indirectly. To work with the flow of thoughts our memory must remember all-important thought and allows to return to them when we needed.

It is obvious that skills of management of flow of thoughts can be enhanced with help of the special training or some special tools described below.

Self-talk. A kind of flow of thoughts is a conversation with itself. Self-talk is basically inner voice of any person, the voice in his mind that says the things which this person don't necessarily say out loud. Self-talk can be s a stream of chatter, positive or negative. Anyway it is very important self-regulatory mechanism for cognitive control (Kross, 2014). For any person, items of these self-talks can be different.

Willpower charging. Willpower and attention are two main tools for cognitive control. Most of people possess some harmful habits or addictions. Many people wish to overcome such habits and addiction as smoking, excessive eating, laziness, procrastination or lack of assertiveness. A strong willpower and self-discipline brings an inner strength to overcome any negative habit. These traits can make a great difference in the life of any individual. They can make him a winner in whatever he does. If a person is tired, overwhelmed, uninspired, afraid, hurting, or simply lazy, to overcome this condition he needs to charge willpower according to the theory of McGonigal (McGonigal, 2012).

At present many methods and practices developed to enhance self-control and willpower, including meditation, yoga, and even diet (Bukosky, 2014). Above, in the section 9.6 was demonstrated a simple method of willpower charging.

Multitasking, described in section 7.4, is very important skill that is needed today for many professions. Many entrepreneurs assume that multitasking is the best way to increase productivity of workers. Although multitasking is commonplace, relatively little is known about when and why people perform more than one attention-demanding task at a time. Because multitasking enables people to achieve more goals and to experience more activities- this skill is the object of many researches that took place at nowadays. It is obvious that developed multasking ability demands the serious cognitive control.

Mind training. At present for mind training created a huge number of different methods, practices and tricks, made as computer programs, games, courses, described in the different books, video or film.

10.7. Creativity control. Algorithms and methods

In addition to what is said in section 1.7, it should be added that the creative process can be formalized, algorithmized, and therefore managed in one way or another. For any individual creative abilities are determined by two main components – an ability to generate new ideas and the ability to select and develop the best of them. If the first component is part of the intellect, the second component concerning choosing decision has the special algorithms, described below. These algorithms can be a part of the special tools of reasoning, described, for example, in the book of Richard Nisbett "Mindware: Tools for Smart Thinking" (2015).

Choosing decision. In real life, usually, no serious decision is ever made with complete information. It concerns, for example, the design of complex systems, the problem of spouse choosing, decision making in business in the competition conditions, etc.

On the other hand, we must choose something in the limited time diapason. From time to time, we are making choices in our life - always with incomplete information. Even postponing to make a decision is also a decision, and has its own consequences. Therefore, we need to have the tools or instrument for choosing as an example of the decision making.

Using math methods and computer science, humankind in the last century created such methodologies like Systemic (systematic) approach, Brainstorming, Synectics, Morphological analysis, nonlinear programming, to name just a few.

In 1941 Alex Osborn, an advertising executive, described brainstorming as "a conference technique by which a group attempts to find a solution for a specific problem by amassing all the ideas spontaneously by its members". The rules he came up with are the following:

- No criticism of ideas
- Go for large quantities of ideas
- Build on each other's ideas
- Encourage wild and exaggerated ideas

Alex Osborn together with Sidney Parnes formulated a Creative Problem Solving Process - complex strategy, also known as Osborn-Parnes-process or CPS Model. The model is usually presented as five steps, but sometimes a preliminary step is added called mess-finding which involves locating a challenge or problem to which to apply the model.

The total six stages are:

- Mess-finding (Objective Finding)

- Fact-finding
- Problem-Finding
- Idea-finding
- Solution finding (Idea evaluation)
- Acceptance-finding (Idea implementation).

Generally, any genius had his special method of innovation, based on his mind and character abilities. In many respects, these methods are unique, but it turned out that they have some common algorithms. In the 1950s, the famous Sovjet inventor and science fiction author Genrich Altshuller based on the analysis of more than 200,000 patents discovered that technological systems can be changed according to specific laws, and found the patterns that predict breakthrough solutions to problems. Genrich Altshuller developed a special methodology calling it TRIZ, acronym that means in Russian "Theory of Inventive Problem Solving".

The TRIZ was based on the hypothesis that many problems can be structured according to several basic inventive principles. Using these principles allows to make the process of creativity more predictable and successful.

One of the central ideas of TRIZ is a design analysis to find conflicts between improving and worsening features. For example, if you want to make a more powerful engine for the car, you must increase the engine's weight and therefore increase mass of a car, this fact decreases the real power of the car's engine. Theoretically here can be a moment when increasing power of engine will not give any effect, because more heavy car demands more power to drive itself. Among the thirty-nine sources of conflicts that Altshuller defined in the mechanical systems were strength, stability, brightness, volume, and temperature. After that, he found about forty principles to solve these conflicts, including segmentation (dividing an object into independent parts), prior action (perform changes to an object in advance), and so on.

TRIZ converts the front-end of innovation to an organized and structured process, enrich idea-generation sessions with systematic methods, helps with generating and selecting best ideas to enable faster and better innovative problem solving, helps with increasing the degree of innovative productivity, accelerate new intellectual property and value creation, and generally significantly increases the effectiveness of creative work and reduce the time to solution.

The main feature of this methodology is that it allows making a normal person with needed level of knowledge be able to do that can be done before only skilled masters or geniuses.

In all, Altshuler wrote fourteen books and close to 500 articles. In the 1980's and later, in the Soviet Union it has been created about two hundred TRIZ centers (Altshuller, 1998, 2007).

During the 1990's, many TRIZ pioneers from the former Soviet Union immigrated to the USA. One group from Kishinev, Moldova, led by Boris Zlotin and Alla Zusman, founded the Michigan-based company, Ideation International (www.ideationtriz.com), The Altshuller Institute for TRIZ Studies (www.aitriz.org), and The Technical Innovation Center (www.triz.org).

Ideation/TRIZ Methodology. Ideation International Inc. created Ideation/TRIZ Methodology (I-TRIZ), which is the result of the history of technological and social evolution, including an analysis of over 3 million worldwide patents. From this huge volume of information approximately one thousand patterns of invention and more than five hundred patterns and lines of technological, market and organizational evolution have been extracted. I-TRIZ includes integrated tools, problem formulation modules, training systems and a special knowledge base. It comprises the following systems:

- Creative Education — a learning system to solve creative problems in any area (technology, marketing, management, etc.)
- Inventive problem Solving System — a tool to solve inventive problems in a systematic way
- Anticipatory Failure Determination System — tools for failure analysis and prediction, to eliminate existing or potential system failures
- Directed Evolution System — tools to develop future generations of a system, and to control system evolution.

The Inventive Problem Solving System, based on the Innovation Workbench software, conducts a five-step process (see Fig.10-1) to solve inventive problems in a systematic way (please visit http://www.ideationtriz.com/IPS.asp).

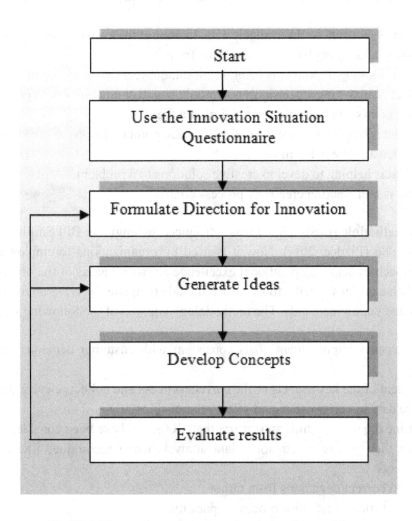

Fig.10-1 Five-step process to solve inventive problems

The resulting semantic indexing technology has been dubbed "Semantic TRIZ". Its software searches deep into the Internet and then analyzes large volumes of text from technical journals, scientific forums, and other sources, including a complete index of U.S. patents. By breaking down sentence structure, the software reorganizes the content of documents into a "problems and solutions" format.

At present based on TRIZ method is creating Theory of Developing the Creative Personality (TDCP) (Orloff, 2012).

Beside TRIZ are appears many tools and methodologies to support creativity. Here are some from them.

Method "Six Thinking Hats" created by Edward de Bono and described him in the book (Bono,1999). This method is based on the principles of a systematic approach, which can help to look at problems from different perspectives, but one at a time, to avoid confusion from too many angles crowding your thinking. In 2005, the tool found some use in the United Kingdom innovation sector, where it was offered by some facilitation companies and had been trialed within the United Kingdom's civil service. The Six Thinking Hats Model represent thinking strategies identified by Edward de Bono, consciously applied in techniques to enhance creativity:

- White Hat is focusing on the available data, looking at the available information, analyzing past trends, and seeing what you can learn from it
- Red Hat is looking at problems using intuition, gut reaction, and emotion
- Black Hat is analyzing a decision's potentially negative outcomes with goal to highlights the weak points and find solution to eliminate them
- Yellow Hat is helping to have the optimistic viewpoint that helps to see all the benefits of the decision and the value in it
- Green Hat is helping to develop creative solutions to a problem
- Blue Hat is representing creativity process control.

Six sigma methodology. Six Sigma was introduced by engineer Bill Smith while working at Motorola in 1986 (Pizdec, 2014). Now it is used by organizations to employ a structured, systematic approach to achieve operational excellence across all areas of the organization. The Six Sigma initiative utilizes specific tools, which include templates, process analysis techniques, statistical tests and deployment aids. The methodology consists of the following stages:

- Define process improvement goals consistent with customer demands and enterprise strategy
- Assessment of the key aspects of the current process and collect relevant data
- Analyze data to verify cause-and-effect relationships
- Determine the relationships and ensure that all factors have been considered
- Improve the process based upon data analysis using techniques like the design of experiments
- Control to correct deviations from target
- Set up pilot runs to establish process capability.
- Finally move onto production, set up control mechanisms and continuously monitor the process.

10.8. Control emotions

In the section 5.7 we considered definitions and importance of emotion in the human life.

Why it is so important to control emotions? First of all, as was proved by scientific studies, negative energy, attracted by negative emotions (anger, envy, grievance, resentment, etc.), is often the cause of all sorts of physical and mental diseases. Conversely, positive energy, attracted by positive emotions, strengthens both psychological and physical health.

Secondly, people who are not able to control emotions, can fall into a state of short-term insanity (affect). A person who is unable to cope with unordinary circumstances, being in a state of affect, can make rash actions, make decisions that in a normal state would not acceptable for him. Needless to say, that frequent staying in such a state can cause disorders of the psyche - schizophrenia or a split personality.

Thirdly, being in the condition of lack of emotional control, a person can destroy interpersonal relationships inside of family or working team, or with friends. Even a little inability to control emotions sometimes could not allow a person to properly perform office and family duties.

At fourth, emotional control provides for individual the ability to be resistant to the influence of other people's emotions. Emotional control closely connects with emotional culture, which can be developed by nurture in young age and training later.

Generally, control emotions is the ability to consciously choose how to live and express emotions by right way. If a person does not learn how to control emotions, he can become a "hostage" of his emotions.

The ability to control emotions effectively is a key function of intellect. Regulating emotions, responding appropriately and responding to the emotions of others are all important aspect of emotional management. Since emotions contain information and influence the thinking, it makes sense to take them into account when constructing logical chains, solving various problems, making and realizing decisions.

The possibilities of emotional control depend on many personal traits (age, gender, temperament, willpower, type of character (introvert or extravert), emotionality, etc.), culture and mindset. As the literature shows, this problem is still far from solution (de Sousa, 2015), so we can talk about some directions that have emerged in the way of solving this problem.

Emotions are an evolutionary earlier mechanism for regulating behavior than the mind. Therefore, they choose even simpler ways of solving life situations. To those who follow their "advice", emotions add energy, because they are directly related to physiological processes, in contrast to the mind to which not all systems of the organism obey. Under the strong impact of emotions, the body organs can mobilize additional forces, what cannot be done by any mind orders.

In the real life, a person too often has to suppress emotions both for individual reasons and following traditions. Using such a mechanism for managing emotions, he acts reasonably to maintain normal relations with others, and at the same time his actions are unreasonable, as they damage the health and psychological state.

In this regard, the control emotions should be carried out in accordance with certain rules that take into account the individual characteristics of the person and the situation. Generally, the control of emotions is part of the control of the subconscious and can therefore be carried out using the same methods and practices. For example, meditations, which may reduce stress, improve concentration, encourage a healthy lifestyle, increase self-awareness, increases happiness, slows aging and so forth (Crane, 2017).

At the present time, considerable experience has already been accumulated in solving emotion control problems. For example Darlene Mininni created and describe in her book (Mininni, 2006) "The Emotional Toolkit", which includes seven tools to help managing emotional life. Justin Mars created "The Best Guide on Mastering Emotions" (Mars, 2016), which allows to increase self-esteem and eliminate anxiety. Kirk Saugareli described one more Guide for Control Emotions (Saugarelli, 2015).

According to Lynn Clark, emotional culture as an ability to understand, feel and manage emotions, can be much more important for successful and joyful life than other intellectual abilities (Clark Lynn, 2015).

This culture that can be varied in the different countries allows to identify which emotions are good or bad, when emotions are appropriate to be expressed, and even how they should be displayed.

Effective managing emotions involve the following skills:
- Ability to control emotions in a reasonable range in the different situation including prevention of negative emotions
- Ability to control impulses
- Ability to cope with the inevitable defeats and failures
- Resistance to alien emotional impact.

At its core, emotional control has much in common with the very popular concept "emotional intelligence" proposed by Dr. Daniel Coleman in his book, first published in 1995 (Goleman, 2005).

According to this author, there are four parts to emotional intelligence – self-awareness, self-management, social awareness and relationship management.

Self-awareness is the ability to recognize own emotions and how they affect thoughts and behavior, know own strengths and weaknesses, and have self-confidence. This is the ability to step outside yourself with focused observation of how feel and react in various situations.

Self-management is ability to control impulsive feelings and behaviors, manage emotions in healthy ways, easily adapt to changing circumstances, and so forth.

In a nutshell, social awareness looks like as empathy, the ability to observe and understand the emotions, needs, and concerns of other people, pick up on emotional cues, feel comfortable socially in any situations.

Relationship management is the capability to develop and maintain in a team or other part of society good relationships, inspire and influence others, understand their needs, and work well without conflicts with other people.

Today, one of the main problems of emotional control is the reduction of stress. A little stress is actually healthy in today's world, but the truth is the average adult (and now kids) contain high levels of stress, which can lead to hypertension, heart disease, cancers, and metabolic problems. Among many methods of stress reduction music can be the best, if remember that in the medical practice widely used the method of musical psychotherapy to correct stress and other emotional states.

About power of emotion is written tons of literature. For example Flower Darby write "Emotions are powerful. Emotions grab our attention and keep our interest. Emotions motivate us, inspire us, and move us to action" (Darby, 2018). "Emotions have immense power. This power can propel you towards your dreams and goals, or sabotage and ruin your life. Choose wisely how to use the power of your emotions." – write Stan Jacobs in the book " The Dusk And Dawn Master: A Practical Guide to Transforming Evening and Morning Habits, Achieving Better Sleep, and Mastering Your Life" (2016).

Using described in previous chapter method of optimal tuning, it possible, first of all, to measure this power for different kind of emotion, at second to find an effective way for minimizing the impact of negative emotions. As an example of such an opportunity, stress was considered, which was minimized in 5 days with the help of music, because many studies proved the serious effectivity of music hearing or playing (Collingwood, 2016).

In experiments, represented on Fig. 10-2, demonstrated results of minimization of stress level by the impact of some sessions of stress healing soundtrack (classic music of Mozart) hearing by a person during 20-30 minutes ones a day during 5 days. Stress was measured by pendulum with R-scale for stress measurement on Fig. 3-14.

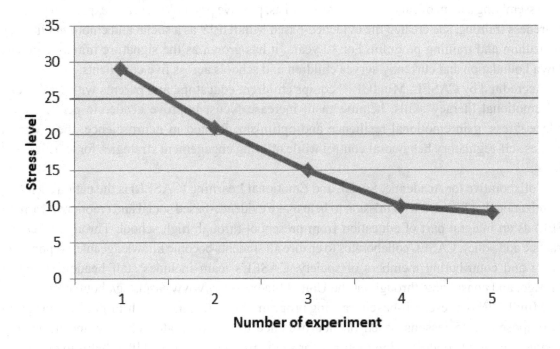

Fig. 10-2. Minimization of the stress level. RES = 400%

This example opens wide possibilities for optimization of emotional control by using the method of measuring the strength of emotions and effectivity of impact via different kind of practices.

10.9. Brain training as a part of intellect control

Brain training must start from the childhood. In an interview with Anderson Cooper, Professor Stephen Hawking said: "The brain is like a muscle, you must work it out and use supplements just like body builders use, but for your brain, and that's exactly what I've been doing to enhance my mental capabilities". The experience of many people demonstrates that brain training is needed, not only in the childhood but during whole life, especially for creative and talented people.

Numerous studies and results presented in the previous chapter convincingly prove the possibilities and effectiveness of brain training at any age. According to Jaušovec, those who did brain training saw improvements on all dimensions of their intellect (Jaušovec, 2012). The most encouraging findings show that training does not have to continue permanently as intensive as in the beginning, but cognitive training should be integrated into our lifestyle.

The famous and beautiful Goldie Hawn, an Academy Award-winning actress, producer, director, best-selling author and children's advocate is the founder of The Hawn Foundation, a public charity with a mission to equip children with the social and emotional skills they need to lead smarter, healthier, and happier lives.

Goldie began The Hawn Foundation in 2003 to apply innovative scientific research to create education programs that support the social and emotional development of children.

Assembling a team of educators, neuroscientists, positive psychologists and experts in mindful awareness training, she created the evidence-based MindUP™ as a social and emotional literacy curriculum and training program. For 10 years, it has grown as the signature initiative of The Hawn Foundation and currently serves children and schools across five continents.

Accredited by CASEL, MindUP™ equips children, educators, and parents with vital social and emotional literacy skills, helping them increase focus, improve academic performance, reduce stress, gain emotional resilience and optimism. Rooted in neuroscience, the program teaches self-regulatory behavioral control while offering engagement strategies for learning and living.

Collaborative for Academic, Social, and Emotional Learning (CASEL) is the nation's leading organization that founded with mission to help make evidence-based social and emotional learning (SEL) as an integral part of education from preschool through high school. Through research, practice and policy, CASEL collaborates to ensure all students become knowledgeable, responsible, caring and contributing members of society. CASEL's team includes staff headquartered in Chicago, and consultants throughout the United States (http://www.casel.org/about/)

MindUP™ is a research-based training program for educators and children. This program is composed of 15 lessons based on neuroscience to self-regulate behavior and mindfully engage in focused concentration required for academic success. MindUP™ helps to eliminate aggression, increases empathy and compassion while resolving peer conflicts in schools (Schawbel, 2013).

Located in San-Francisco company Lumos Labs developed some programs realized as online games, that allow training the main components of the intellect: memory, attention, speed of actions, mental flexibility and skill of problem-solving (www.lumosity.com).

To train their brains many people use in their everyday practice the following methods, which can be used all alone online, with help of literature or via courses:

- Meditation
- Brain Teasers and Games. Free online games can be found, for example, on the websites www.mindgames.com, freebrainagegames.com, www.lumosity.com, etc.
- Writing and reading. Writing and reading are closely related and inseparable.
- Learning a foreign language. Such websites as Babbel.com, duolingo.com allow free learning many European languages: Spanish, French, German, etc.
- Super Brain Yoga is a simple series of poses and movements designed to increase mental energy. It was created by Master Choa Kok Sui, founder of Arhatic yoga and Pranic healing, and is based on ear acupuncture points and subtle body energy.
- And so forth.

If a workaholic, or even a busy person will have the opportunity to choose: 1. rest, 2. have fun or 3. train the brain, he/she with a probability of 90% will choose the second option. It's clear. In order for people everywhere to use fitness and body trainings, it took years of propaganda in the schools, colleges and universities and with the huge assistance of media. For many people in terms of their common sense to engage in brain training is meaningless. Today in the American society a tendency appears to solve this problem with the help of mind machines, as some of them allow to work favorably on the brain during physical exercises, hiking, running, reading, drawing, etc.

Brain training sometimes needed to restore mind abilities after exhaustion of brain, for example, as a result of intensive work, or after serious disease. In this case very effective also using mind machines.

10.10. Collective intellect and Apollo Syndrome

The technical progress in fields of information technology allows in many cases to have and share huge volume of information for many people practically not wasting time. Access to the knowledges from different sources through interaction with different kind of media can unite many people in a different form of collective intellect.

Collective intellect (intelligence) - a term that emerged in the mid-1980s in sociology in the study of collective decision-making process. Scientists defined collective intellect as the ability of groups to find solutions to problems more effective than the best individual solution in this group. Structure of group can be different, from classical hierarchical with many levels to different kind of networks. Many fundraising and nonprofit organizations have practically a network structure with given number of employees and with variable number (sometimes-unknown) members (United Nations Children's Fund, American Red Cross, Doctors without Borders, Green Peace, etc.).

Collective intellect (CI) emerges from organized collective efforts, the collaboration and competition of many individuals and appears in consensus decision making. In a narrower sense, CI is the property that arises via interaction between people armed by methods of information processing. Human Collective intellect has much in common with the intellect of some microorganisms, insects (ants and termites), fish and birds, called swarm intelligence (Bonabeau, 1999).

Today any product on the market demands a team for its development. Management of the work of the creative team is a completely new discipline in connection with which at present is a relatively small numbers of studies in this direction. The creative process in the group depends on many factors: the composition of the group, skills of manager, creative climate, funding and work incentives, technical support and much more.

Quantity and professional quality of team members can be different and depend on many reasons. For a startup period, main task of founders and managers to collect an optimal team is to have number of people minimal by quantity (funding) and maximal by creativity and productivity. A dream team must include highly intelligent people (geniuses) that agree to work 7 day per week 24 hours for minimal salary.

At present we can find on the market many methods of team management, including recommendations how to create a team, how to manage a team by the best way in the different situations, how effectively work inside of time, etc. (McGourty, 2001). Below we will try to pay attention on the some trends in developing of team works.

Unfortunately, it is very difficult to compose or collect a dream team, first of all, because in the reality, highly intelligent people too often tend to be uncommunicative, abrasive, arrogant, easily touchy and sensitive to praise, and so on.

Dr. Meredith Belbin in his books, devoted a team management, reported some unexpectedly poor results with teams formed of bright and clever people who had sharp, analytical minds and high mental ability - he called this phenomenon as the Apollo Syndrome (Effect) (Belbin, 2010). His studies demonstrates that Apollo kind teams can be characterized particularly by tendency to active debate, generating too many ideas, diffusion on focus, too many individual agendas, and so on. This can inflict a lot of inconvenience - unneeded conflicts, waste of time and money.

Thanks to Belbin, now, awareness of this problem help professional firms avoid simple mistakes and traps, connected with this phenomenon. Based on studying numerous teams at Henley Management College, he devised to measure preference for nine Team Roles such as: Resource Investigator, Teamworker, Coordinator, Plant, Monitor Evaluator, Specialist, Shaper, Implementer, Completer Finisher. Some roles are understandable, some needs comments. For example, Resource Investigator needed to find ideas to bring back to the team; Implementer - the practical organizer within the team; Plant role to be highly creative and good at solving problems in unconventional ways; Monitor Evaluator provides a logical eye, making impartial judgements where required and weighs up the team's options in a dispassionate way; Shaper provides the necessary drive to ensure that the team keeps moving and does not lose focus or momentum; Completer Finisher used at the end of tasks to polish and scrutinise the work for errors, subjecting it to the highest standards of quality control.

This is obvious, that some members of team, including a manager can execute several roles simultaneously.

The Belbin Team Inventory Method (BTIM) helps to determine the suitability of individuals for team roles. Based on BTIM results, individuals can be assigned appropriate Belbin roles to perform. Above described methods of evaluation of intellect and its component allow enriching the possibilities of the BTIM.

Today, humanity has accumulated extensive experience in solving various problems using huge labor collectives. The creation of new machines such as space crafts, computers and computer systems, new drugs and so on usually correlate with the success of effortful control. This type of controls allow for people working in a team can do much more than individually.

CONCLUSION

This is the third book written by the author in his life, but the first, which was hard to finish. One of the main reasons was that the author took up the study of a very complex system –human brain and body, in which a lot of things are still unclear simply because the problem lies at the junction of many sciences.

On the other hand, due to the lack of a sufficient number of research objects, the author did not include in the book an analysis of some factors affecting the human energy, for example, the rotation of the body and its parts clockwise and counterclockwise, the form of clothing (for example, skirts for women), the form of furniture or even such utensils like a sieve. In this connection, quantitative and qualitative methods for assessing the degree of functioning of the chakras, methods for assessing the degree of people's attachment (affection) to each other, including love and hate, and much more are not shown.

The author hopes to do this in a separate book or in the next edition of this book, if readers will like that.

ABOUT THE AUTHOR

Yuri Iserlis, PhD., graduated in 1961 from St. Petersburg State Polytechnic University (Russia) as mechanical engineer, and completed his postgraduate studies at State Marine Technical University of St. Petersburg (Russia) in 1972.

In the course of his engineering and scientific activity he participated in many projects, concerning intellect development (human and artificial), mathematics (modelling of complex processes, statistics and optimization), medicine (official and alternative) and new methodologies (of systematic designs and automatic control).

Dr. Iserlis has devoted many years to the development of a methodology to optimize the design of complex machines such as marine diesel engines and turbines. He developed also digital control and SCADA systems, and conducted some projects based on theory and methods of Artificial Intelligence. As a result was created one of the first Intellectual Control system for Russian gas compression stations of gas transmission pipelines. He has worked as designer, researcher, analyst and scientist in various industries, education and medicine. Dr. Iserlis founded and operated the company, Intellectual System, Inc., in St. Petersburg, Russia, from 1992 to 1996, where he developed some real-time industrial intellectual systems. In 1996, Dr. Iserlis immigrated to the United States, and in 2002 founded his latest company, Clever Ace, in San Jose, California (closed in 2013).

He published more than 70 books and articles. Amid them during last decades published in the USA an essay "The Formula of Happiness, and Why Fools Rule the World" (in Russian- 1996), and the book:" Artificial Intelligence Around Us" (2009).

REFERENCES

Ackerman Jennifer. The Genius of Birds. 2016

Alexander Eben. Proof of Heaven. Neurosurgion's Journey into the Afterlife. Simon &Schuster Paperbacks. New York 2012

Alberts Bruce et al. Essential Cell Biology. 2014

Albrecht-Buehler Guenter. Cell intelligence. 2013 http://www.basic.northwestern.edu/g-buehler/FRAME.HTM

Alterman T, Luckhaupt SE, Dahlhamer JM, Ward BW, Calvert GM. Prevalence Rates of Work Organization Characteristics Among Workers in the U.S.: Data From the 2010 National Health Interview Survey. Am J Ind Med. 2013

Altshuller G.S. The Algorithm of an Invention. 2007

Altshuller G.S The Innovation Algorithm: TRIZ, Systematic Innovation and Technical Creativity, 1998

Alves Carlos J. S. et al. Optimization in Medicine. 2008.

Amen Daniel G. Change your brain change your body. Three Rivers Press. New York. 2010

Amosov Nikolai. Modeling of thinking and psyche. 1965 (In Russian) Н.М. Амосов Моделирование мышления и психики. 1965

Anagnostopoulos G.C. et al. A study of correlation between seismicity and mental health: Crete, 2008–2010. Geomatics, Natural Hazards and Risk Journal. Volume 6, 2015 - Issue 1

Andrews Synthia.The Path of Emotions. 2013

Armstrong D.M. A Materialist Theory of the Mind. 1993

Atkinson, R.C.; Shiffrin, R.M. "Chapter: Human memory: A proposed system and its control processes". In Spence, K.W.; Spence, J.T. The psychology of learning and motivation. 2. New York: Academic Press. 1968

Atkinson William Walker. Suggestion and Auto-Suggestion. 1909

Atkinson William Walker. Mind Power: The Secret of Mental Magic. Advanced Thought Publishing Co., Chicago.1912

Atkinson, William Walker; Dumont, Theron Q.. How to heal oneself and others: mental therapeutics.1918

Axmour, J.A. Anatomy and Junction of the intrathoracic neurons regulating the mammalian heart. In: I.H. Zucker and J.P.Gilmore, eds. Reflex Control of the Circulation. Boca Raton, Fl, CRC Press. 1991.

Аугустинавичюте Аушра. Соционика: Введение. - 1998.

Backstein Karen. The Blind Men and the Elephant. 1992

Baddeley Alan et al. Memory. 2015

Baker Douglas, de la Barr G.W. et al. Biomagnetism. 2011

Balter Michael. What Does IQ Really Measure? Science. Vol 364, Issue 6435. Apr. 25, 2011

Bandler R. and Grinder J. Frogs into Princes: Neuro Linguistic Programming. 1979

Bandler R., Fitzpatrick O. How to Take Charge of Your Life: The User's Guide to NLP, 2014

Barbey A.K. et al., An integrative architecture for general intelligence and executive function revealed by lesion mapping, Brain: A Journal of Neurology, 2012.

Baum, Peter. Thought Forms - The Structure, Power, and Limitations of Thought: Volume 1 - Introduction to the Theory. 2012

Baumeister Roy F., Tierney John. Willpower: Rediscovering the Greatest Human Strength. The Pinguins Press. New-York, 2011

Beichler James E. The Biofield: Life as a Hyper-Dimensional Quantity. Annual Meeting of the U.S. Psychotronics Association 16 to 18 July 2004, Columbus, OH

Begley Sharon. Train Your Mind, Change Your Brain: How a New Science Reveals Our Extraordinary Potential to Transform, 2007

Begley Sharon. Buff your brain. Newsweek 1/1/2012

Belyaeva E.A et al. The physiological value of different vibrations and rhythms, Электронный журнал. Вестник новых медицинских технологий-2015-N1

Bekhtereva Natalia. The human brain is over-capacity and prohibitions. Science and life journal, N 7, 2001 (in Russian)

Bender Andrea and Beller Sieghard. Current Perspectives on Cognitive Diversity. Frontiers in Psychology. 2016; April 12

Berekeley School information. New Research: Computer that can identify you by your thoughts. April 3, 2013.

Belbin R. Meredith. Management Teams: Why they succeed or fail. 2010

Biden John. Mind Control, Human Psychology, Manipulation, Persuasion and Deception Techniques Revealed! 2017.

Blackett, Glyn. Mind-Body Intelligence: How to manage your mind using biofeedback and mindfulness. Kindle Edition. 2014

Blakeslee Sandra. Cells That Read Minds. New-York Times. JAN. 10, 2006

Blanke Olaf et al. Leaving body and life behind: out-of body and near-death experience, In the book The Neurology of Consciousness: Cognitive Neuroscience and Neuropathology. 2015

Blanc Sophia (София Бланк. Благославление вселенной. Слово очищает биополе. Вектор. СПб. 2010)

Blau P. Exchange and Power in Social Life. New York: Wiley. 1964.

De Bono Edward. Six Thinking Hats. 1999

De Bono Edward. Lateral thinking: creativity step by step. 2015

Bodian Stephan. Meditation For Dummies. 2016

Bohus Dan, Rudnicky Alexander I. Modeling the Cost of Misunderstanding Errors in the CMU Communicator Dialog System. Carnegie Mellon University. 2001

Bonabeau Eric and Theraulaz Guy. Swarm Intelligence: From Natural to Artificial Systems (Santa Fe Institute Studies on the Sciences of Complexity), 1999

Borsboom Denny. Measuring the Mind: Conceptual Issues in Contemporary Psychometrics. 2009

Bradley Raymond Trevor. The psychophysiology of entrepreneurial intuition: a quantum-holographic theory. Proceedings of the Third AGSE International Entrepreneurship Research Exchange, February 8–10, 2006, Auckland, New Zealand

Brenner Abigail. The Inner Language of the Subconscious.A Picture is Worth a Thousand Words. Psychology Today. Jan 29, 2013.

Brennon Barbara A. Hands of Light: A Guide to Healing Through the Human Energy. Field, 1988

Briggs John. Fire in the crucible understanding the process of creative genius. An Alexandric Book. Los Angeles. 2000

Broadbent, D. Perception and communication. 1958

Brodie Richard. Virus of the Mind: The New Science of the Meme. 2009

Brodsky Joseph. On Grief and Reason. 1997

Bruyere Rosalyn. Wheels of Light: Chakras, Auras, and the Healing Energy of the Body. 1994

Bukalov Alexander V. Socionics: humanitarian, social, political and information intellectual technologies of the XXI century. Journal "Socionics, mentology and personality psychology", N 1, 2000.

Bukowski A. Willpower And Self-Control: Strengthen Your Willpower Today With Mind Performance Hacks To Achieve Your Full Potential, 2014

Buzan Tony. Head First!: 10 Ways to Tap into Your Natural Genius. 2002

Caianiello, E. Outline of a theory of thought-processes and thinking machines. Journal of Theoretical Biology. Volume 1, Issue 2, 1961

Cade, C. Maxwell and Nona Coxhead. The Awakened Mind: Biofeedback and the Development of Higher States of Awareness. Element Books, Shaftesbury, Dorset, England, 1989.

Callahan Roger J. Five Minute Phobia Cure: Dr. Callahan's Treatment for Fears, Phobias and Self-Sabotage. 1985

Callahan Roger J. and Trudo Richard. Tapping the Healer Within: Using Thought- Field Therapy to Instantly Conquer Your Fears, Anxieties, and Emotional Distress. 2001

Cannon B. Walter. The wisdom of the body. 1932

Cardenas Diana and Stout Dale. Personality and Intellectual Abilities as Predictors of Intelligent Behavior. Journal of Interpersonal Relations, Intergroup Relations and Identity. Volume 3, Hiver/Winter. 2010

Castaneda Carlos. The Teachings of Don Juan: A Yaqui Way of Knowledge. University of California Press. 1968

Castaneda Carlos. The Power of Silence. WSP. 1987

Castaneda Carlos. Art of dreaming, HarperPerrenial, 1993

Campbell Don. The Mozart Effect: Tapping the Power of Music. 1997

Carpenter Harry. The Genie Within. Your subconscious mind. Harry Carpenter Publishing, 2011, Fallbrook, CA

Carpenter Harry. The Genie Within. Esses, 2014

Carroll J. B., "The three-stratum theory of cognitive abilities" in D. P. Flanagan, J. L. Genshaft et al., Contemporary intellectual assessment: Theories, tests, and issues, Guilford Press, New York, 1997

Chambers Becky. Whole Body Vibration. 2013

Chassiakos Y.R. et al. Children and Adolescents and Digital Media. From the American Academy of Pediatrics. Technical Report. Pediatrics November 2016, VOLUME 138 / ISSUE 5

Cheek David B. and Rossi Ernest L. Mind-Body Therapy: Methods of Ideodynamic Healing in Hypnosis. 1994

Childre Doc et al. Heart Intelligence: Connecting with the Intuitive Guidance of the Heart. 2017

Chizhevsky Alexander, Space Pulse of Life. (Originally titled The Earth In The Embrace Of The Sun), Moscow: Misl, 1995 (in Russian).

Chopra Deepak. The seven spiritual laws of success. 1994

Chopra Deepak, Tanzi Rufolf E. Super Brain: Unleashing the Explosive Power of Your Mind to Maximize Health, Happiness, and Spiritual Well-Being. 2013

Chun Marvin M. and Turk-Browne Nicholas. Interactions between attention and memory. Science Direct. Current Opinion in Neurobiology 2007, 17

Cipolla Carlo. M. The Basic Laws of Human Stupidity. 2011

Clark Hulda. The Cure for All Diseases. 2010

Clark Lynn. SOS Help For Emotions: Managing Anxiety, Anger & Depression. 2015

Clark Ruth, et al. Efficiency in learning. Evidence-based guidelines to manage cognitive load. 2006

Cobb Jeff. Leading the Learning Revolution: The Expert's Guide to Capitalizing on the Exploding Lifelong Education Market. 2013

Cohen Gene D. The Mature Mind: The Positive Power of the Aging Brain. 2006

Collingwood Jane. The Power of Music To Reduce Stress. Psych Central. 2016

Costa, P.T.,Jr. & McCrae, R.R. (1992). Revised NEO Personality Inventory (NEO-PI-R) and NEO Five-Factor Inventory (NEO-FFI) manual. Odessa, FL: Psychological Assessment Resources.

Connel Lisa. Mind Maps Unlimited: Discover How To Remember Everything, Improve Your Memory and Learning Techniques. Kindle Edition.2014

Cox, W.E. Precognition: An analysis II. J.ASPR, Vol.30. 1956

Cox, T.F.; Cox, M.A.A. Multidimensional Scaling. 2001

Craig Gary. The EFT Manual (Everyday EFT: Emotional Freedom Techniques), 2008

Crane Kristine. 8 Ways Meditation Can Improve Your Life. WELLNESS 09/19/2014 Updated Dec 06, 2017

Crew Bec. Scientists Have Invented a Mind-Reading Machine That Visualizes Your Thoughts. Science Alert. 23 June 2016

Crew Bec. This physicist says consciousness could be a new state of matter 'Perceptronium'. Science Alert. 16 sept. 2016

Csikszentmihalyi, M. Finding flow: The psychology of engagement with everyday life. New York: HarperCollins Publishers. 1997

Csikszentmihalyi, M. Flow and creativity. NAMTA Journal, 1997 22(2), 60-97

Dale Cyndi. The Subtle Body: An Encyclopedia of Your Energetic Anatomy. 2009

Dalai-Lama. Stages of Meditation. 2003

Darby Flower. Harness the Power of Emotions to Help Your Students Learn. Faculty Focus. Higher Ed teaching strategies from Magna publication. January 3, 2018

Danzico Matt. Brains of Buddhist monks scanned in meditation study. BBC News, April 24, 2011

Dawkins Richard. The Selfish Gene: 30th Anniversary Edition–with a new Introduction by the Author. 2006

Dean Liz. Switchwords: How to use one word to get what you want. 2015

Deary IJ, Spinath FM, Bates TC. Genetics of intelligence. Eur J Hum Genet. 2006;14:690–700. http://www.ncbi.nlm.nih.gov/pubmed/16721405

Delaforge, Gaeten, "The Templar Tradition: Yesterday and Today", Gnosis Magazine, #6, 1987. Int J Biometeorol. 1999 Jul;43(1):31-7.

Delyukov A. et al. The effects of extra-low-frequency atmospheric pressure oscillations on human mental activity. International Journal of Biometeorology. 1999 Jul;43(1):31-7.

Demertzi, A., Vanhaudenhuyse, A. et al. Is there anybody in there? Detecting awareness in disorders of consciousness. Expert Review of Neurotherapeutics 8(11), 2008

Densmore Fransis. The use of music in the treatment of the sick by American Indians. The Musical Quarterly. Vol 13, No 4. Oct.1927

De Sousa, Ronald. Emotion. The Stanford Encyclopedia of Philosophy (Winter 2017 Edition)

Deutsch, J. A.; Deutsch, D. "Attention: Some Theoretical Considerations". Psychological Review. 70, 1963

Dispenza Joe. Breaking the habit of being yourself. How to lose your mind and create a new one. 2012

Digman J.M. Personality structure: Emergence of the five-factor model. Annual Review of Psychology 1990,41

Doidge Norman. The Brain That Changes Itself: Stories of Personal Triumph from the Frontiers of Brain Science. 2007

Dolan A. How a child prodigy at Oxford became a £130-an-hour prostitute. Daily Mail. 05 April 2008

Doran G. T. There's a S.M.A.R.T. way to write management's goals and objectives. Management Review. AMA FORUM. 70 (11).1981.

Dorigo Marco, Stützle Thomas. Ant Colony Optimization (Bradford Books). 2004

Dossey L. Power of Premonitions. 2009

Dryden Gordon and Vos Jeanette. The New Learning Revolution 3rd Edition (Visions of Education Series). 2005

Dulnev G.N. Energy-exchange in nature. SP SITMO, St. Petersburg. 2000 (In Russian- Дульнев.Г.Н. Энергоинформационный обмен в природе. СП ГИТМО, Санкт-Петербург. 2000)

Dweck Carol. Mindset: The New Psychology of Success: How We Can Learn to Fulfill Our Potential. 2006

Dyer Wayne W. The Power of Intention: Learning to Co-create Your World Your Way. 2005

Eason Cassandra. The Art Of The Pendulum: Simple techniques to help you make decisions, find lost objects, and channel healing energies. 2005

Eben Alexander. Proof of Heaven: A Neurosurgeon's Journey into the Afterlife 2012

Efroimson V.P. Genetics of the brilliance. 2002 (In Russian)

Eisenbud Jule. The world of Ted Serios: "thoughtographic" studies of an extraordinary mind, 1967

Einstein A. Why Socialism? Monthly Review, New York, May, 1949

Emmons Robert A. Thanks!: How Practicing Gratitude Can Make You Happier.2008

Endres Klaus-Peter. Moon Rhythms in Nature: How Lunar Cycles Affect Living Organisms, 2002

Everitt Brian S and Landau Sabine. Cluster Analysis. 2011

Evtimova Elena. Human Aura, Assemblage point, Field Resonance and Health. Научни трудове на Русенския университет- 2010, том 49, серия 8.1

Farrell Nick. Magic Imagination. The keys to magic. Llewellyn Publications, Minnesota. 2013

Feltman Rachel. Results of NASA's twin study won't actually be released for a year or two. Washington Post. March 4, 2016

Fernandez Alvaro. 10 Neurotechnologies About to Transform Brain Enhancement and Brain Health. SharpBrains. November 10, 2015

Feuillet, et al. Man with tiny brain shocks doctors. New Scientist. 20 July 2007 (The Lancet vol 370, p 262.

Feuerstein, R., Rand, Y., Hoffman, M. B., & Miller, R. Instrumental enrichment: An intervention program for cognitive modifiability. Baltimore, MD: University Park Press.1980.

Fodor Jerry A., The Language Of Thought. Crowell Press, 1975.

Fodor Jerry A. The Modularity of Mind: An Essay on Faculty Psychology, Bradford Books. 1983

Ford Brian J. Single cell intelligence. Mensa at Cambridge. Mensa magazine. February 2010

Fosar Grazyna und Bludorf Franz. Vernetzte Intelligenz. Die Natur geht online. Gruppenbewußtsein, Genetik, Gravitation. Aachen 2001.

Förster Jens et al. Why Love Has Wings and Sex Has Not: How Reminders of Love and Sex Influence Creative and Analytic Thinking. Personality and Social Psychology Bulletin. 2009. Nov;35(11):1479-91.

Fox Mark L. Da Vinci and the 40 Answers. 2008

Fraser Peter H., Massey Harry. Decoding the human body-field/ The new science of information as medicine. Healing Arts Press. Rochester, Vermont, 2008

Franz Alexander. Psychosomatic Medicine: Its Principles and Applications. 1950

Furr R. Michael and Bacharach Verne R. Psychometrics: An Introduction. 2007

Gabrenya Jr., W. K. Culture and Social Class. 2003 http://my.fit.edu/~gabrenya/social/readings/ses.pdf.

Gnezditskiy V.V. The inverse problem of EEG and clinical electroencephalography (in Russian -Гнездицкий В.В. Обратная задача ЭЭГ и клиническая электроэнцефалография). М.: МЕДпресс;информ, 2004. – 624 с.

Galtung Johan. Theories of conflict. Definitions, Dimensions, Negations, Formations Columbia University, 1958-1973

Gardner Howard. Intelligence Reframed: Multiple Intelligences for the 21st Century. Basic Books Inc. 2000

Gariaev, P.P. Wave Genome, Public Profit, Moscow, 1994. [in Russian]

Gariaev, P.P. Linguistic-wave genome: theory and practice. Institute of Quantum Genetics. - Kiev, 2009. (Гаряев П.П. Лингвистико-волновой геном: теория и практика. Институт квантовой генетики. — Киев, 2009)

Gariaev, P.P., M.J. Friedman and E.A. Leonova- Gariaeva, "Principles of Linguistic-Wave Genetics", DNA Decipher Journal 2011 Jan; 1(1):11-24, http://tinyurl.com/76shk6a

Gatehouse Jonathon. America dumbs down. Maclean's. May 15, 2014

Gautam Sandeep. Creativity Components. The Creativity Post. Sep 10, 2012

Gawain Shakti. Creative Visualization: Use the Power of Your Imagination to Create What You Want in Your Life. New World Library, 2010

Georgiou Harris. fMRI Data Reveals the Number of Parallel Processes Running in the Brain. MIT Technology Review. Emerging Technology From the arXiv. November 5, 2014

Gerber Richard. Vibrational Medicine. The #1 Handbook of Subtle-energy Therapies. 2001

Gershon Michael D. The Second Brain: A Groundbreaking New Understanding of Nervous Disorders of the Stomach and Intestine.1998

Glen Rein. Bioinformation within biofield: Beyond Bioelectromagnetics. The Journal of Alternative and Complementary Medicine. Vol 10, No 1, 2004

Godik Edward. Riddle of sensitives: that physicists have seen (in Russian). 2010`

Goleman Daniel, et al. Primal Leadership: Learning to Lead with Emotional Intelligence, 2004

Goleman Daniel. Emotional Intelligence: Why It Can Matter More Than IQ. 2005

Gottfredson Linda S. Mainstream science on intelligence: an editorial with 52 signatories, history, and bibliography. Intelligence 24, 1997

Gottfredson Linda S. The General Intelligence Factor. Scientific American Presents, 9(4),1998

Goffman Erving. The Presentation of Self in Everyday Life. 1959

Gorsuch R. L. Factor Analysis. 1983

Green Linda. Mental biolocation (in Russian. Грин Лю Ментальная биолокация. Воронеж. 1994

Greenberger Daniel M., et al. Going Beyond Bell's Theorem. 1989

Gómez-Pinilla Fernando. Brain foods: the effects of nutrients on brain function. Nat Rev Neurosci. 2008 Jul; 9(7)

Grasland-Mongrain Pol et al. Ultrafast imaging of cell elasticity with optical microelastography. Proceedings of the National Academy of Sciences, 2018

Grush Loren, Modern day telepathy? Scientists develop method for reconstructing thoughts. FoxNews, February 1, 2012

Grof S. "Theoretical and empirical basis of transpersonal psychology and psychotherapy: Observations from LSD research". Journal of Transpersonal Psychology, Vol. 5, 1973

Grof S. The adventure of self-discovery. Dimensions of consciousness and new perspectives in psychotherapy and inner exploration. State University of New York Press, 1988.

Grof S. The Holotropic Mind: The Three Levels of Consciousness and How They Shape Our Lives. San Francisco, CA: Harper Collins.1992

Gott J. Richard. The Cosmic Web: Mysterious Architecture of the UniverseJan 26, 2016

Giulia Enders. Gut: The Inside Story of Our Body's Most Underrated Organ, 2018

Hameroff S. R., & Penrose R. Consciousness in the universe: A review of the 'Orch OR' theory. Physics of Life Reviews Volume 11, Issue 1, March 2014, Pages 39-78

Hameroff S. R., & Penrose R. Orchestrated reduction of quantum coherence in brain microtubules: A model for consciousness. In S. R. Hameroff, A. Kaszniak, & A. C. Scott (Eds.), Toward a science of consciousness the first Tucson discussions and debates (pp. 507-540). Cambridge: MIT Press 1996

Hameroff S.R. Quantum computation in brain microtubules? The Penrose-Hameroff "Orch OR" model of consciousness. Philosophical Transactions Royal Society London, Series A, 1998

Hameroff Stuart and Chopra Deepak. The Quantum Soul: A scientific Hypothesis. Exploring Frontiers of the Mind-Brain Relationship Eds: A Moreira-Almeida and F Santana Santos, Springer, 2011

Hameroff S.R., Craddock T.J., Tuszynski J.A. Quantum effects in the understanding of consciousness. J Integr Neurosci. 2014. Apr 13(2)

Hampshire Adam et al. Fractionating Human Intelligence. Neuron, Volume 76, Issue 6, 2012

Haxeltine Michael. Bovis Units in Everyday Life: A Discussion Book - Dowsing Spiral-bound. 2008

Hartshorne Joshua K. and Germine Laura T. When Does Cognitive Functioning Peak? The Asynchronous Rise and Fall of Different Cognitive Abilities Across the Life Span. Psychological Science. December 15, 2014

Hassanien (Editor) et al. Applications of Intelligent Optimization in Biology and Medicine: Current Trends and Open Problems. 2016

Hawkins David R. Power vs. Force, 2002

Hawkins Jeff with Sandra Blakeslee. On Intelligence. Times Books, New-York, 2004

Hay Loiuse. You Can Heal Your Life. 1984

Henriques Gregg. 10 Problems With Consciousness. Psychology Today. Dec 05, 2018

Hermann, N., What is the Function of the Various Brainwaves? Scientific American, Dec. 22, 1997

Herrnstein Richard J. and Murray Charles. Bell Curve: Intelligence and Class Structure in American Life. 1996

Heylighen Francis. Gifted people and their problems. Personal website http://pespmc1.vub.ac.be/HEYL.html

Homans G.C. Social Behavior as Exchange. American Journal of Sociology,1958. 63

Huang T.L., & Charyton, C. A comprehensive review of the psychological effects of brainwave entrainment. Alternative Therapies in Health and Medicine, 14(5), 2008

Hutchison Michael. The Book of Floating: Exploring the Private Sea. 1984, 2003

Hutchison, Michael. Mega Brain: New Tools And Techniques For Brain Growth And Mind Expansion BALLANTINE BOOKS – NEW YORK, 2014

Harris Bill. Thresholds of the mind. Centerpointe Press. Beaverton, Oregon, 2007

Harris Bill. The New Science of Super Awareness. Centerpointe Press. Beaverton, Oregon, 2015

Hellige Joseph. Hemispheric Asymmetry: What's Right and What's Left (Perspectives in Cognitive Neuroscience) 2001

Horn J. & Cattell, R. "Refinement and test of the theory of fluid and crystallized general intelligences.". Journal Of Educational Psychology. 57. 1966

Horn Terry, Wootton Simon. Train Your Brain, Hodder EducationTrain. 2007

Hölzel B., Carmody J., Vangel M., Congleton, C., Yerramsetti, S., Gard, T. & Lazar, S. (2011) Mindfulness practice leads to increases in regional brain gray matter density. Neuroimaging. 191. 36-43.

Horowitz Janice M. The Man with Magic Fingers. Web.archive.org. 2011-01-13.

Hunt Valerie. Infinite Mind. Science of the Human Vibrations of Consciousness. Malibu Publishing. 2000

Hofstadter Richard. Anti-Intellectualism in American Life. 1966

Hunt V. - The Promise of Bioenergy Fields. An End to Au Disease. An Interview with Dr. Valerie Hunt by Susan Barber, V. 2, 2000. http://www.spiritofmaat.com/archive/nov1/vh.htm

Iserlis Yu. E., Lojko V.I. To question of optimization individual therapeutic treatment. Transactions of Len. San. hygienic Medical Institute, 1981, vol. 138, p. 93-100. In Russian.

Iserlis Yuri. The formula of happiness and why fools rule the world.(In Russian: Формула счастья и почему дураки правят миром. Доместик. Сан Хозе, 1996, №5,6)

Iserlis Yu.E., Ur'ev A.M. "Modeling of the magnetic field of arbitrary shape in a body with the aid of a set of elementary sources system", "Electrichestvo" (Electricity), 1985, No 6. In Russian. Abstract: Electrical & Electronics Abstracts (Science Abstracts Series B), November 1985, No 62727.

Iserlis Yu.E., Malin V.A. Optimizing load distribution between gas-pumping assembles in the compressor unit of a main gas pipeline". "Soviet Energy Technology "(English translation of "Energomashinostroenie"), 1986, No 7, p.1-9.

Iserlis Y. Artificial intelligence around us. Bookstandpublishing. 2009

Izard C. E. Human emotions. New York: Plenum Press 1977

Jantz Gregory L. Brain Differences Between Genders. Psychology Today. 2014

Jaruševičius G. et al. Correlation between Changes in Local Earth's Magnetic Field and Cases of Acute Myocardial Infarction. International Journal Environmental Research and Public Health. 2018 Mar; 15(3): 399.

Jaušovec, N., Jaušovec, K. Working memory training: Improving intelligence – Changing brain activity. Brain and Cognition; 79(2).2012

Jabr Ferris. How brainless slime molds redefine intelligence. Nature. 13 November 2012

Jahn, R.G.; Dunne B.J.. "On the Quantum Mechanics of Consciousness with Application to Anomalous Phenomena". Foundations of Physics 18 (6): 721–772. 1986

Jahn Robert and Dunne Brenda. Science of the Subjective. Journal of Scientific Exploration, Vol. 11, No. 2, 2007

Jahn Robert and Dunne Brenda. Consciousness and the Source of Reality. 2011

Jaeggi, S. M., Buschkuehl, M., Jonides, J., & Perrig, W. J. Improving fluid intelligence with training on working memory. Proceedings of the National Academy of Sciences of the United States of America, 105(19), 2008

Johnson Jerry Alan, "Chinese Medical Qigong Therapy: A Comprehensive Clinical Text", 2000. The International Institute of Medical Qigong, Pacific Grove, Ca

Josipovic Zoran et al. Influence of meditation on anti-correlated networks in the brain. Frontiers in Human Neuroscience. 03 January 2012

Kahneman, Daniel. Attention and Effort. Prentice Hall. 1973

Kahneman, Daniel. Thinking, Fast and Slow Paperback. 2013

Kahneman, Daniel and Tversky Amos. Judgment Under Uncertainty: Heuristics and Biases. Cambridge University Press. 1982

Kain Paul. Positive Thinking: The Secret To Reprogramming Your Mind For Maximum Happiness. 2016

Kafatos Menas C, et al. Biofield Science: Current Physics Perspectives. Biofield Science and Healing: Toward a Transdisciplinary Approach. Global Advances in Health and Medicine. 2016.

Karagulla Shafica. Breakthrough To Creativity.1967

Kancharla. Veera Raghavaiah. Happiness and Sadness. International Journal of Psychology and Counseling (IJPC) Volume 4, Number 1.2014.

Kaufman Scott Barry. Intelligence Is Still Not Fixed at Birth. Psychology Today. Oct 21, 2011

Kaufman Alan S. and Raiford Susan Engi. Intelligent Testing with the WISC-V, 2016

Kegan Robert, Lahey Lisa Laskow. Immunity to Change: How to Overcome It and Unlock the Potential in Yourself and Your Organization. 2009

Kehoe J. Mind Power into the 21st century. Zoetic Inc. Vancover. 2011

Keith, T. & Reynolds, M. Cattell-Horn-Carroll abilities and cognitive tests: What we've learned from 20 years of research. Psychology in the Schools, 2010, 47(7).

Kelly Kevin. The Inevitable: Understanding the 12 Technological Forces That Will Shape Our Future. 2016

Khabarova O.V (in Russian -. Хабарова О. В. Биоэффективные частоты и их связь с собственными частотами живых организмов. «Биомедицинские технологии и радиоэлектроника». 2002, №5

Khalsa Dharma Singh with Stauth Cameron. Brain Longevity. Warner Books. 1997

Kiefer Charles F. and Constable Malcolm.The Art of Insight: How to Have More Aha! Moments. 2013

Kilner, Walter J., The Human Atmosphere, or the Aura Made Visible by the aid of Chemical Screens, 1911

Kiseleva et al. The psychological portrait as a tool to improve the subjective well-being of the client in the context of personal sales. SHS Web of Conferences 28, 2016. https://core.ac.uk/download/pdf/53095632.pdf

Kitamura T. et al. Engrams and circuits crucial for systems consolidation of a memory. Science 2017. Vol. 356, Issue 6333

Knapton Sarah. Intelligence genes discovered by scientists. The Telegraph. 21 Dec 2015

Koch Rudolf. The book of signs. 493 symbols used from earliest times to the middle ages by primitive peoples & early christians. 1965.

Kobayashi Masaki et al. Imaging of Ultraweak Spontaneous Photon Emission from Human Body Displaying Diurnal Rhythm. Journal PLOS. July 16, 2009

Koposov Nicolay. To assess the scale of Stalin's repressions. (In Russian – Николай Копосов. К оценке масштаба сталинских репрессий. 11 декабря 2007. http://polit.ru/article/2007/12/11/repressii/

Korotkov K.G., Krizhanovsky E.V. et al. The Dynamic of the Gas Discharge around Drops of Liquids. In book: Measuring Energy Fields: State of the Science, Backbone Publ.Co., Fair Lawn, USA, 2004.

Korotkov K., Korotkin D. Concentration Dependence of Gas Discharge around Drops of Inorganic Electrolytes, Journal of Applied Physics, 89, 9, 2001

Korotkov Konstantin, et al. Assessing Biophysical Energy Transfer Mechanisms in Living Systems: The Basis of Life Processes. The Journal of Alternative and Complementary Medicine. Jun 2004

Koyokina O.N. Managed by the consciousness of space-time structuring of the active environment. (In Russian- Коёкина О.И. - Управляемое сознанием пространственно-временное структурирование активной среды. (Нейрофизиологические исследования), Ж. «Традиционная медицина» №1, 2003)

Koyokina O.N Ability to integral bodily perception and clairvoyance of traditional healers (In Russian- Коёкина О.И. Способности к интегральному телесному восприятию и ясновидению у народных целителей (Нейрофизиологические исследования). Парапсихология и психофизика. - 2000. - №2)

Kosslyn Stephen M. and Miller G. Wayne. A New Map of How We Think: Top Brain/Bottom Brain. The Wall Street Journal. Oct. 20, 2013

Kross Ethan et al. Self-Talk as a Regulatory Mechanism: How You Do It Matters. Journal of Personality and Social Psychology.2014, Vol. 106, No. 2.

Kruger, Justin; Dunning, David. "Unskilled and Unaware of It: How Difficulties in Recognizing One's Own Incompetence Lead to Inflated Self-Assessments". Journal of Personality and Social Psychology. 1999. 77 (6)

Larson Christina. Gene-editing Chinese scientist kept much of his work secret. The Washington Times. November 27, 2018

Law, Larry. Willpower Instinct Guide: Proven Methods to Increase Willpower with Self Control and Self Discipline 2013

Lacey J.I, and Lacey B.C. Two-way communication between the heart and the brain. Significance of time within the Cardiac Cycle. American Psychologist. 1978, February

Lakhovsky Georges. The Secret of Life: Cosmic Rays and Radiations of Living Beings. 1935.

Leadbeater C.W. Thought-Forms: A Record of Clairvoyant Investigation. Written in collaboration with Annie Besant. London: Theosophical Publishing Society. 1901

Lehar Steven M. The World in Your Head: A Gestalt View of the Mechanism of Conscious Experience. 2013

Lebedev et al. Editorial: Augmentation of Brain Function: Facts, Fiction and Controversy. Frontier in Systems Neuroscience, 12 Sep. 2018

Leonovich A.S., Mazur V.A. Resonance excitation of standing Alfven waves in an axisymmetric magnetosphere (monochromatic oscillations) // Planet. Space Sci. - 1989. V. 37. – P. 1095 -1107.

Levitin Daniel J. This Is Your Brain on Music: The Science of a Human Obsession. 2007

Levitin Daniel J. The Organized Mind: Thinking Straight in the Age of Information Overload Penguin Group US. 2015

Liang P. et al. CRISPR/Cas9-mediated gene editing in human tripronuclear zygotes. Protein Cell. 2015 May;6(5):363-372

Lilly John C. Programming and Metaprogramming in the Human Biocomputer: Theory and Experiments. Coincidence Control Publishing, Portland, 2014

Lynes Barry. Rife's World of Electromedicine: The Story, the Corruption and the Promise. 2009

Lipton B.H. The Biology of Belief: Unleashing the Power of Consciousness, Matter & Miracles. 2005

Lipton B.H, Nature, Nurture and Human Development, 2012 https://www.brucelipton.com/resource/article/nature-nurture-and-human-development

Liu Tianjun. The scientific hypothesis of an "energy system" in the human body.Journal of Traditional Chinese Medical Sciences. Volume 5, Issue 1, January 2018, Pages 29-34

Longley, Peter. Auras: The complete guide to auras, seeing auras, feeling auras, sensing auras, and understanding auras and astral colors! 2014

Longley, Peter. Pendulum Dowsing: The complete guide to pendulum dowsing, divination, and more!. Kindle Edition. 2015

Looi, C.Y. et al. The Use of Transcranial Direct Current Stimulation for Cognitive Enhancement (published in Cognitive Enhancement: Pharmacologic, Environment and Genetic Factors). Elsevier. January 5, 2015

Looi, C.Y. et al. Combining brain stimulation and video game to promote long-term transfer of learning and cognitive enhancement. Scientific Reports. Nature Publishing Group. 2016

Ludwig A.W. Altered states of consciousness // Altered states of consciousness: A book of reading. — N.Y., 1969.

Ludwig, Arnold M. Altered States of Consciousness (presentation to symposium on Possession States in Primitive People). Archives of General Psychiatry September 1966, 15 (3)

Lugavere Max and Grewal Paul. Genius Foods: Become Smarter, Happier, and More Productive While Protecting Your Brain for Life 2018

Lulfityanto, G., Donkin, C. & Pearson, J. Measuring Intuition: Non-conscious Emotional Information Boosts Decision Accuracy and Confidence. Psychological Science. April 2016

Lutz, A., Slagter, H. A., Dunne, J. D., & Davidson, R. J. Attention regulation and monitoring in meditation. Trends Cogn Sci, 12(4), 2008.

Maltby J. et al. Personality, Individual Differences and Intelligence, 2017

Mangan James T. The Secret of Perfect Living. 1963

Mårtensson Johan, Lovden Martin. Do Intensive Studies of a Foreign Language Improve Associative Memory Performance? Frontiers in Psychology.V.2; 2011

Mars Justin. Control Emotions: The Best Guide On Mastering Emotions – Build Self Confidence, Increase Self Esteem & Eliminate Anxiety. 2016

Martin Wendy Bruce. Breaking Through a Mental Block: The Athlete's Guide to Becoming Fearless. 2016

Magaldi Kristin. The Sleep-Memory Connection And All The Ways We Can Learn In Our Sleep. Medical Daily. Aug 21, 2015

Mayer John D., Brackett Marc A., Salovey Peter. Emotional Intelligence: Key Readings on the Mayer and Salovey Model. 2004

McGonigal Kelly. The Willpower instinct. Aviry, New York, 2012

McGourty, J & DeMeuse, KP. Team developer: an assessment and skill building program, John Wiley & Sons, 2001

McGreal Scott A. The Illusory Theory of Multiple Intelligences. Psychology Today, November 23, 2013

McCrae, Robert R., and Oliver P. John. An introduction to the five-factor model and its applications. Journal of Personality 1992. 60.2

McCraty Rollin, et al. Electrophysiological Evidence of Intuition: Part.1 The Surprising Role of the Heart. The Journal of alternative and complementary medicine. Volume10, Number 1, 2004.

McCraty Rollin, Mike Atkinson, and Raymond Trevor Bradley. Electrophysiological Evidence of Intuition: Part 2. A System-Wide Process? The Journal of alternative and complementary medicine. Volume 10, Number 2, 2004

McCraty Rollin. The Coherent Heart: Heart-Brain Interactions, Psychophysiological Coherence, and the Emergence of System-Wide.2006

McCraty, R., R.T. Bradley, & D. Tomasino (2004-2005), The resonant heart, Shift, special issue—"The Science of Fields," No. 5, Dec. 2004-Feb.2005

McCrea Simon M. Intuition, insight, and the right hemisphere: Emergence of higher sociocognitive functions. Psychology Research and Bahavior Management. 2010, 3.1-39

McGonigal Kelly. The Willpower Instinct: How Self-Control Work. 2012

McLendon Russel. Why Ping-Pong is good for your brain. Mother Nature network. April 18, 2016

Meier Dave. The Accelerated Learning Handbook: A Creative Guide to Designing and Delivering Faster, More Effective Training Programs. 2000

Merrick Christina, et al. External control of the stream of consciousness: Stimulus-based effects on involuntary thought sequences. Consciousness and Cognition. Volume 33, May 2015

Mermet Abbe and Clement Mark. Principles and Practice of Radiesthesia: A Textbook for Practitioners and Students. 1991

Merzenich, Michael. Soft-Wired: How the New Science of Brain Plasticity Can Change Your Life. 2013

Mientka Matthew. Adulthood Extended To Age 25 By Child Psychologists In UK. Medical Daily. Sep 24, 2013

Milbradt David. Bonghan Channels in Acupuncture. Acupuncture Today. April, 2009, Vol. 10, Issue 04

Michalko M. Craking Creativity: The Secrets of Creative Genius. Crown Publishing Group.New York, 2001

Michalko M. Creative Thinkering: Putting Your Imagination to Work. New World Library, Novato, 2011

Miller, I., Mill, R. A. & Webb, B., Quantum Bioholography. DNA Decipher Journal. March 2011. Vol. 1. Issue 2

Mininni Darlene. The Emotional Toolkit: Seven Power-Skills to Nail Your Bad Feelings, 2006

Miyake Akira and Shah Priti. Models of Working Memory: Mechanisms of Active Maintenance and Executive Control. 1999

Moir Anna and Jessel David. Brain Sex: The real difference between men and women. 2015

Monroe Robert. Journeys Out of the Body Harmony, New-York, 1971

Morcan, James & Lance. Genius intelligence: Secret Techniques and Technologies to Increase IQ, 2014

Moody Raymond. Life after life. 1975

Muhl Anita M. Automatic Writing: An Approach to the Unconscious. 1963

Murphy Josef. The Power of Your Subconscious Mind, 2008

Murphy Brendan D. The Grand Illusion: A Synthesis of Science and Spirituality. Book One/ Balboa Press. 2012.

Muto Akira, et al. Real-Time Visualization of Neuronal Activity during Perception. Current Biology. Volume 23, Issue 4, 18 February 2013.

Montagnier L., et al. DNA waves and water, 2011

Mishlove Jeffrey and Saul-Paul Sirag. The Roots of Consciousness: The Classic Encyclopedia of Consciousness Studies Revised and Expanded. 1997

Myers, Isabel Briggs with Myers Peter B. Gifts Differing: Understanding Personality Type. Mountain View, CA: Davies-Black Publishing. 1980,1995

Narby Jeremy. Intelligence in Nature. 2006

Nelder, John A.; Mead R. "A simplex method for function minimization". Computer Journal. 7:, 1965

Nelson J, et al. The Effects of transcranial direct current stimulation (tDCS) on multitasking throughput capacity. Frontiers in Human Neuroscience. 29 November 2016

Nelson, R. D. 'Princeton Engineering Anomalies Research (PEAR)'. Psi Encyclopedia. London: The Society for Psychical Research. 2017

Nims, Larry P., and Sotkin Joan. Be Set Free Fast!: A Revolutionary New Way to Eliminate Discomforts, 2003

Nisbett Richard E. Mindware: Tools for Smart Thinking. 2015

North, A. C. and Hargreaves, D. J. (2008). The social and applied psychology of music. Oxford: Oxford University Press. 2008

Norman, D. A. Toward a theory of memory and attention. Psychological Review. 75 (6): 1968

Oreshkin P. «Reading the muscles», not the thoughts. //Journal "Technics of Youth". Moscow, 1961. № 1

Orloff Michael A. Modern TRIZ: A Practical Course with EASyTRIZ Technology, 2012

Osborn Alex. Wake Up Your Mind; 101 Ways to Develop Creativeness.1952

Oster Gerald. Auditory Beats In The Brain. Scientific American. Mt. Sinai Medical Center. October 1973

Ostrander Sheila and Schroeder Lynn. Superlearning 2000: New Triple Fast Ways You Can Learn, Earn, and Succeed in the 21st Century. 2012

Palmary Tomas J. et al. What is synesthesia? Science. September 11, 2006

Pandya Sunil K. Understanding Brain, Mind and Soul: Contributions from Neurology and Neurosurgery. Mens Sana Monograph. 2011

Panksepp J. The emotional sources of "Chills" induced by music. Music Perception, 13, 1995

Perkins D. His Outsmarting IQ: The Emerging Science of Learnable Intelligence. The Free Press, 1995

Personality and Intelligence. Sternberg R. (editor) and Ruzgis P. (Editor). 1994

Peter, Laurence J.; Hull, Raymond. The Peter Principle: Why Things Always Go Wrong. New York: William Morrow and Company. 1969

Peters R. Ageing and the brain. Postgrad Med Journal. Feb 2006; 82(964): 84–88.

Piaget J. Piaget's theory. In P. Mussen. Handbook of Child Psychology. 1983.

Pinker Steven. How the Mind Works. W.W. Norton & Company. New-York, London. 2009

Plotkin Henry. Evolution in Mind: An Introduction to Evolutionary Psychology. Blackwell Publishing, 2004

Proctor Robert, Schiebinger Londa. Agnotology: The Making and Unmaking of Ignorance. 2008

Puchko L.G. Discovery of the Future. Multidimensional medicine in questions and answers. Issue 1. 2008 (In Russian -Пучко Л.Г. Открытия будующего. Многомерная медицина в вопросах и ответах. Выпуск 1. 2008)

Puchko L.G. Biolocation for all. System of self-diagnosis and self-healing of man. ANS. 2010. (In Russian -Пучко Л. Г. Биолокация для всех. Система самодиагностики и самоисцеления человека. АНС. 2010)

Puchko L.G. Discoveries of the future. Multidimensional person. A new highly effective algorithm for human self-healing and treatment of animals. 2008 (In Russian -Пучко Л.Г. Открытия будущего. Многомерный человек. Новый высоко-эффективный алгоритм самоисцеления человека и лечения животных. 2008)

Puchko L.G. Multidimensional medicine. 2015 (In Russian Пучко Л.Г. Многомерная медицина. 2015)

Putnam, Hilary, "Brains and Behavior", originally read as part of the program of the American Association for the Advancement of Science, Section L (History and Philosophy of Science), December 27, 1961

Quevli Nels. Cell Intelligence: The Cause of Growth, Heredity, and Instinctive Actions, Illustrating That the Cell Is a Concious, Intelligent Being, And, by Reason... Manner That Man Contructs Houses, Railroads. 1916.

Radin D.I. Effects of a-priori probability on psi perception: does precognition predict actual or probable futures? JP 52, 1988

Radin Dean. The Conscious Universe: The Scientific Truth of Psychic Phenomena. HarperCollins Publishers NY. 2010

Radin Dean. Supernormal: Science, Yoga, and the Evidence for Extraordinary Psychic Abilities, Deepak Chopra Books. NY. 2013

Radin Dean. Real Magic: Ancient Wisdom, Modern Science, and a Guide to the Secret Power of the Universe. 2018

Raichle Marcus E. et al. A default mode of brain function. Mallinckrodt Institute of Radiology and Departments of Neurology and Psychiatry, Washington University School of Medicine, St. Louis, 2000. (http://www.pnas.org/content/98/2/676.full.pdf)

Raichle Marcus E. and Abraham Z. Snyder. A default mode of brain function: A brief history of an evolving idea. NeuroImage. Washington University School of Medicine, St Louis, 6 March 2007

Reamer Andrew. The Impacts of Technological Invention on Economic Growth – A Review of the Literature. The George Washington University. Washington, DC. February 28, 2014

Regalado Antonio. China's CRISPR twins might have had their brains inadvertently enhanced. MIT Technology Review. Feb 21, 2019

Reese Debbie Denise, et al. Inspiration Brief 1. Concept Paper: Defining Inspiration, the Inspiration Challenge, and the Informal Event. Center for Educational Technologies. Wheeling Jesuit University. 2005

Rescorla Michael. The Computational Theory of Mind. Stanford Encyclopedia of Philosophy. 2015

Restak Richard. The Modular Brain, Bantam, 1995

Restak Richard. Mozart's Brain and the Fighter Pilot: Unleashing Your Brain's, 2007

Restack Richard. The big questions Mind, 2012

Rich Mark. Energetic Anatomy. An Illustrated Guide to Understanding and Using the Human Energy System. ZKife Align, Texas. 2004

Robinson Ken. Out of Our Minds: Learning to be Creative. Capston Publishing Ltd., 2011

Robson David. The surprising downsides of being clever. BBC, 14 April, 2015

Rossi E.L. et al. Twenty Minute Break C. 1991

Rossi Ernest. The Psychobiology of Mind-Body Healing: New Concepts of Therapeutic Hypnosis, 1993

Rubik, B. The Biofield Hypothesis: Its Biophysical Basis and Role in Medicine. Journal of Alternative and Complementary Medicine, December 2002, 8(6)

Ruigrok A. et al. A meta-analysis of sex differences in human brain structure. Neuroscience & Biobehavioral Reviews. Volume 39, February 2014.

Rummel R.J. Lethal Politics: Soviet Genocide and Mass Murder Since 1917. New Brunswick, N.J.: Transaction Publishers, 1990

Salovey, P., & Mayer, J. Emotional intelligence. Imagination, cognition, and personality, 9(3). 1989

Saey Tina Hesman. A recount of human genes ups the number to at least 46,831.ScienceNews Vol. 194, No. 7, October 13, 2018, p. 5

Sartini de OliveiraI Lidiane and SaramagoII Sezimária F. P. Multiobjective optimization techniques applied to engineering problems. Journal of the Brazilian Society of Mechanical Sciences and Engineering. vol.32 no.1 2010

Saugareli Kirk. How to Control Emotions: An Essential Guide to Controlling Your Emotions, Behaving Calmly, and Exuding Emotional Stability and Maturity, 2015

Shannahoff-Khalsa David S, The Ultradian Rhythm of Alternating Cerebral Hemispheric Activity, International Journal of Neuroscience, 1993, 70

Schawbel Dan. Goldie Hawn: How Her Foundation Is Supporting Our Youth. Forbes. 7/31/2013

Scheele Paul R. Photoreading: Read with Greater Speed, Comprehension, and Enjoyment to Absorb Complete Books in Minutes, 2007

Scherer Jiri. Kreativitätstechniken: In 10 Schritten Ideen finden, bewerten, umsetzen. 2007.

Schwartz Gary E. Human Energy Systems Laboratory at the University of Arizona with GDV and others instruments documented the measurement of bioelectromagnetic signals. In Ionas, W.B. et al. Proceeding: Bridging worlds and filing gabs in the science of healing. Samueli Institute for informational biology. Nov.29-Dec.3, 2001

Schwartz J.M. Begley S. The Mind and the Brain: Neuroplasticity and the Power of Mental Force, 2003

Schwarz Jack, Human Energy Systems: A Way of Good Health Using Our Electromagnetic Fields. 2001

Schumann, G Single nucleotide polymorphism in the neuroplastin locus associates with cortical thickness and intellectual ability in adolescents. Molecular Psychiatry, 2014; DOI: 10.1038

Sieger Dave. Audio-Visual Entrainment: History and Physiological Mechanisms- as published in the Association for Applied Psychophysiology and Biofeedback (AAPB) publication, "Biofeedback Magazine" Volume 31, Number 2., 2003

Silva Jose and Miele Philip. The Silva Mind Control Method. 1977

Silva Jose and Burt Goldman. The Silva Mind Control Method of Mental Dynamics by Jose Silva, 1988

Simonenko S.I. Intellectual Abilities in Manager's Success: Correlation between Results of Ability Tests and Assessment Centres. Sotsial'naia psikhologiia i obshchestvo [Social Psychology and Society], 2012. no. 1, (In Russ., abstr. in Engl.)

Simonton Dean Keith. Age and Outstanding Achievement: What Do We Know After a Century of Research? Psychological Bulletin 1988, Vol. 104, No. 2

Shcherbak, V.I. Arithmetic inside the Universal Genetic Code. BioSystems, 2003, 70, 187-209. http://dx.doi.org/10.1016/S0303-2647(03)00066-2

Sheldrake Rupert. Morphic Resonance.The Nature of formative causation.Park Street Press. 2009.

Sheldrake Rupert. Dogs That Know When Their Owners Are Coming Home. 2011

Sheldrake Rupert. Science Set Free: 10 Path to New Discovery. Deepak Choppa Books. New York. 2012

Sheldrake Rupert.The Sense of Being Stared At: And Other Unexplained Powers of Human Minds, 2013

Sheremetiev Konstantin. Intellectics. How your brain works.(In Russian). 2013

Shnoll Simon. Cosmophysical factors in random processes. Stockholm. Svenska fysikarkivat, 2009.

Shriram Sharma Acharya. The extrasensory potentials of mind. Shantikunj, Haridwar., India, 1996

Shriram Sharma Acharya. Amrit Chintan. My Long Walk. Akhand Jyoti January-February 2014

Skip Atwater F. Holmes. Induction of Expanded States of Consciousness Using Spatial Angle Modulation™ Audio Support Technology. TMI Journal 2012, № 1

Smart Andrew. Autopilot: The Art and Science of Doing Nothing, 2013

Smith Wesley J. Human Genetic Engineering Begins. Evolution News and Science Today, July 27, 2017

Smolensky Michael and Lamberg Lynne. The Body Clock Guide to Better Health: How to Use your Body's Natural Clock to Fight Illness and Achieve Maximum Health. 2015

Soski Norbert. Only Human: Guide to our internal Human Operating System (iHOS) and Achieving a Better Life. 2019

Sonnenburg Justin, Sonnenburg Erica. Control of Your Weight, Your Mood, and Your Long-term Health. 2015

Spearman, Charles. "General intelligence," objectively determined and measured. American Journal of Psychology, 15, 1904.

Sperry, R. W. "Cerebral Organization and Behavior: The split brain behaves in many respects like two separate brains, providing new research possibilities". Science. 133, 1961.

Sprague Shaw. Below trend: the U.S. productivity slowdown since the Great Recession. Bureau of Labor Statistics. Beyond the Numbers. January 2017 | Vol. 6 / No. 2

Stefanelli Mark. Monaural/binaural beats and brain states. Cymatics. 2014 http://www.thecymartist.com.

Stein Diane. The Women's Book of Healing: Auras, Chakras, Laying On of Hands, Crystals, Gemstones, and Colors Llewellyn Publications. 1998

Sternberg Robert J. Handbook of Human Intelligence, 1982

Sternberg, Robert J. Beyond IQ: A Triarchic Theory of Intelligence. Cambridge: Cambridge University Press.1985

Sternberg Robert J, Detterman Doug. What is Intelligence?: Contemporary Viewpoints on its Nature and Definition, 1986

Sternberg, Robert J. Successful Intelligence: How Practical and Creative Intelligence Determine Success in Life, 1997

Sternberg, Robert J. Contemporary theories of intelligence. In W. M. Reynolds & G. E. Miller (Eds.), Handbook of psychology: Educational psychology (Vol. 7, pp. 23–45). 2003

Stevens, S. S. On the Theory of Scales of Measurement. Science. 103 (2684).7 June 1946.

Stevenson Ian. Twenty Cases Suggestive of Reincarnation: Second Edition, Revised and Enlarged. 1980

Stevenson Ian. Life Before Life: A Scientific Investigation of Children's Memories of Previous Lives. 2005

Svetunkoff S.G. The theory of the dictatorship of the lumpen-proletariat. (In Russian-Светуньков С.Г. Теория диктатуры люмпен-пролетариата). –.2015

Tart, C.T. Card guessing tests: learning paradigm or extinction paradigm? JASPR 60, 1966

Tart Charles T. Discrete States Of Consciousness. Institute for the Study of Human Consciousness, San Francisco. American Association for the Advancement of Science, 140th Meeting, San Francisco, 1974

Tart Charles T. Altered states of consciousness. HerperCollins. Canada, 1990

Tart Charles T. States of consciousness. iUniverse. 2001

Tart Charles. The end of materialism. How evidence of the paranormal is bringing science and spirit together, 2012

Tazkuvel Embrosewyn. Auras: How to see, Feel & Know (Full Color). Kaleidoscope Productions. Ashland, OR, 2013

Terman, Lewis M.; Oden, Melita 1 The Gifted Group at Mid-Life: Thirty-Five Years' Follow-Up of the Superior Child. Genetic Studies of Genius Volume V. Stanford (CA): Stanford University Press. 1959, 2013.

Tesla Nicola. My Inventions: The Autobiography of Nikola Tesla. 2011

Tesla Nicola: My Life, My Research. 2014

The Wiley Handbook of Cognitive Control. Edited by Egner Tobias. 2017

Thorne Michael and Henley Tracy. Connections in the History and Systems of Psychology. 2004

Titchener, E.B. The «feeling of being stared at». Science New Series, 1898

Thurstone L.L. Primary mental abilities. Chicago: University of Chicago Press.1938

Todeschi Kevin J, Liaros Carol Ann. Edgar Cayce on Auras & Colors.A.R.E. Press, Virginia Beach, VA, 2011

Toffler Alvin. Future Shock. 1970

Toffler Alvin. The Third Wave. 1984

Torrance, E. P. The Torrance tests of creative thinking-norms—Technical manual research edition—Verbal tests, Forms A and B—Figural tests, Forms A and B. Princeton, NJ: Personnel Press. 1974

Toshiyuki Nakagaki, Hiroyasu Yamada & Ágota Tóth. Intelligence: Maze-solving by an amoeboid organism. Nature 407, 470 (28 September 2000)

Travis, F. and Shear, J. Focused attention, open monitoring and automatic self-transcending: Categories to organize meditations from Vedic, Buddhist and Chinese traditions. Consciousness and Cognition, 2010.

Travis Fred. Your Brain is a River, Not a Rock. 2012

Treisman A. "Monitoring and storage of irrelevant messages in selective attention". Journal of Verbal Learning and Verbal Behavior 3 (6). 1964

Tulving, E. Episodic and semantic memory. In E. Tulving & W. Donaldson (Eds.), Organization of Memory. New York: Academic Press. 1972.

Turner Matt. DIMON: 'The United States of America is truly an exceptional country,' but 'something is wrong. The Business Insider. Apr. 4, 2017.

Upledger John E. Cell Talk. Transmitting Mind into DNA. North Atlantic Books/ Berkeley 2010

Ukhtomsky, A.A.: The Dominant. Classics of Psychology, Piter Series. St-Petersburg, 2002, M., 1996 (in Russian)

Wagner Ulrich, et al. Sleep inspires insight. NATURE. VOL 427. 22 JANUARY 2004

Walton Chris. Gamma Healing. Amzon Create Space. 2010

Webb James Young. A Technique for Producing Ideas: The simple, five-step formula anyone can use to be more creative in business and in life! 2009

Webster Richard. Dowsing for Beginners. How to Find Water, Wealth & Lost Objects. 2012

Webster Richard. The complete look of Auras. Learn to see, read, strengthen and heal auras. Llewellin Publications Woodbury, Minnesota. 2010

Wenk Gary. Your Brain on Food: How Chemicals Control Your Thoughts and Feelings. 2010

Whale Jon. The Catalyst of Power: The Assemblage Point Of Man.DragonRising Publishing. East Sussex.UK. 2009

Whale Jon. Core energy: shifting the assemblage point. Positive Health Online, issue 17, January 1997.

Whale Jon. Naked Spirit. The Supernatural Odyssey. 2008

Wighton Kate. Intelligence 'networks' discovered in brain for the first time/ Imperial College London. 22 December 2015

Willer Mark. Evidence builds that meditation strengthens the brain. Science Daily. March 14, 2012

Winner Ellen. Gifted Children: Myths and Realities. 1996.

Wise, Anna. The High-Performance Mind: Mastering Brainwaves for Insight, Healing and Creativity. Jeremy P. Tarcher/Putnam. New York, 1997.

Wise, Anna. Awakening the Mind: A Guide to Mastering the Power of Your Brain Waves. Jeremy P. Tarcher/Putnam. New York, 2002.

Wlodarski Rafael and Dunbar Robin I. M. The Effects of Romantic Love on Mentalizing Abilities. Rev Gen Psychol. 2014 Dec 1; 18(4): 313–321.

Wolff Milo. Schrödinger's Universe: Einstein, Waves and the Origin of the Natural Laws. Outskirt Press, 2008

Wood Benjamin. Memory. Train Your Brain.: The Complete Guide on How to Improve Your Memory, Think Faster, Concentrate More and Remember Everything. 2017

Wood Evelyn Nielsen. Reading skills,1958

Wright Margaret H. Nelder, Mead, and the Other Simplex Method. Documenta Mathematica · Extra Volume ISMP.

Ywahoo, D. Voices of our ancestors: Cherokee teachings from the wisdom fire. Boston: Shambhala Publications.Boston.1987

Zaghi Soroush et al. Noninvasive Brain Stimulation with Low-Intensity Electrical Currents: Putative Mechanisms of Action for Direct and Alternating Current Stimulation. Neuroscientist. 2010;16(3)

Zilberman, M. On the training of precognitive ability. JSPR 60, 1995

Printed in the United States
By Bookmasters